Springer Monographs in Mathematics

This series publishes advanced monographs giving well-written presentations of the "state-of-the-art" in fields of mathematical research that have acquired the maturity needed for such a treatment. They are sufficiently self-contained to be accessible to more than just the intimate specialists of the subject, and sufficiently comprehensive to remain valuable references for many years. Besides the current state of knowledge in its field, an SMM volume should ideally describe its relevance to and interaction with neighbouring fields of mathematics, and give pointers to future directions of research.

More information about this series at http://www.springer.com/series/3733

Jan Rataj • Martina Zähle

Curvature Measures of Singular Sets

Springer

Jan Rataj
Mathematical Institute
Charles University
Prague, Czech Republic

Martina Zähle
Mathematisches Institut
Friedrich-Schiller-Universität Jena
Jena, Germany

ISSN 1439-7382 ISSN 2196-9922 (electronic)
Springer Monographs in Mathematics
ISBN 978-3-030-18185-7 ISBN 978-3-030-18183-3 (eBook)
https://doi.org/10.1007/978-3-030-18183-3

Mathematics Subject Classification (2010): Primary: 53C65, 28A75; Secondary: 28A80, 49Q15, 58A25, 53A17

This Springer imprint is published by the registered company Springer Nature Switzerland AG.
The registered company address is: Gewerbestrasse 11, 6330 Cham, Switzerland

Preface

The curvature notions studied in this book have some forerunners in the nineteenth century. Carl Friedrich Gauss (1777–1855) and Bernhard Riemann (1826–1866) essentially stimulated the intrinsic versions in differential geometry. In particular, the Gauss curvature and, more generally, the Riemannian curvature tensor, nowadays, play an important role in many applications. Jacob Steiner (1796–1863) and Hermann Minkowski (1864–1909) started with a parallel development in convex geometry with different tools. There is a long, related history up to now in both fields. The famous milestones are, e.g., the Gauss-Bonnet theorem, the Minkowski-Steiner formula for convex bodies and Weyl's tube formula for compact smooth submanifolds, the Crofton formula, Blaschke, Santalo, and Chern's principal kinematic formula, and Hadwiger's characterization theorem. This concerns the Minkowski quermassintegrals in convex geometry, known as intrinsic volumes, or the integrals of higher-order mean curvatures as differential geometric counterparts, including the scalar curvature. Some background on these topics is summarized in Chaps. 2 and 3.

One particular aim of the present book is to demonstrate that these functionals and their measure-geometric extensions fit into the program of Felix Klein (1849–1925) to study geometric properties that are invariant under certain transformation groups of the underlying space. It turns out that these curvature functionals and measures form a complete system of Euclidean invariants, which are additive, like volume, and continuous in an appropriate sense. In modern algebraic and differential geometry, but also in mathematical physics, the extensions, sometimes, are called Lipschitz-Killing curvatures. We will keep this notation.

Geometric measure theory has been developed since the late 1950s at the beginning mainly stimulated by the higher-dimensional Plateau problem and calculus of variations. (A survey on this topic can be found in [HP16].) Until now, Federer's book [Fed69] is a standard reference for a huge variety of basic notions and relationships. In Chap. 1, we select those which are relevant for our purposes. There, we also refer to the textbook of Krantz and Parks [KP08]. The use of currents with additional properties makes it possible to treat geometric properties of first-order rectifiable sets with powerful analytical and algebraic tools.

Subsets of Euclidean spaces, which do not fit into the context of convex or differential geometry, are not only of interest in calculus of variations. Mathematical models for natural phenomena often lead to singular sets, from piecewise smooth submanifolds up to fractals. The first fundamental paper on curvatures, i.e., second-order properties, of sets with possible singularities was [Fed59]. Therein, Federer unified and extended the abovementioned results from differential geometry and convex geometry for the more general sets with positive reach by methods of geometric measure theory. The curvature measures were introduced implicitly as coefficients of a generalized Steiner-Weyl polynomial. This approach was not included in [Fed69] and was not followed up until the 1980s. We understand the present book as a continuation of [Fed59] and a collection and extension of related investigations from the last 30 years. As a main tool, we use the explicit representation of the generalized Lipschitz-Killing curvatures by integrating certain differential forms over the associated unit normal bundle endowed with a topological index function. This allows us to treat the second-order analysis of the underlying geometric sets with tools of the first-order current theory.

Except for the mentioned approaches from convex and differential geometry, there are also relationships and possible applications to algebraic geometry, geometric analysis, and mathematical physics. Furthermore, methods from this book have been applied to models in stochastic geometry, in particular to random tessellations, which describe real phenomena. The random sets under consideration can be singular in the sense of Chaps. 4, 5, and 9.

We introduce and study the main notions, in particular those of normal cycle and curvature measures, and their relationships for the special cases of sets with positive reach (Chap. 4) and their locally finite unions (Chap. 5). For these classes, the basic definitions and mathematical tools are more accessible and prepare the reader for the study of more general singular sets. (Some extensions and related approaches are discussed in Chap. 9 and the bibliographical notes therein.) The corresponding integral geometric formulas are derived in Chap. 6. One of the main results is the translative integral formula, which leads also to a direct proof of the principal kinematic formula for such sets.

Approximations of the curvature measures of singular sets by those of more regular sets in the present book are mainly used for the purposes of Chap. 8. But they are of independent interest for many theoretical and numerical applications. In Chap. 7, we show approximations of the normal cycles, and hence of the curvature measures, by those of parallel sets of small distances as well as of appropriate polytopes.

Chapter 8 provides the corresponding characterization theorems for the curvature functionals and measures. For the case of sets with positive reach, this question was already asked by Federer. The problem was an appropriate notion of continuity. An answer for unions of sets with positive reach is given in terms of flat convergence of associated normal cycles. By means of the above approximations, we can reduce the characterization to the convex cases considered by Hadwiger [Had57] and Schneider [Sch78]. It is shown that the Lipschitz-Killing curvature functionals (or measures) form a basis for all Euclidean motion invariant continuous (measure-valued) valuations. This gives the interpretation in the spirit of Felix Klein's ideas.

Extensions of normal cycles and curvature measures to more general classes of singular sets are presented in Chap. 9. Moreover, a principal kinematic formula is proved, which includes that for sets with positive reach.

Note that Chaps. 4, 5, 6, 7, 8 and 9 contain the full proofs of the statements, except for some basic tools from the literature.

In Chap. 10, we leave the context of geometric measure theory and present some applications to fractal geometry, i.e., to much more irregular sets. Due to a result of Fu [Fu85], for arbitrary compact sets in Euclidean spaces of dimensions less than four, the closure of the complement to parallel sets of almost all distances has positive reach. In higher dimensions, this property is called tube regularity. By a reflection principle, it allows to introduce the curvature measures for parallel sets with this property, in particular, for those in certain classes of fractals. In the recent literature, appropriate convergences of the rescaled curvature measures for their parallel sets of small distances have been shown. The limits are interpreted as fractal curvature measures. This leads to new parameters for describing the geometry of such fractals. Some of these results with corresponding references are presented in this chapter, partially in a form of a survey. The Minkowski content is included as a special case. The latter has been studied by different authors with various methods.

We are extremely grateful to Andreas Bernig, Ulrich Menne, Werner Nagel, Dušan Pokorný, Thomas Wannerer, Steffen Winter, and Luděk Zajíček, who read parts of earlier versions of the manuscript and made valuable comments. We also thank Christopher Schneider for contributing the computer graphics in Chap. 10.

Finally, we acknowledge the support of the Czech and the German Science Foundations, GAČR and DFG, by some research grants (in particular, GAČR 18–11058S and DFG/ZA 242/5) on related topics during the work on this book.

Prague, Czech Republic | Jan Rataj
Jena, Germany | Martina Zähle

Contents

Chapter 1
Background from Geometric Measure Theory

The role of the first chapter is to collect the notions and results from geometric measure theory which will be needed in the sequel. Most definitions and results from this chapter can be found in the Federer's book [Fed69] and/or in the book of Krantz and Parks [KP08]. Other sources will be cited when needed. Most of these results are presented without proofs.

The basic setting is the d-dimensional Euclidean space, $\mathbb{R}^d$, with standard scalar product $x \cdot y$ and norm $|x| := \sqrt{x \cdot x}$, $x, y \in \mathbb{R}^d$. By $B(x,r)$ ($\mathring{B}(x,r)$) we denote the closed (resp. open) ball with centre $x \in \mathbb{R}^d$ and radius $r \geq 0$. Given an affine subspace L of $\mathbb{R}^d$, p_L denotes the orthogonal projection onto L and $L^\perp$ stands for the orthogonal complement (as a linear subspace) to L. The Lebesgue measure in $\mathbb{R}^d$ is denoted by $\mathcal{L}^d$. All frequently used notations can be found in the List of symbols.

We assume that the reader is familiar with the basic linear algebra, analysis and measure theory.

1.1 Area, Coarea and Rectifiability

1.1.1 Hausdorff Measure

The Hausdorff measures are defined as outer measures on $\mathbb{R}^d$. An *outer measure* on a nonempty set X is a set function μ defined on the family $\mathcal{P}(X)$ of all subsets of X with values in $[0, \infty]$ with the properties:

(i) $\mu(\emptyset) = 0$,
(ii) $A \subset B$ implies $\mu(A) \leq \mu(B)$,
(iii) $\mu(\bigcup_{i=1}^\infty A_i) \leq \sum_{i=1}^\infty \mu(A_i)$.

J. Rataj, M. Zähle, *Curvature Measures of Singular Sets*, Springer Monographs in Mathematics, https://doi.org/10.1007/978-3-030-18183-3_1

To an outer measure μ, the system of μ-measurable sets is assigned:

$$\mathcal{A}_\mu := \{A \subset X : \mu(E) = \mu(E \cap A) + \mu(E \setminus A) \text{ whenever } E \subset X\}.$$

Theorem 1.1 ([KP08, §1.2]) *$\mathcal{A}_\mu$ is a σ-algebra and the restriction of μ to $\mathcal{A}_\mu$ is a (σ-additive) measure.*

There is a plenty of outer measures whose σ-algebra is very poor or even trivial. There exists, however, a simple criterion assuring that the σ-algebra is rich enough. An outer measure μ on a metric space (X, ρ) is called *metric* if $\mu(A \cup B) = \mu(A) + \mu(B)$ whenever

$$\operatorname{dist}(A, B) := \inf\{\rho(a, b) : a \in A, b \in B\} > 0.$$

Theorem 1.2 (Carathéodory's criterion, [KP08, Theorem 1.2.13]) *If μ is a metric outer measure on a metric space (X, ρ) then $\mathcal{A}_\mu$ contains all Borel sets.*

A metric outer measure μ on X is said to be *Borel regular* if to any $A \subset X$ there exists a Borel set $B \supset A$ such that $\mu(A) = \mu(B)$.

Let $s \geq 0$. Denote

$$\omega_s = \frac{\pi^{s/2}}{\Gamma(\frac{s}{2} + 1)}$$

(note that if s is an integer then ω_s is the volume of the unit ball in $\mathbb{R}^s$). We define the *s-dimensional Hausdorff measure* in $\mathbb{R}^d$ as

$$\mathcal{H}^s(A) = \lim_{\delta \to 0+} \inf_{\substack{A \subset \bigcup_i G_i \\ \operatorname{diam} G_i \leq \delta}} \sum_i \omega_s \left(\frac{\operatorname{diam} G_i}{2}\right)^s.$$

(The infimum above is taken over all finite or countable coverings of A with (arbitrary) subsets $G_1, G_2, \ldots$ of $\mathbb{R}^d$ of diameters at most δ.)

Proposition 1.3 ([Fed69, Section 2.10]) *For any $s \geq 0$,*

1. *$\mathcal{H}^s$ is a metric outer measure on $\mathbb{R}^d$;*
2. *$\mathcal{H}^s$ is Borel regular;*
3. *$\mathcal{H}^s$ is translation and rotation invariant;*
4. *$\mathcal{H}^0$ is the counting measure;*
5. *$\mathcal{H}^d$ is the Lebesgue measure ($\mathcal{H}^d = \mathcal{L}^d$);*
6. *$\mathcal{H}^s = 0$ if $s > d$.*

Note that the definition of $\mathcal{H}^s$ would not change if we would consider only coverings by open, or closed, or even compact convex, sets. Covering by balls would, however, produce another measure called *spherical measure* (though its values on "nice" sets would be the same).

Let A be a subset of $\mathbb{R}^d$. The *Hausdorff dimension* of A is defined as

$$\dim_H A := \inf\{s \ge 0 : \mathcal{H}^s(A) < \infty\}.$$

The Hausdorff dimension has the following meaning.

Proposition 1.4 ([KP08, Proposition 2.1.2]) *If $s < \dim_H A$ then $\mathcal{H}^s(A) = \infty$. If $s > \dim_H A$ then $\mathcal{H}^s(A) = 0$.*

Examples

- Any nonempty open set in $\mathbb{R}^d$ has Hausdorff dimension d.
- A nonempty m-dimensional C^1-submanifold has Hausdorff dimension m.
- Any countable set has Hausdorff dimension 0.
- The middle-third Cantor set in $\mathbb{R}^1$ has Hausdorff dimension $\frac{\log 2}{\log 3}$ (see, e.g., [Mat95, §4.10]; cf. Chap. 10).
- The trajectory of Brownian motion in $\mathbb{R}^d$ has almost surely Hausdorff dimension 2 (nevertheless, its two-dimensional Hausdorff measure vanishes; see, e.g., [MP10, Ch. 4]).

Definition 1.5 A set $A \subset \mathbb{R}^d$ is called *Ahlfors s-regular* $(0 \le s \le d)$ if there exist constants $c, C > 0$ such that

$$cr^s \le \mathcal{H}^s(A \cap B(a, r)) \le Cr^s, \quad a \in A,\ 0 < r < \operatorname{diam} A. \tag{1.1}$$

A is said to be *upper* (*lower*) Ahlfors s-regular if the upper (lower, respectively) inequality holds.

1.1.2 Densities of Sets

Let A be a subset of $\mathbb{R}^d$ and $a \in \mathbb{R}^d$ a point. Let $s > 0$. Define

$$\Theta^{*s}(A, a) = \limsup_{r \to 0_+} \frac{\mathcal{H}^s(A \cap B(a, r))}{\omega_s r^s},$$

$$\Theta_*^s(A, a) = \liminf_{r \to 0_+} \frac{\mathcal{H}^s(A \cap B(a, r))}{\omega_s r^s},$$

the *upper and lower s-dimensional density of A at a*. If both the upper and lower densities agree we call the common value *s-dimensional density of A at a* and denote it by $\Theta^s(A, a)$.

Theorem 1.6 ([Fed69, §2.10.19])

1. *If* $A \subset \mathbb{R}^d$ *is Lebesgue measurable then* $\Theta^d(A, \cdot)$ *equals* 1 $\mathcal{L}^d$*-almost everywhere on* A *and equals* 0 $\mathcal{L}^d$*-almost everywhere on the complement of* A *(Lebesgue density theorem).*
2. *If* $\mathcal{H}^s(A) < \infty$ *then* $\Theta^{*s}(A, \cdot) \leq 1$ $\mathcal{H}^s$*-almost everywhere on* A.

1.1.3 Lipschitz Mappings

A mapping $f : A \to \mathbb{R}^n$ defined on a set $A \subset \mathbb{R}^d$ is *Lipschitz* if there exists a number $L \geq 0$ such that

$$|f(y) - f(x)| \leq L|y - x| \quad \text{for all } x, y \in A.$$

The infimum of all constants L with the above property as called the *Lipschitz constant of* f and denoted Lip f. The following result says that we can mostly work with Lipschitz mappings defined on the whole space.

Theorem 1.7 (Kirszbraun, [Fed69, §2.10.43]) *Any Lipschitz mapping from a subset of* $\mathbb{R}^d$ *to* $\mathbb{R}^n$ *can be extended to a Lipschitz mapping defined on the whole* $\mathbb{R}^d$*, with the same Lipschitz constant.*

Lipschitz mappings are often used in geometric measure theory instead of C^1-mappings from the classical calculus. The following two results make this possible.

Theorem 1.8 (Rademacher, [Fed69, §3.1.6]) *A Lipschitz mapping* $f : \mathbb{R}^d \to \mathbb{R}^n$ *is differentiable* $\mathcal{L}^d$*-almost everywhere.*

Theorem 1.9 (Whitney, [Fed69, §3.1.16]) *Let* $f : \mathbb{R}^d \to \mathbb{R}^n$ *be Lipschitz and let* $\varepsilon > 0$*. Then there exists a* C^1 *mapping* $g : \mathbb{R}^d \to \mathbb{R}^n$ *such that*

$$\mathcal{L}^d\{x \in \mathbb{R}^d : f(x) \neq g(x)\} < \varepsilon.$$

The behaviour of Hausdorff measure under Lipschitz mappings is given in the following simple proposition.

Proposition 1.10 ([KP08, Proposition 2.4.7]) *If* $A \subset \mathbb{R}^d$ *and* $f : A \to \mathbb{R}^n$ *is Lipschitz then* $\mathcal{H}^s(f(A)) \leq (\mathrm{Lip}\, f)^s \mathcal{H}^s(A)$, $s \geq 0$.

As a consequence we obtain the Ahlfors regularity of compact Lipschitz submanifolds.

Definition 1.11 A set $M \subset \mathbb{R}^d$ is an *m-dimensional Lipschitz submanifold* ($1 \leq m \leq d-1$) if for any $a \in M$ there exist an open neighbourhood U of a in $\mathbb{R}^d$ and a bi-Lipschitz homeomorphism $f : M \cap U \to \mathrm{im}\, f \subset \mathbb{R}^m$ (called *Lipschitz chart* of M). A collection (f_i) of charts with domains covering M is called an *atlas* of M.

Proposition 1.12 *Any compact m-dimensional Lipschitz submanifold $M \subset \mathbb{R}^d$ is Ahlfors m-regular ($1 \le m \le d-1$).*

Proof Let $a \in M$ be given, let $f : M \cap U \to \operatorname{im} f \subset \mathbb{R}^m$ be a Lipschitz chart, denote $L := \max\{\operatorname{Lip} f, \operatorname{Lip} f^{-1}\}$, and let $\varepsilon > 0$ be such that $B(a,\varepsilon) \subset U$. Then, for any $0 < r \le \varepsilon$,

$$f^{-1}(B(f(a), L^{-1}r)) \subset M \cap B(a,r) \subset f^{-1}(B(f(a), Lr)).$$

Applying Proposition 1.10, we obtain

$$L^{-k}\omega_k(L^{-1}r)^k \le \mathcal{H}^k(M \cap B(a,r)) \le L^k\omega_k(Lr)^k,$$

which shows that the set $A \cap B(a,\varepsilon)$ is Ahlfors k-regular with constants $c = \omega_k L^{-2k}$ and $C = \omega_k L^{2k}$ in (1.1) independent of $a \in M$ and $\varepsilon > 0$. Since M is compact, it has a finite atlas of Lipschitz charts, and it is not difficult to verify that M fulfills (1.1) as well. □

Corollary 1.13 *For any bounded set $F \subset \mathbb{R}^d$ and $r > 0$, the boundary ∂F_r is upper Ahlfors $(d-1)$-regular.*

Proof If $r > 0$ is a regular value of the distance function $x \mapsto \operatorname{dist}(x, F)$ then ∂F_r is a $(d-1)$-dimensional Lipschitz manifold by the inverse function theorem for Lipschitz mappings (see [Cla76]). This appears, in particular, if $r > \operatorname{diam} F$ (cf. [Fu85]). In such a case, ∂F_r is Ahlfors $(d-1)$-regular by Proposition 1.12.

Let now $r > 0$ be arbitrary and partition $F = F^1 \cup \cdots \cup F^n$ into finitely many pieces with $\operatorname{diam} F^i < r$, $i = 1, \ldots, n$. Then, each $\partial(F^i)_r$ is upper Ahlfors $(d-1)$-regular by the above observation, and, since clearly $\partial F_r \subset \partial(F_1)_r \cup \cdots \cup \partial(F_n)_r$, ∂F_r is upper Ahlfors $(d-1)$-regular. □

Let a mapping f with values in $\mathbb{R}^n$ be differentiable at a point $a \in \mathbb{R}^d$ (with differential $Df(a)$) and let $0 \le k \le d$ be an integer. The *k-dimensional Jacobian* of f at a is defined as

$$J_k f(a) := \sup\{\mathcal{H}^k(Df(a)(C)) : C \text{ is a } k\text{-dimensional unit cube in } \mathbb{R}^d\}. \quad (1.2)$$

(Note that $Df(a)$ maps the cube C always to a parallelepiped and we are considering its maximal k-volume.) Another description using the multilinear notation from Sect. 1.2.1 is $J_k f(a) = \|\bigwedge_k Df(a)\|$.

Particular cases (cf. [KP08, pp. 127–8])

1. If $k = d = n$ then $J_d f(a) = |\det Df(a)|$.
2. If $k = d < n$ then

$$J_d f(a) = \sqrt{\det\left(Df(a)^{\mathsf{T}} Df(a)\right)}.$$

Moreover, $J_d f(a) = \mathcal{H}^d(Df(a)(C))$ for any unit cube $C \subset \mathbb{R}^d$, or $J_d f(a) = \mathcal{H}^d(Df(a)(A))/\mathcal{L}^d(A)$ for any measurable subset $A \subset \mathbb{R}^d$ of positive finite Lebesgue measure.

3. If $k = n < d$ then

$$J_n f(a) = \sqrt{\det\left(Df(a)\, Df(a)^{\mathsf{T}}\right)}.$$

If the rank of $Df(a)$ is less than n then $J_n f(a) = 0$. If the rank of $Df(a)$ equals n then $J_n f(a) = \mathcal{H}^n(Df(a)(C))$ for any unit n-cube in $(\ker Df(a))^{\perp}$, the orthogonal complement of the kernel of $Df(a)$ ($\ker Df(a) = \{u \in \mathbb{R}^d : Df(a)u = 0\}$), or $J_n f(a) = \mathcal{H}^n(Df(a)(A))/\mathcal{H}^n(A)$ for any measurable subset $A \subset (\ker Df(a))^{\perp}$ of positive finite n-dimensional Hausdorff measure.

Two basic formulas for change of variables in the setting for Lipschitz functions follow. Notice that we use the notion "integrable function" in the usual sense for a measurable function with finite integral.

Theorem 1.14 (Area formula, [Fed69, §3.2.3]) *Let $f : \mathbb{R}^d \to \mathbb{R}^n$ be Lipschitz, $d \le n$, and let $A \subset \mathbb{R}^d$ be Lebesgue measurable. Then*

$$\int_A J_d f \, \mathrm{d}\mathcal{L}^d = \int_{\mathbb{R}^n} \operatorname{card}(A \cap f^{-1}\{z\})\, \mathcal{H}^d(\mathrm{d}z).$$

If, moreover, h is an $\mathcal{L}^d$-integrable (or nonnegative $\mathcal{L}^d$-measurable) function on A then

$$\int_A h(x) J_d f(x)\, \mathcal{L}^d(\mathrm{d}x) = \int_{\mathbb{R}^n} \sum_{x \in A \cap f^{-1}\{z\}} h(x)\, \mathcal{H}^d(\mathrm{d}z).$$

Theorem 1.15 (Coarea formula, [Fed69, §3.2.11]) *Let $f : \mathbb{R}^d \to \mathbb{R}^n$ be Lipschitz, $d \ge n$, and let $A \subset \mathbb{R}^d$ be Lebesgue measurable. Then*

$$\int_A J_n f \, \mathrm{d}\mathcal{L}^d = \int_{\mathbb{R}^n} \mathcal{H}^{d-n}(A \cap f^{-1}\{z\})\, \mathcal{H}^n(\mathrm{d}z).$$

If, moreover, h is an $\mathcal{L}^d$-integrable (or nonnegative $\mathcal{L}^d$-measurable) function on A then

$$\int_A h(x) J_n f(x)\, \mathcal{L}^d(\mathrm{d}x) = \int_{\mathbb{R}^n} \int_{A \cap f^{-1}\{z\}} h(x)\, \mathcal{H}^{d-n}(dx) \mathcal{H}^n(\mathrm{d}z).$$

1.1.4 Tangent Cones

If $A \subset \mathbb{R}^d$ and $a \in \mathbb{R}^d$, a vector $u \in \mathbb{R}^d$ is said to be *tangent to* A at a if $u = 0$ or if there exist sequences $(a_i) \subset A \setminus \{a\}$ and $(r_i) \subset (0, \infty)$ such that $a_i \to a$ and $r_i(a_i - a) \to u$, $i \to \infty$. Another (equivalent) description is that a nonzero vector u is tangent to A at a if and only if there exist points $a_i \neq a$ from A such that $\frac{a_i - a}{|a_i - a|} \to \frac{u}{|u|}$. We denote by $\mathrm{Tan}(A, a)$ the set of all tangent vectors to A at a. It is always a closed cone with vertex at the origin.

Given an integer $0 \le k \le d$, the cone of *$(\mathcal{H}^k, k)$-approximate tangent vectors of A at a* is defined as

$$\mathrm{Tan}^k(A, a) = \bigcap \{\mathrm{Tan}(E, a) : E \subset A, \Theta^k(A \setminus E, a) = 0\}.$$

Clearly, $\mathrm{Tan}^k(A, a)$ is a closed subcone of $\mathrm{Tan}(A, a)$.

1.1.5 Approximate Differential

A function $f : A \to \mathbb{R}^n$ $(A \subset \mathbb{R}^d)$ is said to be *$(\mathcal{H}^k, k)$-approximatively differentiable* at $a \in A$ if there exists a mapping $g : \mathbb{R}^d \to \mathbb{R}^n$ differentiable at a and such that

$$\Theta^k(\{x \in A : f(x) \neq g(x)\}, a) = 0.$$

The mapping

$$(\mathcal{H}^k, k)\mathrm{ap}\, Df(a) := Dg(a) \mid \mathrm{Tan}^k(A, a)$$

(restriction of $Dg(a)$ to $\mathrm{Tan}^k(A, a)$) is called the *$(\mathcal{H}^k, k)$-approximate differential* of f at a. We often write only ap $Df(a)$, or even only $Df(a)$, for brevity. (Note that the approximate differential coincides with the classical one whenever the latter exists, hence, there is no risk of confusion.)

It can be shown that ap $Df(a)$ does not depend on the choice of the function g.

Assume that f is $(\mathcal{H}^k, k)$-approximatively differentiable at $a \in A$ and that $\mathrm{Tan}^k(A, a)$ is a k-dimensional subspace of $\mathbb{R}^d$. For an integer $0 \le m \le k$, we define the *m-dimensional approximate Jacobian* of f at a as

$$\mathrm{ap}\, J_m f(a) = \sup\{\mathcal{H}^m(\mathrm{ap}\, Df(a)(C)) : C \text{ is a unit } m\text{-cube in } \mathrm{Tan}^k(A, a)\}.$$

We will also use the short notation $J_m f(a)$ for the approximate Jacobian of f, i.e.,

$$J_m f(a) := \mathrm{ap}\, J_m f(a).$$

1.1.6 Rectifiable Sets

Definition 1.16 Let $k \in [0, d]$ be an integer. A set $A \subset \mathbb{R}^d$ is called

- *k-rectifiable* if A is a Lipschitz image of a bounded subset of $\mathbb{R}^k$;
- *countably k-rectifiable* if A is a countable union of k-rectifiable sets;
- *locally k-rectifiable* if for any $a \in A$ there exists $r > 0$ such that $A \cap B(a, r)$ is k-rectifiable;
- *countably $\mathcal{H}^k$-rectifiable* if $A = A_0 \cup A_1 \cup A_2 \cup \dots$ with $\mathcal{H}^k(A_0) = 0$ and A_i k-rectifiable, $i \geq 1$;
- *locally $\mathcal{H}^k$-rectifiable* (*in an open set $U \supset A$*) if it is countably $\mathcal{H}^k$-rectifiable and

$$\mathcal{H}^k(A \cap K) < \infty \quad \text{for any compact } K \subset \mathbb{R}^d \quad (K \subset U, \text{ resp.}).$$

Remark 1.17

1. The reader should be warned that the terminology for rectifiable sets is not unified in the literature. Federer [Fed69] used the term "$(\mathcal{H}^k, k)$-rectifiable set" for a set that we call countably $\mathcal{H}^k$-rectifiable with the additional condition of finite $\mathcal{H}^k$-measure. Krantz and Parks [KP08] use "countably k-rectifiable" for what we call "countably $\mathcal{H}^k$-rectifiable". We adopt here the terminology of Ambrosio et al. [AFP00].
2. The local $\mathcal{H}^k$-rectifiability does not imply $\mathcal{H}^k$-measurability in general. (Consider a non-measurable subset $B \subset \mathbb{R}^k$ and take $A = B \times \{0\} \subset \mathbb{R}^d$.) We will mostly use the rectifiability together with the measurability assumption.

Examples

1. A k-dimensional Lipschitz submanifold of $\mathbb{R}^d$ is locally k-rectifiable.
2. The graph of a Lipschitz function of $d - 1$ variables is locally $(d - 1)$-rectifiable.
3. If $A \subset \mathbb{R}^d$ is countably $\mathcal{H}^k$-rectifiable and $f : A \to \mathbb{R}^n$ Lipschitz then $f(A)$ is countably $\mathcal{H}^k$-rectifiable as well.
4. If $A \subset \mathbb{R}^d$ is locally $\mathcal{H}^k$-rectifiable and $f : A \to \mathbb{R}^n$ Lipschitz and proper (i.e., preimages of compact sets are compact) then $f(A)$ is locally $\mathcal{H}^k$-rectifiable as well.

Theorem 1.18 ([Fed69, §3.2.18]) *Let $A \subset \mathbb{R}^d$ be $\mathcal{H}^k$-measurable and locally $\mathcal{H}^k$-rectifiable and $\gamma > 1$. Then there exist bi-Lipschitz mappings $g_i : \mathbb{R}^k \to \operatorname{im} g_i \subset \mathbb{R}^d$ with $\operatorname{Lip} g_i \leq \gamma$ and $\operatorname{Lip} g_i^{-1} \leq \gamma$ and compact subsets $K_i \subset \mathbb{R}^k$, $i = 1, 2, \dots$, such that $g_i(K_i) \cap g_j(K_j) = \emptyset$ for $i \neq j$ and*

$$\mathcal{H}^k\left(A \setminus \bigcup_i g_i(K_i)\right) = 0.$$

Proposition 1.19 ([Fed69, §3.2.19]) *Let $A \subset \mathbb{R}^d$ be $\mathcal{H}^k$-measurable and locally $\mathcal{H}^k$-rectifiable and $f : A \to \mathbb{R}^n$ be Lipschitz. Then for $\mathcal{H}^k$-almost all $a \in A$, $\Theta^k(A, a) = 1$, $\operatorname{Tan}^k(A, a)$ is a k-dimensional subspace, and f is $(\mathcal{H}^k, k)$-approximately differentiable at a.*

Theorem 1.20 ([Fed69, §3.2.29]) *A set $A \subset \mathbb{R}^d$ is countably $\mathcal{H}^k$-rectifiable if and only if there exist k-dimensional C^1-submanifolds $M_1, M_2, \dots$ of $\mathbb{R}^d$ such that $\mathcal{H}^k(A \setminus \bigcup_i M_i) = 0$.*

We present now a common extension of Theorems 1.14 and 1.15 for rectifiable sets. It is an easy extension of [Fed69, §3.2.22] (where finite global $\mathcal{H}^k$-measure is assumed), see also [KP08, Theorem 5.4.9].

Theorem 1.21 (General Area-coarea theorem) *Let $A \subset \mathbb{R}^d$ be $\mathcal{H}^k$-measurable and locally $\mathcal{H}^k$-rectifiable and $Z \subset \mathbb{R}^n$ be $\mathcal{H}^n$-measurable and locally $\mathcal{H}^m$-rectifiable, $k \geq m$, and let $f : A \to Z$ be Lipschitz. Then*

1. *for $\mathcal{H}^k$-almost all $x \in A$, either* $\operatorname{ap} J_m f(x) = 0$ *or*

$$\operatorname{im} \operatorname{ap} Df(x) = \operatorname{Tan}^m(Z, f(x))$$

is an m-dimensional subspace,

2. *$f^{-1}\{z\}$ is locally $\mathcal{H}^{k-m}$-rectifiable and $\mathcal{H}^{k-m}$-measurable for $\mathcal{H}^m$-almost all $z \in Z$,*
3.

$$\int_A \operatorname{ap} J_m f \, \mathrm{d}\mathcal{H}^k = \int_Z \mathcal{H}^{k-m}(f^{-1}\{z\}) \, \mathcal{H}^m(\mathrm{d}z),$$

4. *for any $\mathcal{H}^k$-integrable (or nonegative $\mathcal{H}^k$-measurable) function h on A,*

$$\int_A \operatorname{ap} J_m f(x) h(x) \, \mathcal{H}^k(\mathrm{d}x) = \int_Z \int_{f^{-1}\{z\}} h(x) \, \mathcal{H}^{k-m}(dx) \, \mathcal{H}^m(\mathrm{d}z).$$

Example 1.22 Given two subspaces L_p, L_q of $\mathbb{R}^d$, of dimension p, q, respectively, we define

$$J(L_p, L_q) = J_r(p_{L_q}|L_p),$$

the r-dimensional Jacobian of the orthogonal projection to L_q restricted to L_p, where $r = \min\{p, q\}$.

Let $A \subset \mathbb{R}^d$ be $\mathcal{H}^k$-measurable and locally $\mathcal{H}^k$-rectifiable, let L be a j-dimensional subspace of $\mathbb{R}^d$ and set $r = \min\{j, k\}$. If $f = p_L|A : A \to L$ is the orthogonal projection from A to L then

$$J_r f(a) = J(\operatorname{Tan}^k(A, a), L)$$

for $\mathcal{H}^k$-almost all $a \in A$. The area-coarea theorem thus yields

$$\int_A J(\mathrm{Tan}^k(A,a),L)\,\mathcal{H}^k(\mathrm{d}a) = \int_L \mathcal{H}^{k-j}(A \cap p_L^{-1}\{z\})\,\mathcal{H}^j(\mathrm{d}z)$$

if $k \geq j$ and

$$\int_A J(\mathrm{Tan}^k(A,a),L)\,\mathcal{H}^k(\mathrm{d}a) = \int_{p_L(A)} \mathcal{H}^{j-k}(A \cap p_L^{-1}\{z\})\,\mathcal{H}^k(\mathrm{d}z)$$

if $k \leq j$. Assume now that $k = j$; integrating the last formula with respect to the invariant probability measure ν_k^d over the Grassmannian $G(d,k)$ of all k-subspaces (for exact definition see Sect. 1.2.2) and using Fubini theorem (provided that the measurability is shown), we get the Crofton formula (consequence of [Fed69, §2.10.15, §3.2.6]):

Theorem 1.23 (Crofton formula for Hausdorff measures) *If $A \subset \mathbb{R}^d$ is $\mathcal{H}^k$-measurable and locally $\mathcal{H}^k$-rectifiable and $0 \leq k \leq d$ then*

$$\mathcal{H}^k(A) = c(d,k) \int_{G(d,k)} \int_L \mathrm{card}(A \cap p_L^{-1}\{z\})\,\mathcal{H}^k(\mathrm{d}z)\nu_k^d(\mathrm{d}L)$$

with

$$c(d,k) = \frac{\Gamma\left(\frac{k+1}{2}\right)\Gamma\left(\frac{d-k+1}{2}\right)}{\Gamma\left(\frac{d+1}{2}\right)\Gamma\left(\frac{1}{2}\right)}.$$

The following theorem is about the rectifiability of level sets of Lipschitz maps. It as a consequence of [Fed69, §3.2.31].

Theorem 1.24 *If $W \subset \mathbb{R}^d$ is $\mathcal{H}^k$-measurable and locally $\mathcal{H}^k$-rectifiable, $f : W \to \mathbb{R}^n$ Lipschitz and $0 \leq m \leq k$, then the set*

$$\{y \in \mathbb{R}^n : \mathcal{H}^{k-m}(f^{-1}\{y\}) > 0\}$$

is countably $\mathcal{H}^m$-rectifiable.

1.1.7 Purely Unrectifiable Sets

A set $E \subset \mathbb{R}^d$ is called *purely k-unrectifiable* if it contains no k-rectifiable subset of positive $\mathcal{H}^k$-measure.

Proposition 1.25 *Any $\mathcal{H}^k$-measurable set $W \subset \mathbb{R}^d$ with $\mathcal{H}^k(W) < \infty$ can be written as disjoint union $W = A \cup E$ of a countably k-rectifiable set A and purely k-unrectifiable set E.*

Theorem 1.26 (Structure theorem, [Fed69, §3.3.13]) *Assume that $\mathcal{H}^k(E) < \infty$. If E is purely k-unrectifiable then $\mathcal{H}^k(p_L(E)) = 0$ for almost all k-subspaces E of $\mathbb{R}^d$. Consequently, the Crofton formula fails for purely unrectifiable sets with positive $\mathcal{H}^k$-measure.*

1.2 Multilinear Algebra and Differential Forms

1.2.1 Multilinear Algebra

Let V be a finite (d-)dimensional vector space over $\mathbb{R}$ (usually $V = \mathbb{R}^d$). If $k \geq 0$ is an integer, we denote by $\bigotimes^k V$ the linear space of all k-linear functions $\alpha : V^k \to \mathbb{R}$. Elements of $\bigotimes^k V$ are called *covariant k-tensors* and $\dim \bigotimes^k V = d^k$. Let $\Sigma(k)$be the set of all permutations of $\{1, \dots, k\}$. A covariant k-tensor α is called *antisymmetric* if for any $\sigma \in \Sigma(k)$ and for all vectors $v_1, \dots, v_k \in V$,

$$\alpha\left(v_{\sigma(1)}, \dots, v_{\sigma(k)}\right) = (\operatorname{sgn} \sigma)\alpha\left(v_1, \dots, v_k\right).$$

The set of all antisymmetric covariant k-tensors will be denoted by $\bigwedge^k V$. It is a linear subspace of $\bigotimes^k V$. Elements of $\bigwedge^k V$ are called *k-covectors* or *multi-covectors* . It follows from the antisymmetry that $\alpha(v_1, \dots, v_k) = 0$ whenever the vectors $v_1, \dots, v_k$ are linearly dependent. Thus, if $k > d$ then $\bigwedge^k V$ is trivial. Given two multi-covectors $\alpha \in \bigwedge^k V$, $\beta \in \bigwedge^m V$, their *exterior* (or *wedge*) *product* is defined by the "shuffle formula"

$$\begin{aligned}(\alpha \wedge \beta)&(v_1, \dots, v_{k+m}) \\ &= \sum_{\sigma \in \Sigma(k,m)} (\operatorname{sgn} \sigma)\alpha(v_{\sigma(1)}, \dots, v_{\sigma(k)})\beta(v_{\sigma(k+1)}, \dots, v_{\sigma(k+m)}),\end{aligned} \tag{1.3}$$

where

$$\Sigma(k, m) = \{\sigma \in \Sigma(k + m) : \sigma(1) < \dots < \sigma(k), \sigma(k + 1) < \dots < \sigma(k + m)\}.$$

It is easy to verify that $\alpha \wedge \beta \in \bigwedge^{k+m} V$. k-covectors of the form $\alpha_1 \wedge \dots \wedge \alpha_k$, where $\alpha_1, \dots, \alpha_k \in V^* = \bigwedge^1 V$, are called *simple*. Any k-covector can be written as a linear combination of simple k-covectors.

The direct sum

$$\bigwedge^* V = \bigoplus_{k=0}^{\infty} \bigwedge^k V$$

is called *exterior algebra* of V; it is obviously an algebra with respect to addition and exterior multiplication.

For a basis $b_1, \ldots, b_n$ in an n-dimensional real vector space the dual basis is denoted by $b_1^*, \ldots, b_n^*$, i.e., $b_i^*(b_j) = \delta_{ij}$. If $\{e_1, \ldots, e_d\}$ is a basis of V then

$$\{e_{i_1}^* \wedge \cdots \wedge e_{i_k}^* : 1 \leq i_1 < \cdots < i_k \leq d\}$$

is a basis of $\bigwedge^k V$.

The space of *k-vectors* or *multivectors* is introduced as

$$\bigwedge_k V := \bigwedge^k V^* ,$$

where $V^* = \bigwedge^1 V$ is the dual space to V. Then the exterior product of multivectors is defined as well. Using the identification $V^{**} = V$ *simple k-vectors* are given by $v_1 \wedge \cdots \wedge v_k$ for $v_1, \ldots, v_k \in V$ and for the above basis in V the k-vectors

$$\{e_{i_1} \wedge \cdots \wedge e_{i_k} : 1 \leq i_1 < \cdots < i_k \leq d\}$$

form a basis of $\bigwedge_k V$.

Example 1.27

1. If $u = u^1 e_1 + \cdots + u^d e_d$ and $v = v^1 e_1 + \cdots + v^d e_d$ then

$$u \wedge v = \sum_{i<j} (u^i v^j - u^j v^i)(e_i \wedge e_j).$$

2. If $u_1, \ldots, u_d \in V$, $u_i = \sum_{j=1}^d u_i^j e_j$, then

$$u_1 \wedge \cdots \wedge u_d = \det \left(u_i^j\right)_{i,j=1}^d (e_1 \wedge \cdots \wedge e_d).$$

3. There exist multivectors which are not simple if $\dim V \geq 4$, e.g., $e_1 \wedge e_2 + e_3 \wedge e_4$ is not simple.

Using the identification $e_i = e_i^{**}$ and the rules of exterior multiplication it follows that

$$e_{i_1} \wedge \cdots \wedge e_{i_k}(e_{j_1}^* \wedge \cdots \wedge e_{j_k}^*) = \delta_{(i_1,\ldots,i_k)(j_1,\ldots,j_k)} .$$

Hence, $e_{i_1} \wedge \cdots \wedge e_{i_k} = (e^*_{i_1} \wedge \ldots \wedge e^*_{i_k})^*$ and consequently,

$$\bigwedge_k V = \bigwedge^k (V^*) = (\bigwedge^k V)^* .$$

Because of this interpretation we can define the *dual pairing* of multivectors $\xi \in \bigwedge_k V$ and k-forms $\alpha \in \bigwedge^k V$

$$\langle \xi, \alpha \rangle := \alpha(\xi) = \xi(\alpha)$$

in the sense of the corresponding mappings. Then one obtains, in particular,

$$\langle v_1 \wedge \cdots \wedge v_k, \alpha \rangle = \alpha(v_1, \ldots, v_k) .$$

With any k-vector $\xi \in \bigwedge_k V$ we associate the subspace

$$L(\xi) := \{u \in V : \xi \wedge u = 0\}.. \tag{1.4}$$

In the sequel the *linear hull* of a family of vectors S is denoted by Lin $\{S\}$.

Proposition 1.28 ([Fed69, §1.6.1])

1. $0 \neq \xi \in \bigwedge_k \mathbb{R}^d$ *is simple if and only if* $\dim L(\xi) = k$. *Moreover,*

$$L(v_1 \wedge \cdots \wedge v_k) = \text{Lin}\,\{v_1, \ldots, v_k\}$$

   for any simple k-vector $\xi = v_1 \wedge \cdots \wedge v_k$.
2. *If* $\xi = v_1 \wedge \cdots \wedge v_k$ *and* $\zeta = w_1 \wedge \cdots \wedge w_k$ *are two non-zero simple k-vectors, then* $L(\xi) = L(\zeta)$ *if and only if* $\xi = c\zeta$ *for some* $c \neq 0$. $c > 0$ *if and only if the vectors* $(v_1, \ldots, v_k)$ *and* $(w_1, \ldots, w_k)$ *have the same orientation.*

A linear mapping $\mathbf{L} : V \to W$ between two vector spaces induces mappings on the spaces of alternating forms and multivectors as follows:
The linear mapping $\bigwedge^k \mathbf{L} : \bigwedge^k W \to \bigwedge^k V$ with

$$(\bigwedge^k \mathbf{L})\alpha(v_1, \ldots, v_k) := \alpha(\mathbf{L}v_1, \ldots, \mathbf{L}v_k),\ \alpha \in \bigwedge_k W , \tag{1.5}$$

is called *pullback* of the alternating k-linear forms under the mapping $\mathbf{L}$. It commutes with exterior multiplication:

$$\bigwedge^{k+l} \mathbf{L}(\alpha \wedge \beta) = \bigwedge^k \mathbf{L}(\alpha) \wedge \bigwedge^l \mathbf{L}(\beta),\ \alpha \in \bigwedge^k W,\ \beta \in \bigwedge^l W . \tag{1.6}$$

The counterpart called *pushforward* of multivectors is the linear mapping $\bigwedge_k \mathbf{L} : \bigwedge_k V \to \bigwedge_k W$ determined by

$$(\bigwedge_k \mathbf{L})(v_1 \wedge \cdots \wedge v_k) := \mathbf{L}v_1 \wedge \cdots \wedge \mathbf{L}v_k . \tag{1.7}$$

(Using multilinearity and Proposition 1.28 one can see that this definition does not depend on the choice of the representing vectors $v_1, \ldots, v_k$ for the k-vector $v_1 \wedge \ldots \wedge v_k$.)
Then one obtains the *duality*

$$\langle \xi, (\bigwedge^k \mathbf{L})\alpha \rangle = \langle (\bigwedge_k \mathbf{L})\xi, \alpha \rangle \,, \xi \in \bigwedge_k V, \ \alpha \in \bigwedge^k W. \tag{1.8}$$

The operations of *interior multiplication* are introduced as follows. Let $\xi \in \bigwedge_k V$ and $\alpha \in \bigwedge^m V$. If $k \le m$ we define $\xi \mathbin{\lrcorner} \alpha \in \bigwedge^{m-k} V$ by

$$\langle \eta, \xi \mathbin{\lrcorner} \alpha \rangle = \langle \eta \wedge \xi, \alpha \rangle, \qquad \eta \in \bigwedge_{m-k} V, \tag{1.9}$$

and if $k \ge m$ we define $\xi \mathbin{\llcorner} \alpha \in \bigwedge_{k-m} V$ by

$$\langle \xi \mathbin{\llcorner} \alpha, \beta \rangle = \langle \xi, \alpha \wedge \beta \rangle, \qquad \beta \in \bigwedge^{k-m} V. \tag{1.10}$$

Let now $V = \mathbb{R}^d$ and $\{e_1, \ldots, e_d\}$ be its canonical orthonormal basis. $u \cdot v$ denotes the basic scalar product. We define a *scalar product* on $\bigwedge_k \mathbb{R}^d$ by means of the values on the induced basis as follows. Assume that $i_1 < \cdots < i_k$ and $j_1 < \cdots < j_k$ and set

$$(e_{i_1} \wedge \cdots \wedge e_{i_k}) \bullet (e_{j_1} \wedge \cdots \wedge e_{j_k}) = \begin{cases} 1 & \text{if } i_1 = j_1, \ldots, i_k = j_k, \\ 0 & \text{otherwise.} \end{cases}$$

The corresponding norm on $\bigwedge_k \mathbb{R}^d$ will be denoted by $\|\cdot\|$. Similarly the dual space $\bigwedge^k \mathbb{R}^d$ gets a Euclidean structure. Here besides the dual norm $\|\cdot\|$, we define another norm (called *comass*)

$$\|\alpha\| = \sup\{|\langle \xi, \alpha \rangle| : \xi \in \bigwedge_k \mathbb{R}^d \text{ simple }, |\xi| = 1\}, \quad \alpha \in \bigwedge^k \mathbb{R}^d \,.$$

Direct calculations lead to the following formula for the scalar product of simple k-vectors:

$$(u_1 \wedge \cdots \wedge u_k) \bullet (v_1 \wedge \cdots \wedge v_k) = \det \left(u_i \cdot v_j\right)_{i,j=1}^k \,. \tag{1.11}$$

From this one can derive that for $v_1 \wedge \cdots \wedge v_k \ne 0$ the norm $|v_1 \wedge \cdots \wedge v_k|$ is equal to the k-volume of the convex hull of the vectors $v_1, \ldots, v_k$, i.e., of the spanned parallelepiped. Because of this and Proposition 1.28 a nonzero simple k-vector in $\bigwedge_k \mathbb{R}^d$ has the *geometric interpretation of an equivalence class of oriented parallelepipeds with the same k-dimensional spanned vector space in* $\mathbb{R}^d$ *and the same k-volume*. This plays a role in applications to geometric integration theory.

Recall that in the general Euclidean setting the dual element v^* of an arbitrary vector v is defined by the relation $v^*(u) = u \cdot v$ for all vectors u, where $\cdot$ means the corresponding scalar product.

Definition 1.29 For any $0 \le k \le d$, the *Hodge star operator* is the linear mapping $\star : \bigwedge_k \mathbb{R}^d \to \bigwedge_{d-k} \mathbb{R}^d$ given by

$$\star : \xi \mapsto (\xi \,\lrcorner\, (e_1^* \wedge \cdots \wedge e_d^*))^*.$$

Note that $\star\xi$ is simple whenever ξ is, and that

$$\langle \star\xi \wedge \xi, e_1^* \wedge \cdots \wedge e_d^* \rangle = |\xi|^2.$$

1.2.2 The Grassmannian

Let $G(d,k)$ denote the set of all k-dimensional linear subspaces of $\mathbb{R}^d$. Due to Proposition 1.28, via the mapping $0 \neq \xi = v_1 \wedge \cdots \wedge v_k \mapsto L(\xi)$ and the orientation of $(v_1, \ldots, v_k)$ we can identify

$$G_0(d,k) := \{\xi \in \textstyle\bigwedge_k \mathbb{R}^d \text{ simple}, |\xi| = 1\}$$

with the set of *oriented* linear k-subspaces of $\mathbb{R}^d$. In [Fed69, §3.2.8] it is shown that G_0(d,k) is a C^∞-submanifold of $\bigwedge_k \mathbb{R}^d$. The *Grassmannian* $G(d,k)$ is then obtained as a factorspace, by identifying k-vectors from $G_0(d,k)$ that differ by sign only:

$$G(d,k) = G_0(d,k)/\text{sgn}\,,$$

and the mapping $\ell : \xi \mapsto L(\xi)$ is the natural projection from $G_0(d,k)$ to $G(d,k)$. We have

$$\dim G_0(d,k) = \dim G(d,k) = k(d-k)$$

(see also Sect. 1.4). (These representations are equivalent to those from classical differential geometry and Lie groups, where $G_0(d,k)$ is identified with the factor group $\mathrm{SO}(d)/(\mathrm{SO}(k) \times \mathrm{SO}(d-k))$ for the special orthogonal groups and $G(d,k)$ with $\mathrm{O}(d)/(\mathrm{O}(k) \times \mathrm{O}(d-k))$ for the orthogonal groups.) The Hausdorff measure of $G_0(d,k)$ is equal to

$$2c_{d,k} := \mathcal{H}^{k(d-k)}(G_0(d,k)) = 2\pi^{\frac{k(d-k)}{2}} \prod_{j=1}^{k} \frac{\Gamma(j/2)}{\Gamma((d-j+1)/2)} \tag{1.12}$$

(see [Fed69, §3.2.28]). Since the Hausdorff measures in Euclidean spaces are motion invariant and $\ell^{-1}(L)$ has two elements for any $L \in G(d,k)$, it follows that

$$\nu_k^d(\cdot) := \left(c_{d,k}^{-1}\mathcal{H}^{k(d-k)} \lfloor_{G_0(d,k)}\right) \ell^{-1}(\cdot) \tag{1.13}$$

is the (unique) O(d)-invariant probability measure on $G(d,k)$. Here we identify an orthogonal mapping $g \in \mathrm{O}(d)$ with the orthogonal mapping $\bigwedge_k g$ on $G(d,k)$ in the above interpretation. Then ν_k^d corresponds to the Haar measure on $\mathrm{O}(d)/(\mathrm{O}(k) \times \mathrm{O}(d-k))$.

Lemma 1.30 *Given any $\xi \in G_0(d,k)$ and $u, v, w \in \mathbb{R}^d$, we have:*

$$(v \wedge \xi) \llcorner u^* = (u \cdot v)\,\xi \quad \text{if } u \perp L(\xi), \tag{1.14}$$

$$\xi \llcorner v^* = \xi \llcorner (p_{L(\xi)}v)^*, \tag{1.15}$$

$$L(\xi \llcorner v^*) = L(\xi) \cap v^\perp \text{ if } v \notin L(\xi)^\perp, \tag{1.16}$$

and

$$(\xi \llcorner v^*) \bullet (\xi \llcorner w^*) = (p_{L(\xi)}v) \cdot (p_{L(\xi)}w). \tag{1.17}$$

Proof Denote for brevity $L := L(\xi)$ (recall (1.4)). If $\psi \in \bigwedge^k \mathbb{R}^d$ and $0 \neq u \perp L$ then

$$\langle (v \wedge \xi) \llcorner u^*, \psi \rangle = \langle v \wedge \xi, u^* \wedge \psi \rangle = \langle u, v^* \rangle \langle \xi, \psi \rangle = (u \cdot v)\langle \xi, \psi \rangle$$

(we have used (1.3) in the second equality). This proves (1.14).

Further, denote $a := p_L v$. If $a = 0$ then it follows from the definition that $\xi \llcorner v^* = 0$ and, hence, (1.15) holds. If $a \neq 0$ we can write ξ in the form $\frac{a}{|a|} \wedge \zeta$ with some $\zeta \in G_0(d, k-1)$, $a, v \perp L$, and we get $\xi \llcorner v^* = \frac{a \cdot v}{|a|}\zeta = |p_L v|\zeta$ by (1.14), and both (1.15), (1.16) follow.

Finally, note that both sides of (1.17) depend linearly on both vectors v, w, and they wanish whenever one of these vectors is perpendicular to L. Thus, it is sufficient to verify (1.17) in the case when v, w are elements of a given orthonormal basis of L, which follows easily using (1.11). □

Definition 1.31 Let L, M be a k, l-dimensional subspace of $\mathbb{R}^d$, respectively, and let $\{a_1, \dots, a_d\}$, $\{b_1, \dots, b_d\}$ be two orthonormal bases of $\mathbb{R}^d$ such that $L = \mathrm{Lin}\{a_1, \dots, a_k\}$ and $M = \mathrm{Lin}\{b_1, \dots, b_l\}$. We define

$$[L, M] := \left| \bigwedge_{i=k+1}^{d} a_i \wedge \bigwedge_{j=l+1}^{d} b_j \right|,$$

where the norm on the right hand side is understood in the space of $(2d-k-l)$-vectors.

Note that $[L, M]$ vanishes unless $L + M$ span the whole $\mathbb{R}^d$. If this is the case, $[L, M]$ equals $J(L, M^\perp)$ from Example 1.22. Geometrically, $[L, M]$ is the volume of the orthogonal projection to $M^\perp$ of a unit $(d-l)$-cube in $L \cap (L \cap M)^\perp$.

The following integral equality will be useful later. For any $0 \le k \le d$,

$$\int_{G(d,k)} [L, M]^2\, \nu_k^d(\mathrm{d}L) = \binom{d}{k}^{-1}, \quad M \in G(d, d-k). \tag{1.18}$$

To see this, note first that the integral clearly does not depend on $M \in G(d, d-k)$ since ν_k^d is rotation invariant. Further, note that $[L(\xi), L(\zeta)^\perp]^2 = (\xi \bullet \zeta)^2$ whenever $\xi, \zeta \in G_0(d, k)$, and that the Parseval's identity in $\bigwedge_k \mathbb{R}^d$ gives

$$\sum_{1 \le i_1 < \cdots < i_k \le d} (\xi \bullet (e_{i_1} \wedge \cdots \wedge e_{i_k}))^2 = 1, \quad \xi \in G_0(d,k),\ |\xi| = 1.$$

Integrating this identity with respect to $(2c_{dk})^{-1}\mathcal{H}^{k(d-k)}(\mathrm{d}\xi)$ and using (1.13), we obtain the result.

1.2.3 Multivector Fields and Differential Forms

A *k-vector field* on $A \subset \mathbb{R}^d$ is a mapping $\xi : A \to \bigwedge_k \mathbb{R}^d$.

A *differential form of degree k* (or briefly a *k-form*) on an open $U \subset \mathbb{R}^d$ is a C^∞-differentiable mapping $\phi : U \to \bigwedge^k \mathbb{R}^d$ (we consider the Euclidean structure on $\bigwedge_k \mathbb{R}^d$). The *support* of a k-form is the closed set

$$\operatorname{spt} \phi := U \setminus \bigcup \{O \subset U : O \text{ open}, \phi = 0 \text{ on } O\}.$$

Example 1.32

1.

$$\mathrm{d}x_i := e_i^*, \quad i = 1, \ldots, d,$$

are the *canonical* constant 1-*forms* on $\mathbb{R}^d$. (They agree with the ordinary differentials of the coordinate functions $x = x_1e_1 + \cdots + x_de_d \mapsto x_i$.)

$$\Omega_d := \mathrm{d}x_1 \wedge \cdots \wedge \mathrm{d}x_d$$

is called *volume form*.

2. If $f_1, \cdots, f_d$ are C^∞-functions then

$$\phi(x) = f_1(x)\mathrm{d}x_1 + \cdots + f_d(x)\mathrm{d}x_d$$

is a 1-form on $\mathbb{R}^d$.
3. A k-form can always be expressed in the form

$$\phi(x) = \sum_{1\le i_1<\cdots<i_k\le d} \phi_{i_1,\dots,i_k}(x)\,\mathrm{d}x_{i_1} \wedge \cdots \wedge \mathrm{d}x_{i_k}$$

with C^∞-functions $\phi_{i_1,\dots,i_k}$ (called coefficients of ϕ).

The *pull-back* of a k-form ϕ on $U \subset \mathbb{R}^d$ by a differentiable mapping $f : G \subset \mathbb{R}^n \to U$ is the k-form $f^\#\phi$ on G defined by $(f^\#\phi)(x) := \bigwedge^k(Df(x))\phi(f(x))$, $x \in G$ (cf. (1.5)). It satisfies

$$\langle \xi(x), f^\#\phi(x)\rangle = \langle (\bigwedge_k Df(x))\xi(x), \phi(f(x))\rangle$$

for any k-vector field ξ on G or on a subset of G (cf. (1.7) and (1.8)). Applying (1.6) pointwise we get

$$f^\#(\phi \wedge \psi) = f^\#\phi \wedge f^\#\psi\,. \tag{1.19}$$

The *exterior derivative* $d\phi$ od a k-form ϕ on U is a $(k+1)$-form on U defined by

$$\langle v_1 \wedge \cdots \wedge v_{k+1}, d\phi(x)\rangle = \sum_{i=1}^{k+1}(-1)^{i-1}\langle v^{(i)}, D\phi(x)v_i\rangle,$$

where $v^{(i)} = v_1 \wedge \cdots \wedge v_{i-1} \wedge v_{i+1} \wedge \cdots \wedge v_{k+1}$ and $D\phi$ is the standard differential of ϕ (obtained by differentiating the coefficients). Using coordinates, one can write

$$d\phi(x) = \sum_{i=1}^{d} \mathrm{d}x_i \wedge \frac{\partial}{\partial x_i}\phi(x).$$

The differential operator d is additive and satisfies

$$d(\phi \wedge \psi) = d\phi \wedge d\psi + (-1)^k\phi \wedge d\psi$$

for k-forms ϕ and l-forms ψ. Furthermore, $d^2 = 0$.

1.3 Currents

1.3.1 Basic Notions

Let U be an open set in $\mathbb{R}^d$ and let $\mathcal{D}^k(U)$ denote the linear space of all k-forms on U with compact support. $\mathcal{D}^k(U)$ is endowed with a topology induced by the seminorms

$$\nu_K^i(\phi) := \sup\{\|D^j\phi(x)\| : 0 \le j \le i,\ x \in K\},$$

where $i = 1, 2, \ldots,$ and $K \subset U$ is compact. (Here $D^j\phi(x)$ stands for the jth differential of ϕ at x, which is a symmetric j-linear mapping from $(\mathbb{R}^d)^j$ to $\bigwedge^k \mathbb{R}^d$, and its norm is understood in the usual functional sense.) The dual space to $\mathcal{D}^k(U)$ is denoted by $\mathcal{D}_k(U)$ and its elements are called *k-dimensional currents* on U. The support of a current $T \in \mathcal{D}_k(U)$ is defined as

$$\operatorname{spt} T = U \setminus \bigcup\{V \subset U : V \text{ open}, T(\phi) = 0 \text{ whenever } \operatorname{spt}\phi \subset V\}.$$

Example 1.33 Let $A \subset U$ be $\mathcal{H}^k$-measurable and locally $\mathcal{H}^k$-rectifiable. We say that $\xi : A \to \bigwedge_k \mathbb{R}^d$ is its *unit orienting k-vector field* if ξ is $\mathcal{H}^k$-measurable and for $\mathcal{H}^k$-almost all $a \in A$, $\xi(a)$ is a unit simple k-vector with associated k-subspace $\operatorname{Tan}^k(A, a)$. Then the mapping

$$T : \phi \mapsto \int_A \langle \xi(a), \phi(a)\rangle\, \mathcal{H}^k(da), \quad \phi \in \mathcal{D}^k(U),$$

is a k-dimensional current in U and is denoted by

$$T = (\mathcal{H}^k \llcorner A) \wedge \xi.$$

This notion will be generalized by allowing ξ to carry multiplicities, see Definition 1.43.

To any current $T \in \mathcal{D}^k(U)$ we attach a functional $\|T\|$ on the space of continuous non-negative functions f on U with compact support

$$\|T\|(f) = \sup\{T(\phi) : \phi \in \mathcal{D}^k(U), \|\phi(x)\| \le f(x) \text{ for all } x \in U\}. \tag{1.20}$$

Definition 1.34 A current $T \in \mathcal{D}_k(U)$ is called *representable by integration* if $\|T\|(f) < \infty$ for any nonnegative continuous function f on U with compact support.

The notion is justified by the following result.

Theorem 1.35 ([Fed69, §4.1.5]) *If a current $T \in \mathcal{D}_k(U)$ is representable by integration then $\|T\|$ defines a Radon measure on U (which we denote by the same symbol) and there exists a $\|T\|$-measurable k-vector field $\xi : U \to \bigwedge_k \mathbb{R}^d$ such that $|\xi| = 1$, $\|T\|$-a.e., and*

$$T(\phi) = \int_U \langle \xi(x), \phi(x) \rangle \, \|T\|(\mathrm{d}x), \quad \phi \in \mathcal{D}^k(U). \tag{1.21}$$

The assertion of the theorem can be reversed in the following sense (see [Fed69, §4.1.5, §4.1.7]):

Proposition 1.36 *If μ is a Radon measure in an open set $U \subset \mathbb{R}^d$ and $\xi : U \mapsto \bigwedge_k \mathbb{R}^d$ is a μ-measurable unit k-vector field then the current $T := \mu \wedge \xi$ defined by*

$$T(\phi) = \int \langle \xi(x), \phi(x) \rangle \, \mu(\mathrm{d}x), \quad \phi \in \mathcal{D}^k(U),$$

is representable by integration and $\|T\| = \mu$.

Remark 1.37

(i) Currents from Example 1.33 are representable by integration.
(ii) If $T \in \mathcal{D}_k(U)$ is representable by integration then $T(\phi)$ is defined whenever $\phi : U \to \bigwedge^k \mathbb{R}^d$ is $\|T\|$-measurable and $\int \|\phi\| \, \mathrm{d}\|T\| < \infty$ (i.e., ϕ has to be neither smooth, nor compactly supported).

Let $T \in \mathcal{D}_k(U)$ be a k-dimensional current and $\phi \in \mathcal{D}^m(U)$ an m-form, $0 \le m \le k$. Then $T \llcorner \phi$ is a $(k-m)$-dimensional current on U defined by

$$(T \llcorner \phi)(\psi) = T(\phi \wedge \psi), \quad \psi \in \mathcal{D}^{k-m}(U). \tag{1.22}$$

If ξ is a C^∞ m-vector field on U, we define the $(k+m)$-dimensional current $T \wedge \xi$ on U by

$$(T \wedge \xi)(\phi) = T(\xi \lrcorner \phi), \quad \phi \in \mathcal{D}^{k+m}(U).$$

If T is representable by integration then $T \wedge \xi$ and $T \llcorner \phi$ are defined even when ξ and ϕ are merely $\|T\|$-integrable (not smooth).

Definition 1.38 If T is a k-dimensional current on U ($k \ge 1$), its *boundary* ∂T is a $(k-1)$-dimensional current on U given by

$$\partial T(\phi) = T(d\phi), \quad \phi \in \mathcal{D}^k(U).$$

If $\partial T = 0$ we call T a *cycle*.

Example 1.39 (Gauss-Green formula) Let $E^d = \mathcal{L}^d \wedge (e_1 \wedge \cdots \wedge e_d)$ denote the d-dimensional current in $\mathbb{R}^d$ defined by Lebesgue integration, with the canonical orientation of $\mathbb{R}^d$. If U is an open subset of $\mathbb{R}^d$ then $T = E^d \llcorner \mathbf{1}_U$ is the d-dimensional current given by integration over U and its boundary fulfills $\operatorname{spt}(\partial T) \subset \partial U$. If, moreover, U is a C^1-domain (i.e., $\overline{U}$ is a d-dimensional C^1-submanifold with boundary) and $n(x)$ is the unit outer normal to U at $x \in \partial U$, then

$$\partial(E^d \llcorner \mathbf{1}_U) = (\partial U) \wedge \eta,$$

where $\eta(x) = \star n(x)$ (recall Definition 1.29; η is the unit simple $(d-1)$-vector field orienting ∂U with orientation given by $\langle \eta(x) \wedge n(x), \Omega_d \rangle = 1$).

Example 1.40 Let $\xi : \mathbb{R}^d \to \bigwedge_k \mathbb{R}^d$ be a C^1 smooth k-vector field in $\mathbb{R}^d$ $(k \geq 1)$ such that $\int_K |\xi| \, d\mathcal{L}^d < \infty$ for any compact set K. Then $\mathcal{L}^d \wedge \xi \in \mathcal{D}_k(\mathbb{R}^d)$ and

$$\partial(\mathcal{L}^d \wedge \xi) = -\mathcal{L}^d \wedge \operatorname{div} \xi,$$

where the divergence of ξ, $\operatorname{div} \xi : \mathbb{R}^d \to \bigwedge_{k-1} \mathbb{R}^d$, is defined through

$$\operatorname{div} \xi(x) = \sum_{i=1}^{d} \left(\frac{\partial}{\partial x_i} \xi(x) \llcorner \mathrm{d}x_i \right).$$

The following result is a weaker variant of the Stokes theorem.

Theorem 1.41 ([Fed69, §4.1.31]) *Let M be an oriented k-dimensional C^1-submanifold of $\mathbb{R}^d$ (with or without boundary ∂M), let ξ be the smooth orienting k-vector field over M, assume that $\mathcal{H}^k(M \cap K) < \infty$ for any $K \subset \mathbb{R}^d$ compact and set $T = (\mathcal{H}^k \llcorner M) \wedge \xi$. Then*

$$\operatorname{spt}(\partial T) \subset (\overline{M} \setminus M) \cup \partial M.$$

If, in particular, $M = \overline{M}$ and M has no boundary then $\partial T = 0$.

1.3.2 Normal and Rectifiable Currents

Given $U \subset \mathbb{R}^d$ open and a current $T \in \mathcal{D}_k(U)$, the *mass of* T is defined as

$$\mathbf{M}(T) = \|T\|(\mathbf{1}),$$

where $\|T\|$ is the functional defined in (1.20) and $\mathbf{1}$ denotes the constant function equal to 1. We will consider also the local variant

$$\mathbf{M}_K(T) = \|T\|(\mathbf{1}_K), \quad K \subset U \text{ compact}$$

($\mathbf{1}_K$ denotes the indicator function of K). We say that T has (locally) finite mass if $\mathbf{M}(T) < \infty$ ($\mathbf{M}_K(T) < \infty$ for all $K \subset U$ compact, respectively). It is not difficult to see that T has locally finite mass if and only if it is representable by integration (cf. Theorem 1.35).

Definition 1.42 A current $T \in \mathcal{D}_k(U)$ ($k \geq 1$) is *normal* if

$$\mathbf{N}(T) := \mathbf{M}(T) + \mathbf{M}(\partial T) < \infty.$$

T is *locally normal* if $\mathbf{N}_K(T) := \mathbf{M}_K(T) + \mathbf{M}_K(\partial T) < \infty$ for all $K \subset U$ compact. If $k = 0$ we set $\mathbf{N}(T) = \mathbf{M}(T)$ and $\mathbf{N}_K(T) = \mathbf{M}_K(T)$.

Definition 1.43 Let $U \subset \mathbb{R}^d$ be open and $k \in \{0, 1, \ldots, d\}$. A current $T \in \mathcal{D}_k(U)$ is called *locally k-rectifiable* if it can be represented in the form

$$T = (\mathcal{H}^k \llcorner A) \wedge \xi$$

for some $\mathcal{H}^k$-measurable and locally $\mathcal{H}^k$-rectifiable set $A \subset U$ and locally $\mathcal{H}^k$-integrable mapping $\xi : A \to \bigwedge_k \mathbb{R}^d$ such that, if $k \geq 1$ then, for $\mathcal{H}^k$-almost all $a \in A$, $\xi(a)$ is a simple k-vector associated with the k-subspace $\mathrm{Tan}^k(A, a)$. If, additionally, $|\xi(a)|$ is an integer for $\mathcal{H}^k$-almost all $a \in A$, we say that T is *integer-multiplicity locally k-rectifiable* (or only *integer-multiplicity locally rectifiable*). (Cf. Example 1.33).

Remark 1.44 The above definition of integer multiplicity rectifiable currents agrees with that in [KP08]. The notion of rectifiable current in [Fed69] is different.

1.3.3 Product of Currents

Let $U \subset \mathbb{R}^d$ and $V \subset \mathbb{R}^n$ be open sets, $\pi_U : (x, y) \mapsto x$, $\pi_V : (x, y) \mapsto y$ be the component projections and let $i_U : x \mapsto (x, 0)$, $i_V : y \mapsto (0, y)$ be the related immersions, $x \in U$, $y \in V$.

Definition 1.45 Given currents $S \in \mathcal{D}_k(U)$ and $T \in \mathcal{D}_l(V)$ ($0 \leq k \leq d, 0 \leq l \leq n$), we define the *Cartesian product* $S \times T \in \mathcal{D}_{k+l}(U \times V)$ of S and T as the unique current satisfying for any $\phi \in \mathcal{D}^i(U)$ and $\psi \in \mathcal{D}^{k+l-i}(V)$:

$$(S \times T)\left((\pi_U)^\# \phi \wedge (\pi_V)^\# \psi\right) = \begin{cases} S(\phi)T(\psi), & i = k, \\ 0, & i \neq k. \end{cases}$$

It can be shown that the differential forms $(\pi_U)^\# \phi \wedge (\pi_V)^\# \psi$ as above form a dense vector subspace of $\mathcal{D}^{k+l}(U \times V)$, hence, $S \times T$ is uniquely determined. The Cartesian product has the following properties (see [Fed69, §4.1.8]):

$$\operatorname{spt}(S \times T) = \operatorname{spt} S \times \operatorname{spt} T, \tag{1.23}$$

$$\partial(S \times T) = (\partial S) \times T + (-1)^k S \times \partial T \text{ if } k, l > 0, \tag{1.24}$$

$$r_\#(T \times S) = (-1)^{kl} S \times T, \tag{1.25}$$

where $r : (x, y) \mapsto (y, x)$. Further, if

$$S = (\mathcal{H}^k \llcorner A) \wedge \xi, \; T = (\mathcal{H}^l \llcorner B) \wedge \zeta$$

are locally rectifiable currents as in Definition 1.43 then

$$S \times T = \Big((\mathcal{H}^k \llcorner A) \otimes (\mathcal{H}^l \llcorner B) \Big) \wedge (\xi \owedge \zeta), \tag{1.26}$$

where

$$(\xi \owedge \zeta)(x, y) := \left(\textstyle\bigwedge_k i_U\right) \xi(x) \wedge \left(\textstyle\bigwedge_l i_V\right) \zeta(y).$$

If, moreover, either A is locally k-rectifiable, or B is locally l-rectifiable, then

$$(\mathcal{H}^k \llcorner A) \otimes (\mathcal{H}^l \llcorner B) = \mathcal{H}^{k+l} \llcorner (A \times B),$$

hence, T is again locally rectifiable in this case (see [Fed69, §3.2.3]).

1.3.4 Push-Forward

Let U, V be open subsets of $\mathbb{R}^d, \mathbb{R}^n$, respectively. The *push-forward* of a current $T \in \mathcal{D}_k(U)$ by a C^∞-mapping $F : U \to V$ is a k-dimensional current on V defined by

$$F_\# T(\phi) = T(F^\# \phi), \quad \phi \in \mathcal{D}^k(V), \tag{1.27}$$

provided that F is proper (i.e., the preimage of any compact set is compact.) The push-forward has the following important properties:

$$\operatorname{spt}(F_\# T) \subset F(\operatorname{spt} T), \tag{1.28}$$

$$\partial(F_\# T) = F_\#(\partial T), \tag{1.29}$$

$$F_\#(T \llcorner F^\# \psi) = (F_\# T) \llcorner \psi, \quad \psi \in \mathcal{D}_l(U). \tag{1.30}$$

Remark 1.46

1. The assumption of properness can be weakened in the sense that only the restriction $F|\operatorname{spt} T$ is proper. In that case, we take any compactly supported real valued function $\gamma \in C^\infty(U)$ which is equal to 1 on $\operatorname{spt} T \cap \operatorname{spt} F^\# \phi$, and set
$$F_\# T(\phi) = T(\gamma \cdot F^\# \phi), \quad \phi \in \mathcal{D}^k(V).$$
This is independent of γ and consistent with the above definition. So, the properness assumption can be relaxed whenever T is compactly supported in U.
2. If T is representable by integration then the definition of push-forward makes sense if F is only C^1 and proper on U. Indeed, we can define
$$(F_\# T)(\phi) = \int \langle (\textstyle\bigwedge_k DF(x))\xi(x), \phi \circ F(x)\rangle \, \|T\|(\mathrm{d}x), \quad \phi \in \mathcal{D}^k(V), \tag{1.31}$$
where ξ is as in Theorem 1.35.

For rectifiable currents, the push-forward can be defined even for locally Lipschitz proper maps. This needs a corresponding extension of the pull-back.

Definition 1.47 Let $W \subset \mathbb{R}^d$ be $\mathcal{H}^k$-measurable and locally k-rectifiable, let $f : W \to \mathbb{R}^n$ be locally Lipschitz, and let $\phi \in \mathcal{D}^m(\mathbb{R}^n)$ $(m \leq k)$. We define the pull-back $\operatorname{ap} f^\# \phi$ of the m-form ϕ to W so that for $\mathcal{H}^k$-almost all $x \in W$ and any $\eta \in \bigwedge_m \operatorname{Tan}^k(W, x)$,
$$\langle \eta, \operatorname{ap} f^\# \phi(x)\rangle = \langle (\textstyle\bigwedge_m \operatorname{ap} Df(x))\eta, \phi(f(x))\rangle.$$

The existence of the approximate differential is guaranteed by Proposition 1.19.

Clearly, if f can be extended to a smooth function F on an open set $U \supset W$, we have
$$\operatorname{ap} f^\# \phi = (F^\# \phi)|W \quad \mathcal{H}^k - \text{a.e.}$$

Now, we can extend the definition of push-forward.

Definition 1.48 Let $T = (\mathcal{H}^k \llcorner W) \wedge \xi$ be a locally k-rectifiable current (see Definition 1.43) and let $f : \overline{W} \to \mathbb{R}^n$ be locally Lipschitz. First, we define the Borel measure $\|f_\# T\|$ on $\mathbb{R}^n$ by
$$\|f_\# T\|(\cdot) := \int_{f^{-1}(\cdot)} \left|(\textstyle\bigwedge_k \operatorname{ap} Df)\xi\right| \, \mathrm{d}\mathcal{H}^k.$$

Provided that $\|f_\# T\|$ is a Radon measure, we define the push-forward $f_\# T \in \mathcal{D}_k(\mathbb{R}^n)$ as the k-dimensional current
$$f_\# T(\phi) = T(\operatorname{ap} f^\# \phi), \quad \phi \in \mathcal{D}^k(\mathbb{R}^n).$$

Of course, $\| f_\# T \|$ is then the variation measure associated with $f_\# T$, so that our notation is consistent.

Properties (1.28) and (1.30) (where $f^\#$ is replaced with ap $f^\#$) remain true. Property (1.29), however, need not hold without properness assumption (cf. [Fed69, p. 359]). It should be emphasized that in both [Fed69] and [KP08], the pushforward is defined only for mapping which are proper on the support of the current.

The Area formula below shows that $f_\# T$ is again locally rectifiable. We refer to [Fed69, §4.1.30], where it is formulated with the additional assumption that W is relatively compact, which can be avoided weakened here, due to our more general notion of push-forward without properness condition.

Theorem 1.49 (Area formula for rectifiable currents) *Let $T = (\mathcal{H}^k \llcorner W) \wedge \xi$ be a locally k-rectifiable current in an open set $U \subset \mathbb{R}^d$, let $f : W \to \mathbb{R}^n$ be locally Lipschitz and assume that $\| f_\# T \|$ is a Radon measure. Then*

$$f_\# T = (\mathcal{H}^k \llcorner f(W)) \wedge \eta,$$

where η is a locally $\mathcal{H}^k$-integrable k-vector field on $f(W)$ fulfilling for $\mathcal{H}^k$-almost all $y \in f(W)$,

$$\eta(y) = \sum_{x \in f^{-1}\{y\}} \frac{(\bigwedge_k \mathrm{ap}\, Df(x))\xi(x)}{\mathrm{ap}\, J_k f(x)} = \sum_{x \in f^{-1}\{y\}} \frac{(\bigwedge_k \mathrm{ap}\, Df(x))\xi(x)}{|(\bigwedge_k \mathrm{ap}\, Df(x))\xi(x)|} |\xi(x)|.$$

Example 1.50 ([Fed69, §4.1.26]) Recall that $E^d = \mathcal{L}^d \wedge (e_1 \wedge \cdots \wedge e_d)$ denotes the d-current in $\mathbb{R}^d$ given by Lebesgue integration (cf. Example 1.39). Let $A \subset \mathbb{R}^d$ be bounded and Lebesgue measurable and let $f : \mathbb{R}^d \to \mathbb{R}^d$ be Lipschitz. Then

$$f_\# (E^d \llcorner A) = E^d \llcorner (\deg f|A),$$

where, for $\mathcal{L}^d$-a.a. $y \in \mathbb{R}^d$,

$$(\deg f|A)(y) = \sum_{x \in A \cap f^{-1}\{y\}} \mathrm{sgn}\, \det Df(x).$$

The degree of $f|A$ – $\deg f|A$ – is zero outside of $f(A)$ and constant on any connected component of $\mathbb{R}^d \setminus f(\partial A)$ (this follows from the Constancy theorem, Theorem 1.51).

1.3.5 *Constancy Theorems*

The (classical) Constancy theorem for currents says in principle that a k-dimensional current without boundary in a k-dimensional linear space equals a multiple of the canonical current given by integration w.r.t. k-dimensional Lebesgue measure:

Theorem 1.51 (Constancy theorem, [Fed69, §4.1.7]) *Let T be a d-dimensional current in an open set $U \subset \mathbb{R}^d$, let G be a connected open subset of U and assume that* $\operatorname{spt} \partial T \subset U \setminus G$. *Then there exists a real number c such that*

$$\operatorname{spt}(T - c(E^d \llcorner U)) \subset U \setminus G.$$

A generalized version is valid for locally k-rectifiable currents without boundary living in a k-dimensional oriented connected C^1-manifold, see [Fed69, §4.1.31]. We will formulate and prove a version for oriented Lipschitz manifolds, the proof being similar to that for C^1-manifolds.

Definition 1.52 Let $M \subset \mathbb{R}^d$ be a k-dimensional Lipschitz submanifold (cf. Definition 1.11). We say that an atlas (f_i) of M *orients* M if for any i, j,

$$\det\left(D(f_j \circ f_i^{-1})(f_i(x))\right) > 0 \quad \text{for } \mathcal{H}^k - \text{a.a } x \in \operatorname{dom} f_i \cap \operatorname{dom} f_j$$

whenever $\operatorname{dom} f_i \cap \operatorname{dom} f_j \neq \emptyset$. The pair $(M, (f_i))$ is then called an *oriented* Lipschitz submanifold. The unit simple k-vector field ξ on M determined for a.e. $x \in \operatorname{dom} f_i$ by

$$\xi(x) := \frac{\bigwedge_k Df_i^{-1}(f_i(x))(e_1 \wedge \ldots \wedge e_k)}{\left|\bigwedge_k Df_i^{-1}(f_i(x))(e_1 \wedge \ldots \wedge e_k)\right|}$$

orients M in the usual way ($(e_1, \ldots, e_k)$ is the canonical basis of $\mathbb{R}^k$).

(Note that $Df_i^{-1}(f_i(x))$ exists for $\mathcal{H}^k$-a.a. $x \in \operatorname{dom} f_i$ by Rademacher's theorem (Theorem 1.8) and Proposition 1.10, hence, ξ is well defined $\mathcal{H}^k$-a.e. on M.)

Theorem 1.53 (Constancy theorem for Lipschitz submanifolds) *Suppose M is a k-dimensional connected Lipschitz submanifold of an open set $U \subset \mathbb{R}^d$ with orienting k-vector field ξ. Then for any integer-multiplicity locally k-rectifiable current T in U with*

$$\operatorname{spt} T \setminus M \text{ closed relative to } U, \quad \operatorname{spt} \partial T \subset U \setminus M,$$

there exists an integer r such that

$$\operatorname{spt}(T - r(\mathcal{H}^k \llcorner M) \wedge \xi) \subset U \setminus M.$$

Corollary 1.54 *Let $T \in \mathcal{D}_k(U)$ be integer-multiplicity locally k-rectifiable and let M be a connected k-dimensional Lipschitz submanifold of U with orienting k-vector field ξ such that* $\operatorname{spt} T \subset \overline{M}$. *If*

$$\operatorname{spt} \partial T \subset \overline{M} \setminus M$$

(in particular, if T is a cycle), then

$$T = r(\mathcal{H}^k \llcorner M) \wedge \xi$$

for some $r \in \mathbb{Z}$.

Proof (of Theorem 1.53*)* We present a proof which uses the idea of [Fed69, §4.1.31]) for C^1-submanifolds.

Let $(f_i)_{i=1}^{\infty}$ be an atlas of bi-Lipschitz charts orienting M. Choose C^{∞}-functions $\gamma_i : \mathbb{R}^d \to [0, 1]$ with compact supports fulfilling $M \cap \operatorname{spt} \gamma_i \subset \operatorname{dom} f_i$ and such that the open sets $W_i := \operatorname{int}\{x : \gamma_i(x) = 1\}$ cover M (it is not difficult to see that such functions always exist).

Then, $T \llcorner \gamma_i \in \mathcal{D}_k(\mathbb{R}^d)$ is locally k-rectifiable, and

$$\operatorname{spt} \partial(T \llcorner \gamma_i) \subset M \setminus W_i.$$

Since f_i is bi-Lipschitz, it is proper and we can define the push-forward (cf. Sect. 1.3.4)

$$(f_i)_\#(T \llcorner \gamma_i) \in \mathcal{D}_k(\operatorname{im} f_i)$$

with $\operatorname{spt} \partial(f_i)_\#(T \llcorner \gamma_i) \subset \operatorname{im} f_i \setminus f_i(M \cap W_i)$. Applying Theorem 1.51, we get that for some $r_i \in \mathbb{R}$,

$$\operatorname{spt}((f_i)_\#(T \llcorner \gamma_i) - r_i(\mathbb{E}_k \llcorner \operatorname{im} f_i)) \subset \operatorname{im} f_i \setminus f_i(M \cap W_i)$$

(here $\mathbb{E}_k$ denotes the k-current $\mathcal{L}^k \wedge (e_1 \wedge \cdots \wedge e_k)$ in $\mathbb{R}^k$). By the area formula for currents, Theorem 1.49, and the choice of the orienting k-vector field ξ,

$$(f_i^{-1})_\#(\mathbb{E}_k \llcorner \operatorname{im} f_i) = (\mathcal{H}^k \llcorner \operatorname{dom} f_i) \wedge (\xi | \operatorname{dom} f_i).$$

Thus,

$$\operatorname{spt}((T \llcorner \gamma_i) - r_i(\mathcal{H}^k \llcorner M) \wedge \xi) \subset M \setminus W_i.$$

If $\mathcal{H}^k(M \cap W_i \cap W_j) > 0$ for two indices i, j then it follows that $r_i = r_j$. Thus, since M is connected and covered by the W_i's, we infer that all the r_i's equal a constant $r \in \mathbb{R}$ and $\operatorname{spt}(T - r(\mathcal{H}^k \llcorner M) \wedge \xi) \subset U \setminus M$. Since T has integer multiplicity, it is clear that r is an integer. □

1.3.6 Integral Currents and Compactness

Definition 1.55 A current $T \in \mathcal{D}_k(U)$ is called *locally integral* if both T and ∂T are integer-multiplicity locally rectifiable. (Of course, if $k = 0$ only the condition on T applies, since ∂T is not defined in this case.) Let $\mathbf{I}_k^{\mathrm{loc}}(U)$ denote the set of all locally integral k-currents in $U \subset \mathbb{R}^d$, and let

$$\mathbf{I}_k(U) = \{T \in \mathbf{I}_k^{\mathrm{loc}}(U) : \operatorname{spt} T \text{ is compact}\}$$

be the family of *integral k-currents.*

Definition 1.56 We say that a sequence (T_i) of currents from $\mathcal{D}_k(U)$ converges *weakly* to a current $T \in \mathcal{D}_k(U)$ if $\lim_{i\to\infty} T_i(\phi) = T(\phi)$ whenever $\phi \in \mathcal{D}^k(U)$.

Theorem 1.57 (Compactness Theorem, [KP08, Theorem 7.5.2]) *Let $(T_i) \subset \mathbf{I}_k^{\mathrm{loc}}(U)$ be a sequence of locally integral k-currents such that*

$$\sup_i \mathbf{N}_K(T_i) < \infty \textit{ for all } K \subset \mathbb{R}^d \textit{ compact.}$$

Then there exist a subsequence $(T_{i_j}) \subset (T_i)$ and a locally integral k-current $T \in \mathbf{I}_k^{\mathrm{loc}}(U)$ such that $T_{i_j} \to T$ weakly.

Definition 1.58 (Flat convergence) Let $U \subset \mathbb{R}^d$ be open and $T \in \mathcal{D}_k(U)$ a current representable by integration with compact support. We define

$$\mathbf{F}(T) = \sup\{T(\phi) : \phi \in \mathcal{D}^k(U), \|\phi(x)\| \le 1, \|d\phi(x)\| \le 1\}$$

and call it *flat norm* of T provided it is finite (cf. [Mor88, §4.5]). If $T_i, T \in \mathcal{D}_k(U)$ are currents representable by integration with compact supports, the *flat convergence* of T_i to T, denoted

$$(\mathrm{F}) \lim_{i\to\infty} T_i = T,$$

is defined by $\mathbf{F}(T_i - T) \to 0$, $i \to \infty$.

Remark 1.59 If $\operatorname{spt} T$ is compact, we can express the flat norm equivalently as

$$\mathbf{F}(T) = \inf\{\mathbf{M}(T - \partial S) + \mathbf{M}(S) : S \in \mathcal{D}_{k+1}(U), \operatorname{spt} S \text{ compact}\}. \tag{1.32}$$

In fact, if $\operatorname{spt} T \subset K \subset U$ and K is compact then $\mathbf{F}(T)$ agrees with the seminorm $\mathbf{F}_K(T)$ from [Fed69, §4.1.12]. Further, if $T_1, T_2 \in \mathcal{D}_k(U)$ are representable by integration and their supports are contained in a compact set $K \subset U$ then $d_K(T_1, T_2) = \mathbf{F}(T_1 - T_2)$, where d_K is the pseudometric defined in [KP08, Section 8.2].

Lemma 1.60 ([KP08, Theorem 8.2.1]) *Let T_i, T be locally integral k-currents with compact supports and such that* $\sup_i \mathbf{N}(T_i) < \infty$. *Then, the flat convergence of T_i to T is equivalent to the weak convergence $T_i \to T$.*

The **F**-distance of two push-forwards of a locally rectifiable current can be estimated as follows.

Lemma 1.61 ([Fed69, §4.1.13, §4.1.14]) *Let T be a k-dimensional locally integral current with compact support in an open set $U \subset \mathbb{R}^d$, and let f, g be Lipschitz proper mappings from U to $\mathbb{R}^d$. Then we have*

$$\mathbf{F}(g_\# T - f_\# T) \leq (\max\{\operatorname{Lip} f, \operatorname{Lip} g\})^k \int |g - f| \, \mathrm{d}\|T\| \\ +(\max\{\operatorname{Lip} f, \operatorname{Lip} g\})^{k-1} \int |g - f| \, \mathrm{d}\|\partial T\|$$

(the second summand vanishes if $k = 0$).

1.3.7 *Slicing of Currents*

We start with an implicit definition of slices of rectifiable currents. Recall Definition 1.47 of pull-back for Lipschitz maps.

Theorem 1.62 *Let T be a locally k-rectifiable current in $U \subset \mathbb{R}^d$ and let f :* $\operatorname{spt} T \to \mathbb{R}^n$ *be locally Lipschitz, $n \leq k$. Then, for $\mathcal{L}^n$-almost all $y \in \mathbb{R}^n$ there exists a locally $(k - n)$-rectifiable current*

$$\langle T, f, y \rangle \in \mathcal{D}_{k-n}(U)$$

so that $\operatorname{spt} \langle T, f, y \rangle \subset f^{-1}\{y\} \cap \operatorname{spt} T$ *and for any $g \in C^\infty(\mathbb{R}^n)$,*

$$T \llcorner \operatorname{ap} f^\#(g \wedge \Omega_n)(\cdot) = \int g(y) \langle T, f, y \rangle (\cdot) \, \mathcal{L}^n(\mathrm{d}y), \tag{1.33}$$

$$\mathbf{M}(T \llcorner \operatorname{ap} f^\#(g \wedge \Omega_n)) = \int |g(y)| \mathbf{M} \langle T, f, y \rangle \, \mathcal{L}^n(\mathrm{d}y). \tag{1.34}$$

Definition 1.63 The current $\langle T, f, y \rangle$ from the above theorem is called *slice of T through f at y.*

Remark 1.64 The definition of slicing can be generalized in a straightforward way for mappings $f : \operatorname{spt} T \to M$, where M is an n-dimensional C^1-submanifold of $\mathbb{R}^N$. In the formulas above, $\mathcal{L}^n$ is replaced with $\mathcal{H}^n|M$ and Ω_n with an n-volume form of M (determining the orientation of M).

The slices can be expressed explicitly as follows.

Theorem 1.65 ([Fed69, §4.3.8]) *Let T be a locally k-rectifiable current on $U \subset \mathbb{R}^d$ of the form*

$$T = (\mathcal{H}^k \llcorner W) \wedge \xi,$$

let $f : W \to \mathbb{R}^n$ be locally Lipschitz, $n \leq k$. Then, for $\mathcal{H}^n$-almost all y,

$$\langle T, f, y\rangle = (\mathcal{H}^{k-n} \llcorner f^{-1}\{y\}) \wedge \zeta,$$

where

$$\zeta(x) = \xi(x) \llcorner (\mathrm{ap}\, f^{\#}\Omega_n)/J_n f(x)$$

whenever $J_n f(x)$ is defined and positive, and 0 *otherwise.*

The next proposition lists some basic properties of slices that can be found in [Fed69, §4.3.1].

Proposition 1.66 *Le $T_i, T \in \mathbf{I}_k(U)$, $U \subset \mathbb{R}^d$ open, $f : U \to \mathbb{R}^n$ locally Lipschitz, $k \geq n$, and assume that* (F) $\lim_{i\to\infty} T_i = T$. *Then*

1. $\operatorname{spt}\langle T, f, y\rangle \subset f^{-1}\{y\} \cap \operatorname{spt} T$,
2. $\partial\langle T, f, y\rangle = (-1)^n\langle \partial T, f, y\rangle$ *if* $k > n$,
3. (F) $\lim_{i\to\infty}\langle T_i, f, y\rangle = \langle T, f, y\rangle$ *for $\mathcal{L}^n$-almost all $y \in \mathbb{R}^n$.*

1.3.8 Indecomposable Integral Currents

Definition 1.67 A current $T \in \mathbf{I}_k(\mathbb{R}^d)$ is called *indecomposable* if there exists no $S \in \mathbf{I}_k(U)$ with $S \neq 0 \neq T - S$ and $\mathbf{N}(T) = \mathbf{N}(S) + \mathbf{N}(T - S)$.

Proposition 1.68 ([Fed69, §4.2.25]) *For any $T \in \mathbf{I}_k(\mathbb{R}^d)$ there exists a sequence of indecomposable currents $T_i \in \mathbf{I}_k(\mathbb{R}^d)$ such that*

$$T = \sum_{i=1}^{\infty} T_i \text{ and } \mathbf{N}(T) = \sum_{i=1}^{\infty} \mathbf{N}(T_i).$$

The following result can be found in a paper by Hardt [Har77], see a reformulation of Theorem 1 on page 1 therein. In fact, the result is formulated for real functions, but an extension to vector-valued mappings is obvious.

Theorem 1.69 (Hardt [Har77]) *Let $T = (\mathcal{H}^k \llcorner W) \wedge \xi \in \mathbf{I}_k(\mathbb{R}^d)$ be indecomposable and let $h : \mathbb{R}^d \to \mathbb{R}^n$ be a C^1-mapping such that $\xi \llcorner Dh = 0$ $\mathcal{H}^k$-almost everywhere on W. Then $h|\operatorname{spt} T$ is constant.*

1.4 Computations on Grassmannians

In this section we derive several formulas which will be useful later.

For computations of differentials and Jacobians of functions on the Grassmannian, we will need to work with tangent vectors to $G_0(d, k)$. These can be described as follows. First, consider a unit simple multivector $\xi \in G_0(d, k)$ be given, let $L := L(\xi)$ be the associated k-subspace, and let $v \in L$ and $w \in L^\perp$ be unit vectors. Then,

$$(\xi \llcorner v^*) \wedge w \in \mathrm{Tan}(G_0(d, k), \xi)$$

(recall (1.10) for the definition of interior multiplication). To see this, consider the multivectors

$$\xi_i := (\xi \llcorner v^*) \wedge \left(i^{-1}w + \sqrt{1 - i^{-2}}v\right) \in G_0(d, k), \quad i = 1, 2, \ldots,$$

and note that $\xi = (\xi \llcorner v^*) \wedge v$ and

$$i(\xi_i - \xi) = (\xi \llcorner v^*) \wedge w + (\xi \llcorner v^*) \wedge i\left(\sqrt{1 - i^{-2}} - 1\right)v \to (\xi \llcorner v^*) \wedge w$$

as $i \to \infty$. Consequently, we can obtain on orthonormal basis of $\mathrm{Tan}(G_0(d, k), \xi)$ in the following way. Choose an orthonormal basis $\{u_1, \ldots, u_d\}$ of $\mathbb{R}^d$ such that $\xi = u_1 \wedge \cdots \wedge u_k$ and denote

$$\eta_{i,j} = u_1 \wedge \cdots \wedge u_{i-1} \wedge u_j \wedge u_{i+1} \wedge \cdots \wedge u_k \in \textstyle\bigwedge_k \mathbb{R}^d, \quad 1 \le i \le k,\ k+1 \le j \le d.$$

Then,

$$\{\eta_{i,j},\ 1 \le i \le k, k+1 \le j \le d\}$$

is an orthonormal basis of $\mathrm{Tan}(G_0(d, k), \xi)$. Note that the given basis vectors of $\mathrm{Tan}(G_0(d, k), \xi)$ are unit simple k-vectors (but, of course, not all tangent vectors are simple).

Lemma 1.70 *Let $v \in S^{d-1}$ be a fixed unit vector and consider the mapping*

$$g : \xi \mapsto p_{L(\xi)}v, \quad \xi \in G_0(d, k).$$

Fix a $\xi \in G_0(d, k)$ such that $v \notin L := L(\xi)$ and $v \notin L^\perp$. Then, the differential of g at ξ is given by

$$Dg(\xi)((\xi \llcorner u^*) \wedge w) = (w \cdot p_{L^\perp}v)u + (u \cdot p_L v)w,$$

$v \in L$, $w \in L^\perp$. The Jacobian of g is

$$J_{d-1}g(\xi) = \sin^{k-1} \angle(v, L) \cos^{d-1-k} \angle(v, L).$$

Proof Note that we can describe g equivalently by

$$g(\xi) = \xi \llcorner (v \lrcorner \xi^*),$$

hence, ξ is clearly differentiable. Let $u_i, \eta_{i,j}$ be as above and assume, moreover, that $u_1 = p_L v/|p_L v|$ and $u_d = p_{L^\perp} v/|p_{L^\perp} v|$. The differential of g fulfills

$$\begin{aligned}
Dg(\xi)(\eta_{1,j}) &= |p_L v| u_j, && k < j < d,\\
Dg(\xi)(\eta_{i,d}) &= |p_{L^\perp} v| u_i, && 1 < i \le k,\\
Dg(\xi)(\eta_{1,d}) &= |p_{L^\perp} v| u_1 + |p_L v| u_d, &&\\
Dg(\xi)(\eta_{i,j}) &= 0, && 1 < i \le k, k < j < d.
\end{aligned}$$

Hence, the form for the differential follows by linearity and the Jacobian is

$$\begin{aligned}
J_{d-1}g(\xi) &= \left| \bigwedge_{i=2}^{k} |p_{L^\perp} v| u_i \wedge \bigwedge_{i=k+1}^{d-1} |p_L v| u_i \wedge (|p_{L^\perp} v| u_1 + |p_L v| u_d) \right|\\
&= |p_{L^\perp} v|^{k-1} |p_L v|^{d-1-k}\\
&= \sin^{k-1} \angle(v, L) \cos^{d-1-k} \angle(v, L).
\end{aligned}$$

□

In the sequel the normalized orthogonal projection onto a linear subspace L of $\mathbb{R}^d$ will be denoted by π_L , i.e.,

$$\pi_L w := \frac{p_L w}{|p_L w|}, \quad w \in \mathbb{R}^d.$$

Lemma 1.71 *Let the mapping h be given as*

$$h(\xi) := \pi_{L(\xi)} v, \quad \xi \in G_0(d, k), L(\xi) \not\perp v$$

(as above, v is a fixed unit vector). Then its Jacobian equals

$$J_{k-1}h(\xi) = \tan^{k-1} \angle(v, L(\xi)).$$

Proof We will use the same basis of the tangent space as in the proof of Lemma 1.70. Now we have

$$Dh(\xi)\eta = p_{g(\xi)^\perp} \left(\frac{Dg(\xi)\eta}{|p_L v|} \right).$$

Hence,

$$\begin{aligned}
Dg(\xi)(\eta_{1,j}) &= u_j, && k < j < d,\\
Dg(\xi)(\eta_{i,d}) &= (\tan\angle(v, L))u_i, && 1 < i \le k,\\
Dg(\xi)(\eta_{1,d}) &= u_d,\\
Dg(\xi)(\eta_{i,j}) &= 0, && 1 < i \le k, k < j < d,
\end{aligned}$$

and the assertion follows. □

Lemma 1.72 *Given $v \in S^{d-1}$ fixed, consider the mapping*

$$q : \xi \mapsto \frac{\xi \llcorner v^*}{|\xi \llcorner v^*|}, \quad \xi \in G_0(d,k), L(\xi) \not\perp v$$

(Note that $L(q(\xi)) = L(\xi) \cap v^\perp$, see (1.15).*) Then,*

$$J_{(k-1)(d-k)}q(\xi) = \cos\angle(v, L(\xi))^{-(k-1)}.$$

Proof We have

$$Dq(\xi)\eta = p_{q(\xi)^\perp}\frac{\eta \llcorner v^*}{|\xi \llcorner v^*|}.$$

Using the same basis vectors as in the previous proofs, and the fact that $|\xi \llcorner v^*| = |p_L v|$, we obtain

$$\begin{aligned}
Dq(\xi)(\eta_{1,j}) &= 0, && k < j \le d,\\
Dg(\xi)(\eta_{i,j}) &= \pm(u_2 \wedge \cdots \wedge u_{i-1} \wedge u_j \wedge u_{i+1} \wedge \cdots \wedge u_k), && 1 < i \le k < j < d,\\
Dg(\xi)(\eta_{i,d}) &= \pm|p_L v|^{-1}(v' \wedge u_2 \wedge \cdots \wedge u_k), && 1 < i \le k,
\end{aligned}$$

where v' is a unit vector in Lin $\{u_1, u_d\}$ perpendicular to v, and the assertion follows. □

In the following considerations, we will consider the special orthogonal group SO(d) in $\mathbb{R}^d$, which is known to be a smooth manifold of dimension $\frac{d(d-1)}{2}$ and we consider it to be embedded into the Euclidean space $\mathcal{L}(\mathbb{R}^d, \mathbb{R}^d)$ of linear mappings from $\mathbb{R}^d$ to $\mathbb{R}^d$ (see [Fed69, §3.2.28]). We construct a particular mapping from the sphere to SO(d) which will be useful later.

Fix a unit vector $e \in \mathbb{R}^d$ and consider the mapping

$$F_e : S^{d-1} \setminus \{-e\} \to \mathrm{SO}(d)$$

defined as follows. Given a unit vector $u \neq \pm e$, $F_e(u)$ is the orthogonal mapping acting as the rotation in the plane Lin (e, u) sending e to u, and leaving the orthogonal complement to this plane unchanged. Formally, we can write

$$F_e(u)z = \frac{-(z\cdot u)+(e\cdot u)(z\cdot e)}{1+(e\cdot u)}(e+u)+(z\cdot e)(u-e)+z, \quad z\in\mathbb{R}^d. \tag{1.35}$$

Clearly, F_e is differentiable, and its differential can be computed as follows. For $z\in\mathbb{R}^d$ fixed and direction $v\perp u$, we have

$$\begin{aligned} D[F_e(u)z](v) = {} & \frac{-(z\cdot v)(1+e\cdot u)+(v\cdot e)(z\cdot(e+u))}{(1+e\cdot u)^2}(e+u) \\ & + \frac{-(z\cdot u)+(z\cdot e)(1+2e\cdot u)}{1+e\cdot u}v. \end{aligned} \tag{1.36}$$

Lemma 1.73 *Let $W\subset(\mathbb{R}^d\times S^{d-1})^2$ satisfy $\mathcal{H}^p(W)=0$. Then $\mathcal{H}^q(A)=0$, where*

$$A=\{(u_1,v_1,u_2,v_2,\rho)\in W\times \mathrm{SO}(d):\ v_1+\rho v_2=0\}$$

and $q=p+\frac{(d-1)(d-2)}{2}$.

Proof We will use the locally Lipschitz mapping F_e from (1.35) and the following two easy facts:

1. The mapping $\rho\mapsto\rho^{-1}$ is an isometry on $\mathrm{SO}(d)$.
2. Given a fixed rotation $\sigma\in\mathrm{SO}(d)$, the mappings $\rho\mapsto\rho\circ\sigma$ and $\rho\mapsto\sigma\circ\rho$ are isometries on $\mathrm{SO}(d)$.

Given $e\in S^{d-1}$, we fix some isometry $\psi: e^\perp\to\mathbb{R}^{d-1}$. and, to any $\sigma\in\mathrm{SO}(d-1)$ we attach $\sigma^e\in\mathrm{SO}(d)$ which leaves e fixed and acts as $\sigma\circ\psi$ on $e^\perp$. From the above statements, we see that

$$(u_1,v_1,u_2,v_2,\sigma)\mapsto\left(u_1,v_1,u_2,v_2,F_e(-v_1)\circ\sigma^e\circ(F_e(v_2))^{-1}\right)$$

is a locally Lipschitz mapping from $(W\cap Q_e)\times\mathrm{SO}(d-1)$ onto $A\cap(Q_e\times\mathrm{SO}(d))$, where

$$Q_e=\{(u_1,v_1,u_2,v_2):\ v_1\neq -e,\ v_2\neq -e\}.$$

Since $\mathcal{H}^p(W)=0$ and $\mathrm{SO}(d-1)$ is a smooth manifold of dimension $\frac{(d-1)(d-2)}{2}$, we have (see [Fed69, §2.10.45]

$$\mathcal{H}^q(W\times\mathrm{SO}(d-1))=0.$$

Hence, also $\mathcal{H}^q(A\cap(Q_e\times\mathrm{SO}(d)))=0$. Choosing three different unit vectors e, we cover the whole set A and get $\mathcal{H}^q(A)=0$. □

In the following result, we associate with a locally rectifiable "unit normal bundle" $W \subset \mathbb{R}^d \times S^{d-1}$ its "m-flag space" by adding all m-subspaces orthogonal to the normal vectors, show its rectifiability and describe the approximate tangent vectors. This will be needed later for some integral-geometric formulas.

Lemma 1.74 *Let W be a locally $\mathcal{H}^k$-rectifiable Borel subset of $\mathbb{R}^d \times S^{d-1}$, $0 \leq m \leq d-1-k$, and consider the set*

$$X = \{(x, n, \zeta) : (x, n) \in W, \zeta \in G_0^{n^\perp}(d-1, m)\}$$

($G_0^{n^\perp}(d-1, m)$ denotes the space of unit simple m-vectors in the $(d-1)$-dimensional subspace $n^\perp$). Then, X is locally $\mathcal{H}^p$-rectifiable with

$$p = k + m(d-1-m)$$

and for any $(x, n, \zeta) \in X$, the approximate tangent space $\mathrm{Tan}^p(X, (x, n, \zeta)$ *is spanned by the multivectors*

$$(0, 0, \eta), \quad \eta \in \mathrm{Tan}(G_0^{n^\perp}(d-1, m)),$$

and

$$\big(u, v, n \wedge (\zeta \llcorner v^*)\big), \quad (u, v) \in \mathrm{Tan}^k(W, (x, n)).$$

Proof Let F_e be the mapping from (1.35) ($e \in S^{d-1}$) and consider the mapping

$$\Phi_e : (x, n, \xi) \mapsto \Big(x, n, \big(\textstyle\bigwedge_k F_e(n)\big)\xi\Big)$$

defined on $(W \cap (\mathbb{R}^d \times (S^{d-1} \setminus \{-e\})) \times G_0^{e^\perp}(d-1, m)$. It is clear that, since F_e is locally Lipschitz, Φ_e is locally Lipschitz as well. Further, since W is locally $\mathcal{H}^k$-rectifiable and $G_0^{e^\perp}(d-1, m)$ is a smooth manifold of dimension $m(d-1-m)$, the product $(W \cap (\mathbb{R}^d \times (S^{d-1} \setminus \{-e\})) \times G_0^{e^\perp}(d-1, m)$ is locally p-rectifiable (see [Fed69, §3.2.23]) and its locally Lipschitz image

$$\mathrm{im}\, \Phi_e = \{(x, n, \zeta) \in X : n \neq e\}$$

retains the same property. Repeating the same construction for another unit vector e, we can cover the whole set X.

Let now $(x, n, \zeta) \in X$ be given and let V be the m-subspace associated with ζ. We will find a basis of $\mathrm{Tan}^p(X, (x, n, \zeta))$. Note that for given $(x, n, \xi) \in W \times G_0^{e^\perp}(d-1, m)$, $(u, v) \in \mathrm{Tan}^k(W, (x, n))$ and $\eta \in \mathrm{Tan}(G_0^{e^\perp}(d-1, m), \xi)$ we have

$$D\Phi_e(x, n, \xi)(u, v, \eta) = \Big(u, v, \big(D\big(\textstyle\bigwedge_k F_e(n)\big)v\big)\xi + F_e(n)\eta\Big).$$

Clearly, the vectors $(0, 0, \eta)$ with $\eta \in \operatorname{Tan}(G^{n^\perp}(d-1, m)$ are tangent to X. They form a subspace of dimension $m(d-1-m)$. The remaining k basis tangent vectors can be obtained as follows. Assume that $e = n$ and let $\xi = \zeta = w_1 \wedge \cdots \wedge w_k$ be a representation such that the vectors $w_1, \ldots, w_k$ are orthonormal and $w_1 = p_V v / |p_V v|$ if $|p_V v| > 0$ (otherwise, we choose w_1 arbitrarily). Then we have (see (1.36))

$$D[F_n(n)w_i](v) = |v|(w_1 \cdot n)n = \begin{cases} |p_V v| n, & i = 1, \\ 0, & i = 2, \ldots, k. \end{cases}$$

Thus we get

$$\begin{aligned} \big(D\big(\textstyle\bigwedge_k F_n(n)\big)v\big)\Big(\bigwedge_{i=1}^{k} w_i\Big) &= \big(DF(n)v\big)w_1 \wedge \bigwedge_{i=2}^{k} F(n)w_i \\ &= |p_V v|\, n \wedge \big(\zeta \llcorner (F(n)w_1)^*\big) \\ &= n \wedge \big(\zeta \llcorner (p_V v)^*\big) \\ &= n \wedge \big(\zeta \llcorner v^*\big) \end{aligned}$$

(see (1.15)), which implies the assertion. □

Chapter 2
Background from Convex Geometry

2.1 Intrinsic Volumes

Let K be a convex body (i.e., a nonempty, convex and compact set) in $\mathbb{R}^d$. We use the notation for the *r-parallel body* ($r \geq 0$)

$$K_r := \{z \in \mathbb{R}^d : \operatorname{dist}(z, K) \leq r\}. \tag{2.1}$$

(Equivalently, we can write using Minkowski summation $K_r = K + rB^d$, where B^d is the unit ball centred in the origin in $\mathbb{R}^d$.) The *Steiner formula* expresses the volume of K_r as a polynomial:

$$\mathcal{L}^d(K_r) = \sum_{k=0}^{d} \omega_k r^k V_{d-k}(K), \quad r \geq 0. \tag{2.2}$$

The coefficient $V_k(K)$ is called the *k-th intrinsic volume* of K ($k = 0, 1, \ldots, d$).

Another representation of intrinsic volumes is

$$V_k(K) = \binom{d}{k} \frac{\omega_d}{\omega_k \omega_{d-k}} \int_{G(d,k)} \mathcal{L}_k(\pi_L K)\, \nu_k^d(\mathrm{d}L).$$

This means that $V_k(K)$ is, up to a constant factor, the mean volume of the projections of K into the k-dimensional subspaces of $\mathbb{R}^d$. From classical convex geometry, the re-indexed and re-normed versions

$$W_k^d(K) = \frac{\omega_k}{\binom{d}{k}} V_{d-k}$$

J. Rataj, M. Zähle, *Curvature Measures of Singular Sets*, Springer Monographs in Mathematics, https://doi.org/10.1007/978-3-030-18183-3_2

are known as the *quermassintegrals* or *Minkowski functionals* of K. One advantage of the intrinsic volumes is their independence of the dimension of the embedding space. In particular, if $K \subset \mathbb{R}^d$ is k-dimensional then $V_k(K) = \mathcal{H}^k(K)$.

For any convex body K we have $V_0(K) = 1$ and $V_d(K) = \mathcal{L}^d(K)$. If K has inner points then also $V_{d-1}(K) = \frac{1}{2}\mathcal{H}^{d-1}(\partial K)$. The number $\frac{2\omega_{d-1}}{d\omega_d} V_1(K)$ is known as the *mean width* of K.

It is easy to show that the intrinsic volumes are *valuations* on the space of convex bodies, i.e., they are *additive* in the sense that for two convex bodies K, L such that $K \cup L$ is convex as well,

$$V_k(K \cup L) + V_k(K \cap L) = V_k(K) + V_k(L).$$

Further, V_k is invariant with respect to all Euclidean motions, and it is continuous with respect to the *Hausdorff metric*. The latter is defined for nonempty closed sets A and B in $\mathbb{R}^d$ by

$$d_H(A, B) := \min\left(\inf_{a \in A} \operatorname{dist}(a, B), \inf_{b \in B} \operatorname{dist}(b, A)\right). \tag{2.3}$$

It follows from the general theory of valuations that V_k can be extended to an additive and motion invariant functional on the convex ring (the space of finite unions of convex bodies) denoted again by V_k. (Here one defines additionally $V_k(\emptyset) = 0$, $k = 0, 1, \dots, d$.) V_d still coincides with the Lebesgue measure and $V_{d-1}(R) = \frac{1}{2}\mathcal{H}^{d-1}(\partial R)$ whenever $R = K_1 \cup \dots \cup K_N$ is a set from the convex ring with full-dimensional convex bodies $K_1, \dots, K_N$. The zeroth intrinsic volume, V_0, is the Euler-Poincaré characteristic. (The V_k's are, however, no more continuous on the convex ring with respect to the Hausdorff distance; as a counterexample, consider two approaching parallel segments.) First proofs were given in Hadwiger [Had57], Groemer [Gro78], Schneider [Sch80], see also the survey McMullen and Schneider [MS83].

The following famous Hadwiger's characterization theorem [Had57] says that the intrinsic volumes form a basis of all additive, continuous and motion invariant functionals on the family of convex bodies. Simpler proofs were given in Klain [Kla95], Klain and Rota [KR97], and Chen [Che04]. Here and in the sequel we will use the notation $\mathcal{C}'$ for the *space of convex bodies* (i.e., noempty convex compact sets in $\mathbb{R}^d$), and $\mathcal{P}'$ for the subspace of nonempty convex polytopes (i.e., convex hulls of nonempty finite sets), both being equipped with the Hausdorff distance. Note that the exclusion of the empty set is necessary in order that the Hausdorff distance is well defined.

Theorem 2.1 (Hadwiger [Had57]) *If ψ is an additive, continuous and motion invariant functional on $\mathcal{C}'$ then there exist real constants $c_0, \dots, c_d$ such that*

$$\psi = \sum_{k=0}^{d} c_k V_k.$$

It is an open problem, whether this characterization remains valid if $\mathcal{C}'$ is replaced by $\mathcal{P}'$.

2.2 Curvature and Area Measures

Let Π_K denote the *metric projection* onto the convex body K, i.e., for $z \in \mathbb{R}^d$, $\Pi_K(z)$ is determined by the property

$$|z - \Pi_K(z)| = \operatorname{dist}(z, K).$$

The following local Steiner formula is due to Federer [Fed59] who proved it even for more general sets, see Chap. 4. If B is a Borel subset of $\mathbb{R}^d$ then

$$\mathcal{L}^d(K_r \cap \Pi_K^{-1}(B)) = \sum_{k=0}^{d} \omega_k r^k C_{d-k}(K, B), \quad r \geq 0, \tag{2.4}$$

with nonnegative coefficients $C_{d-k}(K, B)$ which depend σ-additively on B and, hence $C_k(K, \cdot)$ is a finite Borel measure on $\mathbb{R}^d$ called *kth curvature measure of* K. Clearly, we have $C_k(K, \mathbb{R}^d) = V_k(K)$.

Regarding additionally the directions of the metric projection onto K, the Steiner formula has the following generalization. Consider the mapping

$$\tilde{\Pi}_K : z \mapsto \left(\Pi_K(z), \frac{z - \Pi_K(z)}{|z - \Pi_K(z)|}\right), \quad z \in \mathbb{R}^d \setminus K,$$

and let E be a Borel subset of $\mathbb{R}^d \times S^{d-1}$. Then

$$\mathcal{L}^d\big((K_r \setminus K) \cap \tilde{\Pi}_K^{-1}(E)\big) = \sum_{k=1}^{d} \omega_k r^k \widetilde{C}_{d-k}(K, E), \quad r \geq 0, \tag{2.5}$$

with nonnegative coefficients $\widetilde{C}_{d-k}(K, E)$ which again depend σ-additively on E and, hence $\widetilde{C}_k(K, \cdot)$ is a finite Borel measure on $\mathbb{R}^d \times S^{d-1}$; it is called sometimes kth support measure of K in convex geometry. Of course, the image of $\widetilde{C}_k(K, \cdot)$ under the first projection mapping $\pi_0 : (x, n) \mapsto x$ is $C_k(K, \cdot)$. The image under the second projection $\pi_1 : (x, n) \mapsto n$ is a finite Borel measure on the sphere, denoted $S_k(K, \cdot) = \widetilde{C}_k(K, \mathbb{R}^d \times \cdot)$, and called *kth area measure of* K. Curvature and area measures can again be extended additively to the convex ring.

Considering also the dependence on the convex body K, we can consider the curvature measures as functionals $C_k : \mathcal{C}' \times \mathcal{B}(\mathbb{R}^d) \to [0, \infty)$, where $\mathcal{B}(\mathbb{R}^d)$ denotes the family of Borel subsets of $\mathbb{R}^d$. The curvature measures possess the following properties. (Again we define $C_k(\emptyset, \cdot) = 0, k = 0, \ldots, d$.)

Additivity: $C_k(K \cup L, \cdot) + C_k(K \cap L, \cdot) = C_k(K, \cdot) + C_k(L, \cdot)$ whenever K, L and $K \cup L$ are convex bodies.

Motion covariance: $C_k(gK, gA) = C_k(K, A)$ for any convex body K, Borel subset A of $\mathbb{R}^d$ and Euclidean motion $g : \mathbb{R}^d \to \mathbb{R}^d$.

Weak continuity: If a sequence of convex bodies K_n converges to K as $n \to \infty$ in the Hausdorff distance then $C_k(K_n, \cdot) \to C_k(K, \cdot)$ weakly.

Local determinancy: If $K \cap G = L \cap G$ for convex bodies K, L and an open set $G \subset \mathbb{R}^d$ then the restrictions of $C_k(K, \cdot)$ and $C_k(L, \cdot)$ to G coincide.

The following generalization of Hadwiger's representation theorem was shown by Schneider [Sch78, Theorem 6.1]. An analysis of the proof shows that here the corresponding statement remains valid for the space of convex polytopes.

Theorem 2.2 *Let $\Phi : \mathcal{C}' \times \mathcal{B}(\mathbb{R}^d) \to [0, \infty)$ (or $\Phi : \mathcal{P}' \times \mathcal{B}(\mathbb{R}^d) \to [0, \infty)$) be a functional such that $\Phi(K, \cdot)$ is a finite Borel measure for any $K \in \mathcal{C}'$ (for any $K \in \mathcal{P}'$, respectively). Assume further that Φ is additive, motion covariant, weakly continuous and locally determined, in the sense given above. Then there exist nonnegative constants $c_0, \dots, c_d$ such that*

$$\Phi(K, \cdot) = \sum_{k=0}^{d} c_k C_k(K, \cdot), \quad K \in \mathcal{C}' \quad (K \in \mathcal{P}').$$

An analogous result holds for the area measures $S_k(K, \cdot)$, $K \in \mathcal{C}'$. Note that these measures are motion covariant in a slightly different sense: we have $S_k(gK, g_0 B) = S_k(K, B)$ whenever K is a convex body, B a Borel subset of S^{d-1} and g a Euclidean motion in $\mathbb{R}^d$ with linear component g_0. Also, a different locality condition applies. Given a convex body K, we use the notation for the *support function* of K

$$h(K, u) := \sup\{x \cdot u : x \in K\}, \quad u \in \mathbb{R}^d,$$

and for the *support set of K and $u \in S^{d-1}$*

$$K(u) := \{x \in K : x \cdot u = h(K, u)\}.$$

The following representation result was shown by Schneider [Sch75, Satz 2]. Note that here the functional considered need not be nonnegative.

Theorem 2.3 *Let $\Psi : \mathcal{C}' \times \mathcal{B}(S^{d-1}) \to \mathbb{R}$ be a functional such that $\Psi(K, \cdot)$ is a finite signed Borel measure for any $K \in \mathcal{C}'$. Assume further that Ψ is additive, motion covariant and weakly continuous in the sense given above, and that it satisfies the following locality condition:*

$$\text{If} \bigcup_{u \in B} K_1(u) = \bigcup_{u \in B} K_2(u) \text{ then } \Psi(K_1, B) = \Psi(K_2, B), \quad B \in \mathcal{B}(S^{d-1}),$$

$K_1, K_2 \in \mathcal{C}'$. *Then there exist real constants* $c_0, \ldots, c_d$ *such that*

$$\Psi(K,\cdot) = \sum_{k=0}^{d} c_k S_k(K,\cdot), \quad K \in \mathcal{C}'.$$

2.3 Integral-Geometric Formulas

The above representation theorems make it possible to find simple proofs of main integral-geometric results as Principal kinematic theorem and Crofton formula (see [Sch78]). Let $\mathcal{G}_d$ be the group of all rigid motions (i.e., isometries preserving orientation) in $\mathbb{R}^d$. Any $g \in \mathcal{G}_d$ is an affine mapping of the form

$$g(\cdot) = b + \rho(\cdot)$$

with a shift $b \in \mathbb{R}^d$ and linear mapping ρ from the special orthogonal group SO(d). (We will often call ρ a "rotation", though it need not be a rotation in the usual narrower sense.) Let ϑ_d denote the normalized Haar measure on SO(d). Then, the measure Θ_d on $\mathcal{G}_d$ given by

$$\int_{\mathcal{G}_d} f(g)\,\mathrm{d}\Theta_d(g) = \int_{\mathrm{SO}(d)} \int_{\mathbb{R}^d} f(b+\rho)\,\mathcal{L}^d(\mathrm{d}b)\vartheta_d(\mathrm{d}\rho)$$

for any nonnegative measurable function f on $\mathcal{G}_d$, is a (both left and right) Haar measure on $\mathcal{G}_d$. We will often write only $\mathrm{d}g$ instead of $\Theta_d(\mathrm{d}g)$ in the integrals in order to save space.

Theorem 2.4 (Principal kinematic formula) *If* K, L *are convex bodies,* $0 \le k \le d$ *and* A, B *Borel subsets of* $\mathbb{R}^d$ *then*

$$\int_{\mathcal{G}_d} C_k\big(K \cap gL, A \cap gB\big)\mathrm{d}g = \sum_{\substack{1\le r,s\le d\\ r+s=d+k}} \gamma(d,r,s) C_r(K,A) C_s(L,B),$$

where

$$\gamma(d,r,s) = \frac{\Gamma\left(\frac{r+1}{2}\right)\Gamma\left(\frac{s+1}{2}\right)}{\Gamma\left(\frac{r+s-d+1}{2}\right)\Gamma\left(\frac{d+1}{2}\right)}.$$

Let further $\mathcal{A}(d,j)$ be the space of all affine j-subspaces of $\mathbb{R}^d$, $0 \le j \le d$. Any affine subspace $E \in \mathcal{A}(d,j)$ can be represented uniquely as $E = V + z$ with

$V \in G(d, j)$ and $z \in V^{\perp}$. The motion invariant measure μ_j^d on $\mathcal{A}(d, k)$ is defined by $\mu_j^d(\mathrm{d}E) = \mathcal{L}^{d-j}(\mathrm{d}z)\nu_j^d(\mathrm{d}V)$, using this correspondence.

Theorem 2.5 (Crofton formula) *Let K be a convex body, $0 \leq k \leq j \leq d$, and $A \subset \mathbb{R}^d$ a Borel set. Then*

$$\int_{\mathcal{A}(d,j)} C_k(K \cap E, A \cap E)\, \mu_j^d(\mathrm{d}E) = \gamma(d, j, k) C_{d+k-j}(K, A).$$

2.4 Polar Cones

A cone (with centre at the origin) is a set C with the property that if $x \in C$ then also $tx \in C$ for any $t > 0$.

Definition 2.6 The *polar cone* to a set $C \subset \mathbb{R}^d$ is given by

$$C^o := \{v \in \mathbb{R}^d : u \cdot v \leq 0 \text{ for all } u \in C\}.$$

It is clear from the definition that C^o is always a closed convex cone.

If C is already a cone then its second polar satisfies

$$C^{oo} = \overline{\operatorname{conv} C}. \tag{2.6}$$

If $C, D \subset \mathbb{R}^d$ are closed convex cones then

$$(C \cap D)^o = C^o + D^o, \quad (C + D)^o = C^o \cap D^o \tag{2.7}$$

(see [Roc70, Sect. 14, Corollary 16.4.2]).

2.5 Simplicial Complexes

A *k-simplex* is the convex hull $\sigma := \operatorname{conv}\{x_0, \ldots, x_k\}$ of $k+1$ affinely independent points $x_0, \ldots, x_k \in \mathbb{R}^d$. A *$j$-face of σ* ($j = 0, 1, \ldots, d$) is any j-simplex given as the convex hull of $j + 1$ points (vertices) of $\{x_0, \ldots, x_k\}$. The $(d-1)$-faces of d-simplices are called *facets*, the 1-faces are *edges* and the 0-faces are *vertices* of σ. If σ is a d-simplex and F is a facet, we denote by $\nu(F) \in S^{d-1}$ its unit outer normal vector.

A *simplicial k-complex* Σ in $\mathbb{R}^d$ is a family of k-simplices with the following property: If σ, σ' belong to Σ then $\sigma \cap \sigma'$ is either empty or a face of both σ and σ'. A simplicial k-complex Σ is *closed* (or *face-to-face*) if for any $\sigma \in \Sigma$, any $(k-1)$-face μ of σ is a face of another k-simplex $\sigma' \in \Sigma$.

Finally, a *simplicial d-polytope* is the union of a locally finite simplicial d-complex, $P = \bigcup \Sigma$. If a j-face (facet, edge, vertex) of a simplex $\sigma \in \Sigma$ lies in ∂P, it as called a j-face (facet, edge, vertex) of P. We denote by $\mathcal{F}_k(P)$ the set of all k-faces of P, $k = 0, \ldots, d-1$. Of course, the notion of a face depends not only on the union P, but on the d-complex Σ. Therefore, whenever speaking about a simplicial d-polytope, we consider the union P together with the simplicial decomposition Σ. When speaking about a d-polytope, we consider only the union P.

Fatness is the crucial property of "good" triangulations (we use the term *triangulation* for simplicial decomposition in any dimension). Given a k-simplex $\sigma = \operatorname{conv}\{x_0, \ldots, x_k\}$ in $\mathbb{R}^d$, its *size* is

$$\rho(\sigma) := \max_{0 \le i, j \le k} |x_i - x_j|,$$

its *altitude*

$$h(\sigma) := \min_{0 \le i \le k} \operatorname{dist}(x_i, \operatorname{aff}(\sigma_i)),$$

where $\sigma_i := \operatorname{conv}\{x_j : 0 \le j \le k,\ j \ne i\}$, and and its *fatness*

$$\Theta(\sigma) := \frac{h(\sigma)}{k\rho(\sigma)}.$$

Remark 2.7 The following properties can be easily verified.

1. If μ is a face of a simplex σ then $h(\mu) \ge h(\sigma)$, $\rho(\mu) \le \rho(\sigma)$ and $\Theta(\mu) \ge \Theta(\sigma)$.
2. If all edges of a d-simplex σ have lengths at least ε and μ is a k-face of σ then $\mathcal{H}^k(\mu) \ge \Theta(\sigma)^{k-1}\varepsilon^k$.

In order to construct polytopal approximations (Lemma 1.19) we will use *Delaunay triangulations* and its weighted variant.

Let $\varepsilon > 0$, $C > 0$ and $S \subset \mathbb{R}^d$ be given and let M be a locally finite point set in S such that any two different points of M have distance at least $\varepsilon > 0$, and $\operatorname{dist}(x, M) \le C\varepsilon$ for all $x \in S$. Let us call such a set M an (ε, C)*-net in* S. Note that a nonempty set $S \subset \mathbb{R}^d$ always contains an $(\varepsilon, 1)$-net (consider a maximal subset of points having any two points at least ε apart).

Let now M be an (ε, C)-net in $\mathbb{R}^d$. If the points of M lie in general mutual position, meaning that no $d+2$ points of M lie on a common $(d-1)$-sphere, we define the *Delaunay d-complex of M* as the simplicial d-complex Σ_M consisting of all d-simplices σ with vertices in M and such that

$$(\operatorname{int} B_\sigma) \cap M = \emptyset, \tag{2.8}$$

where B_σ denotes the circumscribed ball to σ; its radius is called *circumradius of* σ and will be denoted by R_σ. Note that

$$R_\sigma \le C\varepsilon, \quad \sigma \in \Sigma_M,$$

since M is an (ε, C)-net. If the mutual position assumption is violated a Delaunay d-complex of M (i.e., a d-complex with vertices in M whose each d-simplex satisfies (2.8)) still exists but need not be uniquely determined. We refer to [BCY18] and to the references given therein.

If $d \le 2$, it is not difficult to show that such a Delaunay complex Σ_M consists of uniformly fat triangles (i.e., $\inf_{\sigma\in\Sigma_M} \Theta(\sigma) > 0$). This is, however, not true in higher dimension, where "bad" simplices may occur: Consider a 3-simplex whose four vertices lie on a 2ε-sphere, close to a common great sphere, and have mutual distances greater that ε. Such a simplex can have fatness arbitrarily close to 0; these simplices are called "slivers" in the literature. There exist, however, methods modifying slightly the Delaunay construction and removing such slivers. In particular, we can use a *weighted* Delaunay triangulation, see e.g. [BCY18, Chapter 5], known also as *Laguerre Delaunay tessellation* or *power Delaunay tessellation*.

Assume that to any point $x \in M$ a weight $w(x) \in [0, \varepsilon/2)$ is assigned. Given a k-simplex $\sigma = \operatorname{conv}\{x_0, \dots, x_k\}$ with vertices $x_i \in M$, its *orthocentre* c_σ and *orthoradius* $\hat{R}_\sigma$ are defined by the conditions

$$|x_i - c_\sigma|^2 = \hat{R}_\sigma^2 + w(x_i)^2, \quad i = 0, \dots, k,$$

and by the minimality of $\hat{R}_\sigma$. The ball $B(c_\sigma, \hat{R}_\sigma)$ is called *orthoball* of σ; this name comes from the fact that its boundary sphere intersects the boundary spheres of $B(x_i, w(x_i))$ orthogonally. The *weighted Delaunay d-complex* is a d-complex $\hat{\Sigma}_M$ fulfilling

$$|x - c_\sigma|^2 \ge \hat{R}_\sigma^2 + w(x)^2, \quad x \in M, \quad \sigma \in \hat{\Sigma}_M$$

(note that this is condition (2.8) in the case $w(x) = 0$, $x \in M$). Again, the weighted Delaunay triangulation is not unique, unless we assume that no orthoball contains more than $d + 1$ points on its boundary.

It can be easily seen that the orthoradius $\hat{R}_\sigma$ of any simplex $\sigma \in \hat{\Sigma}_M$ cannot be larger than $C\varepsilon$, thus we get

$$R_\sigma \le (C + \frac{1}{2})\varepsilon, \quad \sigma \in \hat{\Sigma}_M. \tag{2.9}$$

Boissonnat et al. [BCY18, Theorem 5.25] guarantees that there exist weights $w(x)$, $x \in M$, such that the resulting weighted Delaunay d-simplex $\hat{\Sigma}$ has uniformly fat simplexes. It is formulated for $\mathbb{Z}^d$-periodic point patterns (or, equivalently, for point patterns in the torus $\mathbb{T}^d = \mathbb{R}^d/\mathbb{Z}^d$) (Fig. 2.1).

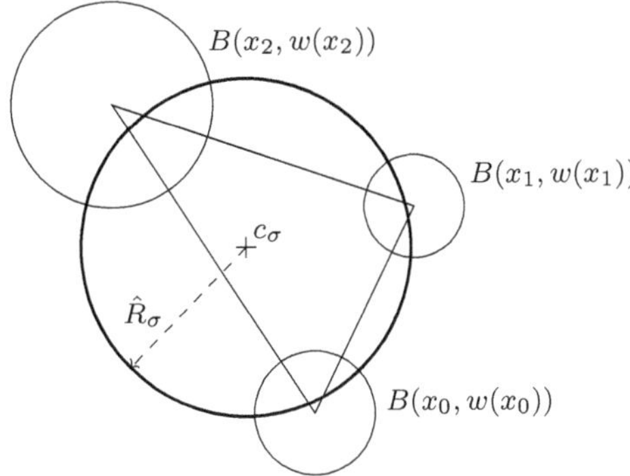

Fig. 2.1 The orthoball of $\sigma = \text{conv}\{x_0, x_1, x_2\}$

Theorem 2.8 ([BCY18, Theorem 5.25]) *Let M be an (ε, C)-net in $\mathbb{R}^d$, $\varepsilon, C > 0$, and assume that M is $\mathbb{Z}^d$-periodic. Then there exist a constant $\theta = \theta(d, C) > 0$ (independent of ε) and weights $w(x) \in [0, \varepsilon/2)$, $x \in M$, such that the weighted Delauynay d-complex $\hat{\Sigma}_M$ fulfills the fatness condition*

$$\Theta(\sigma) \geq \theta, \quad \sigma \in \hat{\Sigma}_M.$$

Chapter 3
Background from Differential Geometry and Topology

3.1 Curvature Measures of Smooth Submanifolds

There exist close analogues to the curvature measures for convex bodies in classical differential geometry of smooth submanifolds. Basic notions have been developed nearly at the same time starting from the 1930th, mainly within the so-called integral geometry.

Assume that a convex body K has a C^2-boundary ∂K and $\nu_K(x)$ denotes the outer unit normal at $x \in \partial K$; the mapping

$$\nu_K : \partial K \to S^{d-1}$$

is called the *Gauss map* of K. The differential $D\nu_K(x)$ exists everywhere on ∂K and it is a self-adjoint linear map from $\mathrm{Tan}(K, x) \cong \nu_K(x)^\perp$ to $\mathrm{Tan}(S^{d-1}, \nu_K(x)) \cong \nu_K(x)^\perp$. The (real) eigenvalues

$$\kappa_1(x), \dots, \kappa_{d-1}(x) \geq 0$$

of $D\nu(x)$ are called *principal curvatures* of K at x; we assume that the corresponding orthonormal eigenvectors $b_1(x), \dots, b_{d-1}(x)$ have the property that

$$(b_1(x), \dots, b_{d-1}(x), \nu(x))$$

is a positively oriented basis of $\mathbb{R}^d$.

For $0 \leq k \leq d-1$, we denote the *elementary symmetric function* of the principal curvatures of order k by

$$s_k(K, x) = \sum_{1 \leq i_1 < \cdots i_k \leq d-1} \prod_{j=1}^{k} \kappa_{i_j}(x), \quad x \in \partial K,$$

J. Rataj, M. Zähle, *Curvature Measures of Singular Sets*, Springer Monographs in Mathematics, https://doi.org/10.1007/978-3-030-18183-3_3

and its normalized variant

$$H_k(K,x) = \binom{d-1}{k}^{-1} s_k(K,x),$$

called also the *mean curvature of order* k of K at point x. In particular, we obtain:

$$H_1(K,x) = \frac{1}{d-1}\sum_{i=1}^{d-1} \kappa_i(x) \qquad \textit{mean curvature}$$

$$H_2(K,x) = \frac{2}{(d-1)(d-2)} \sum_{1\le i<j\le d-1} \kappa_i(x)\kappa_j(x) \qquad \textit{scalar curvature}$$

$$H_{d-1}(K,x) = \kappa_1(x)\kappa_2(x)\ldots\kappa_{d-1}(x) \qquad \textit{Gauss curvature.}$$

Then we have the following relationship to the curvature measures from the preceding section:

$$C_k(K,B) = \frac{1}{(d-k)\omega_{d-k}} \int_{\partial K\cap B} s_{d-1-k}(K,x)\,\mathcal{H}^{d-1}(\mathrm{d}x)\,. \tag{3.1}$$

For $B = \mathbb{R}^d$ the integrals

$$V_k(K) = \mathbf{C}_k(K) := \frac{1}{(d-k)\omega_{d-k}} \int_{\partial K} s_{d-1-k}(K,x)\,\mathcal{H}^{d-1}(\mathrm{d}x) \tag{3.2}$$

are called in classical differential geometry *total mean curvatures* of order $d-1-k$.

Let us drop now the convexity assumption and assume that A is a compact C^2-*domain* in $\mathbb{R}^d$, i.e., a compact set with nonempty interior whose boundary ∂A is a $(d-1)$-dimensional C^2-submanifold of $\mathbb{R}^d$. Then, we can differentiate the Gauss map $D\nu_A$ exactly as described above, introduce the principal curvatures and their elementary symmetric functions, and we can define the curvature measures $C_k(A,\cdot)$ and total mean curvatures $\mathbf{C}_k(A)$ by (3.1) and (3.2), with the only difference that the principal curvatures may be negative and, hence, $C_k(A,\cdot)$ is a signed Radon measure. Also, the Steiner formula holds as in the convex case (Eq. (2.2))

$$\mathcal{L}^d(A_r) = \sum_{k=0}^{d} \omega_k r^k \mathbf{C}_{d-k}(A),$$

but only for sufficiently small $r > 0$.

Closely related is the famous "tube formula" of H. Weyl [Wey39]. If M_m is an m-dimensional, compact, C^2-smooth submanifold of $\mathbb{R}^d$, then

$$\mathcal{L}^d((M_m)_r) = \sum_{k=d-m}^{d} \omega_k r^k \mathbf{C}_{d-k}(M_m)$$

for sufficiently small $r > 0$, where the kth total curvature $\mathbf{C}_k(M_m)$ of M_m can be expressed in modern notation as follows:

$$\mathbf{C}_k(M_m) = \frac{1}{(m-k)\omega_{m-k}} \int_{M_m} s_{m-k}(M_m, x)\, \mathcal{H}^m(\mathrm{d}x) \tag{3.3}$$

if $m - k$ is even. Here $s_{m-k}(M_m; x)$ agrees with a constant multiple of the *trace of the* $(m-k)/2$-th *power of the Riemannian curvature tensor* $\mathcal{R}$ of the manifold at point x, which is determined intrinsically. ($\mathcal{R}^{(m-k)/2}(x)$ is considered as an element of $\bigwedge^{m-k}(T_x M_m) \otimes \bigwedge^{m-k}(T_x M_m)$, where $\mathrm{trace}(a \otimes b) := a \bullet b$ for the inner product in $\bigwedge^{m-k}(T_x M_m)$. For details see also Federer [Fed59, Remark 5.21].) If $m - k$ is odd or $k > m$ one obtains $\mathbf{C}_k(M_m) = 0$.

When restricting the integrals of the mean curvature functions to Borel subsets B of $\mathbb{R}^d$ one obtains the corresponding *curvature measures*

$$C_k(M_m, B) := \frac{1}{(m-k)\omega_{m-k}} \int_{M_m \cap B} s_{m-k}(M_m; x)\, \mathcal{H}^m(\mathrm{d}x)\,.$$

One of the basic results connecting the local differential geometry of a compact C^2-submanifold M_m, $m \le d$, with its *global topology* is the *Gauss-Bonnet theorem* telling that the total Gauss curvature agrees with the *Euler-Poincaré characteristic*:

$$\mathbf{C}_0(M_m) = \chi(M_m)\,. \tag{3.4}$$

This can be shown by means of *Morse theory* in the more general context of Riemannian geometry.

Let us turn back to the case of a compact C^2-domain. Then ∂A is a C^2-submanifold of dimension $d - 1$ and, calculating the contribution to the Steiner formula of ∂A, we have to consider both normal vectors at each $x \in \partial A$, $\nu_A(x)$ and $-\nu_A(x)$. The principal curvatures will change the sign, which means that the elementary symmetric function associated with $-\nu_A(x)$ equals $(-1)^k$ times the elementary symmetric function associated with $\nu_A(x)$. Consequently, the curvature measures of ∂A satisfy

$$C_k(\partial A, \cdot) = (1 + (-1)^{d-1-k}) C_k(A, \cdot).$$

This means, in particular, that for even $d-1-k$ the curvature measures $C_k(A,\cdot)$ are objects of the *intrinsic geometry* of ∂A. In general, the latter does not hold true for the mean curvatures $H_{d-1-k}(A,x)$ of the domain A, and thus for their curvature integrals $C_k(A,\cdot)$, if $d-1-k$ is odd. More generally, in classical differential geometry and Riemannian geometry the above notions and partial results have been extended to arbitrary compact C^2-manifolds with boundary.

3.2 Euler Characteristic

Under the *Euler-Poincaré characteristic* $\chi(A)$ of a set $A \subset \mathbb{R}^d$ we will always mean the Euler characteristic with respect to the singular homology (see [Dol72, Ch. III]). (Nevertheless, we will mainly work with assumptions assuring the equivalence of all homology classes.)

Definition 3.1 A set $A \subset \mathbb{R}^d$ is *contractible* to a point $a \in A$ if there exists a continuous mapping $h : A \times [0,1] \to A$ such that $h(x,0) = x$ and $h(x,1) = a$ for any $x \in A$. The set A is *locally contractible* if for any $a \in A$ and $\varepsilon > 0$ there exists a neighbourhood $a \in V \subset U(a,\varepsilon)$ such that $A \cap V$ is contractible to a.

Definition 3.2 A continuous mapping $f : A \to B \subset A$ is a *retraction* (onto B) if $f(b) = b$ for any $b \in B$. A set $A \subset \mathbb{R}^d$ is a *neighbourhood retract* if there exists an open set $U \supset A$ and a retraction $f : U \to A$.

Theorem 3.3 ([Hat02, Theorem A.7]) *If $A \subset \mathbb{R}^d$ is compact and locally contractible then A is a neighbourhood retract.*

Theorem 3.4 ([Hat02, Corollary A.8]) *If $A \subset \mathbb{R}^d$ is a compact neighbourhood retract then its homology groups and fundamental group are finitely generated. Hence, the Euler-Poincaré characteristic $\chi(A)$ is defined.*

Theorem 3.5 *Assume that $X, Y \subset \mathbb{R}^d$ are compact and locally contractible, and that at least one of the Euler-Poincaré characteristics $\chi(X \cap Y)$, $\chi(X \cup Y)$ exists. Then, the other exists as well, and we have the additivity formula*

$$\chi(X \cup Y) + \chi(X \cap Y) = \chi(X) + \chi(Y).$$

Proof Since X, Y are compact, $(\mathbb{R}^d, X, Y)$ is an excisive triad with respect to the Čech cohomology (see [Dol72, VIII.6.13,15]). Further, since both X and Y are compact and locally contractible, they are neighbourhood retracts by Theorem 3.3 and their Čech cohomologies coincide with the usual cohomologies [Dol72, VIII.6.12]. The existence and additivity of the Euler characteristic follows now by [Dol72, Propositions V.4.11 and V.5.8]. □

3.3 Integral Geometry

Integral geometry as a branch of classical differential geometry studies kinematic relationships, as e.g. the famous Principal kinematic formula of Santaló, Blaschke and Chern for the integral of the Euler-Poincaré characteristic of the intersection of two bodies as above, one fixed and the other moving. H. Federer [Fed59] extended it to a more general situation including the total curvatures of all orders for compact C^2-submanifolds and S.S. Chern [Che66] gave a direct differential geometric proof for the latter case (see also Sulanke and Wintgen [SW72] for a different approach to this version). In the above notations Chern proved the following *Principal kinematic formula*:

$$\int_{\mathcal{G}_d} \mathbf{C}_k(M_m \cap gM_n)\, dg = \sum_{\substack{0 \le r \le m,\, 0 \le s \le n \\ r+s=d+k}} \gamma(d, r, s)\mathbf{C}_r(M_m)\mathbf{C}_s(M_n) \tag{3.5}$$

where $1 \le m, n \le d$ and the constants γ are as in the convex case.
Moreover, the differential geometric analogue of the measure version of the Principal kinematic formula from the convex case (Theorem 2.4) remains valid:

$$\int_{\mathcal{G}_d} C_k(M_m \cap gM_n, A \cap gB)\, dg = \sum_{\substack{0 \le r \le m,\, 0 \le s \le n \\ r+s=d+k}} \gamma(d, r, s)C_r(M_m, A)C_s(M_n, B)$$

for Borel sets A and B. The same holds true for the Crofton formula (Theorem 2.5). Note, that even in the convex case both measure versions go back to Federer [Fed59], although some of the global cases were developed much earlier.

3.4 Morse Theory

Let A be a compact C^2-domain in $\mathbb{R}^d$. As above, we denote by $\mathrm{Tan}(\partial A, x)$ the tangent space and by $\nu_A(x) \perp \mathrm{Tan}(A, x)$ the unit outer normal vector to A at $x \in \partial A$. The *second fundamental form* of A at $x \in \partial A$ is a symmetric bilinear form defined on $\mathrm{Tan}(\partial A, x)$ as

$$II_x(u, v) := u \cdot (D\nu_A(x)v), \quad u, v \in \mathrm{Tan}(\partial A, x).$$

Let further f be a C^2-smooth function defined on an open neighbourhood of A without stationary points (i.e., its gradient is nonzero everywhere). We define the *Hessian form* of f relative to A as the symmetric bilinear form

$$H_A f(x) := D^2 f|\mathrm{Tan}(\partial A, x) - (\nabla f(x) \cdot \nu_A(x)) II_x.$$

Definition 3.6 We say that a point $x \in \partial A$ is *critical for* f if $-\nabla f(x)$ is a positive multiple of $\nu_A(x)$. The value $f(x)$ is then called a *critical value* of f. Any real number c which is not a critical value is called a *regular value* of f.

Further, we say that a critical point $x \in \partial A$ of f is *nondegenerate* if $H_A f(x)$ has full rank $(d-1)$. At a nondegenerate critical point $x \in \partial A$, the *index* of F relative to A, denoted $\operatorname{ind}_{f,A}(x)$, is defined as the number of negative eigenvalues of $H_A f(x)$.

Finally, we say that a smooth function f without stationary points is a *Morse function for A* if

1. f has only nondegenerate critical points on ∂A,
2. for each $c \in \mathbb{R}$ there is at most one critical point $x \in \partial A$ with $f(x) = c$.

Note that the critical points of a Morse function must be isolated. Thus, any Morse function of a compact C^2-domain has only finitely many critical points.

The classical Morse theory says that given a Morse function for A, the domain A can be represented as a CW complex with each k-cell corresponding to a nondegenerate critical point of index k (see [MC69] or [Hir76]). This yields the following formula for the Euler-Poincaré characteristic of A.

Theorem 3.7 *If A is a compact C^2-domain in $\mathbb{R}^d$, f a Morse function for A and $c \in \mathbb{R}$ then*

$$\chi(A \cap \{f \le c\}) = \sum_{x:\, f(x) \le c} (-1)^{\operatorname{ind}_{f,A}(x)}$$

(the summation is carried out over critical points of f). In particular,

$$\chi(A) = \sum_x (-1)^{\operatorname{ind}_{f,A}(x)}.$$

We will need a generalization of the above formula to a more general framework where ∂A is only C^1 smooth, but the unit outer normal ν_A is Lipschitz on ∂A.

Definition 3.8 A closed $C^{1,1}$*-domain* in $\mathbb{R}^d$ is a C^1-domain (i.e., a d-dimensional C^1-submanifold with boundary) such that the Gauss map $x \mapsto \nu_A(x)$ is Lipschitz.

By the well-known Rademacher theorem, a Lipschitz mapping (in a finite-dimensional space) is differentiable almost everywhere. Hence, if A is a closed $C^{1,1}$-domain, its Gauss map ν_A is differentiable $\mathcal{H}^{d-1}$-almost everywhere on ∂A. We adopt Definition 3.6 for closed $C^{1,1}$-domains (with the only difference that we understand the nondegeneracy condition so that it presumes the existence of $D\nu_A(x)$).

Fu [Fu89a] showed that the Morse formula for the Euler-Poincaré characteristic holds for compact $C^{1,1}$-domains as well.

Theorem 3.9 *If A is a compact $C^{1,1}$-domain in $\mathbb{R}^d$, f a Morse function for A and $c \in \mathbb{R}$ then*

$$\chi(A \cap \{f \le c\}) = \sum_{x:\, f(x) \le c} (-1)^{\mathrm{ind}_{f,A}(x)}.$$

In particular,

$$\chi(A) = \sum_{x} (-1)^{\mathrm{ind}_{f,A}(x)}.$$

Fu also showed that given a compact $C^{1,1}$ domain in $\mathbb{R}^d$, the "height function"

$$h_v : x \mapsto x \cdot v, \quad x \in \mathbb{R}^d,$$

is Morse for A for $\mathcal{H}^{d-1}$-almost all $v \in S^{d-1}$ [Fu89a, §5]. Note further that $D^2 h_v = 0$ and $\nabla h_v(x) = v$, hence x is critical for $h_v|A$ if and only if $v = -\nu_A(x)$ and $H_A h_v(x) = -II_x$ at critical points $x \in \partial A$. Thus, the index $\mathrm{ind}_{h_v,A}(x)$ agrees with

$$\lambda_A(x) := \text{ the number of negative principal curvatures at } x \in \partial A.$$

Consequently, we get the following.

Corollary 3.10 *If A is a compact $C^{1,1}$-domain in $\mathbb{R}^d$ then for $\mathcal{H}^{d-1}$-almost all $v \in S^{d-1}$ and all $t \in \mathbb{R}$,*

$$\chi(A \cap \{x : x \cdot v \le t\}) = \sum_{x \in \nu_A^{-1}\{-v\}:\, x \cdot v \le t} (-1)^{\lambda_A(x)}.$$

In particular,

$$\chi(A) = \sum_{x \in \nu_A^{-1}\{-v\}} (-1)^{\lambda_A(x)}.$$

Chapter 4
Sets with Positive Reach

4.1 Geometric Properties

The metric projection to a set $\emptyset \neq X \subset \mathbb{R}^d$ is not determined everywhere unless X is convex and closed. Sets with positive reach are sets with the property that the metric projection is defined on some open neighbourhood of X.

Given a set $\emptyset \neq X \subset \mathbb{R}^d$, we define its *distance function* as

$$d_X : x \mapsto \operatorname{dist}(x, X) = \inf\{|x - a| : a \in X\},$$

and $\operatorname{Unp} X$ denotes the set of all $x \in \mathbb{R}^d$ for which there exists a unique point $a \in X$ nearest to x. We write then $a =: \Pi_X(x)$, and the mapping

$$\Pi_X : \operatorname{Unp} X \to X$$

is called the *metric projection* to X.

Lemma 4.1 *Π_X is continuous.*

Proof Assume for the contrary that Π_X is not continuous at a point $x \in \operatorname{Unp} X$, i.e., there exist points $x_i \in \operatorname{Unp} X$, $x_i \to x$, $\inf_i |\Pi_X(x_i) - \Pi_X(x)| > 0$. Turning to a subsequence, we may assume that $\Pi_X(x_i) \to y \in X$. Clearly $y \neq \Pi_X(x)$, but $|y - x| = d_X(x)$, which contradicts the assumption $x \in \operatorname{Unp} X$. □

The *reach function* of X is defined as

$$\operatorname{reach}(X, a) := \sup\{r \geq 0 : \mathring{B}(a, r) \subset \operatorname{Unp} X\}$$

and the reach of X is set as

$$\operatorname{reach} X := \inf_{a \in X} \operatorname{reach}(X, a)$$

J. Rataj, M. Zähle, *Curvature Measures of Singular Sets*, Springer Monographs in Mathematics, https://doi.org/10.1007/978-3-030-18183-3_4

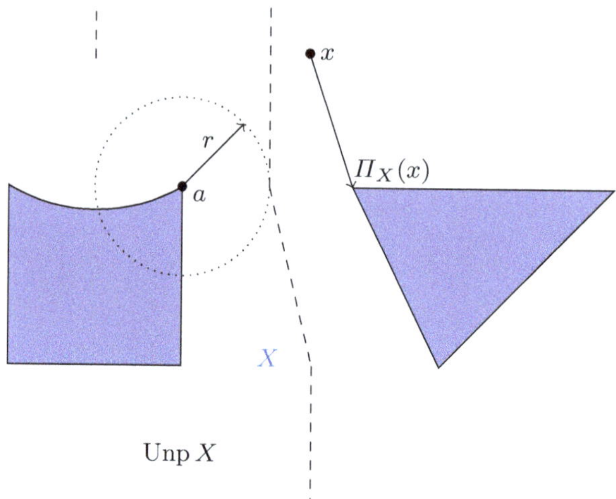

Fig. 4.1 A set X with positive reach, the metric projection Π_X to X and the reach $r = \text{reach}(X, a)$. The dashed lines do not belong to Unp X

(i.e., reach X is the supremum of all $r \geq 0$ such that any $z \in X_r$ has its unique nearest neighbour in X). Note that reach $\emptyset = \infty$ and that any set with positive reach is closed (Fig. 4.1).

Examples of sets with positive reach:

1. If K is closed and convex then (and only then) reach $K = \infty$.
2. If F is finite nonempty then reach $F = \min\{\frac{1}{2}|x - y| : x, y \in F, x \neq y\}$.
3. If X is a compact C^2-domain then reach $X > 0$. (Without compactness, the assertion is not true.)
4. The union of two disjoint compact sets with positive reach has positive reach.

First we show a differentiability property of the distance function.

Lemma 4.2 *If the distance function d_X is differentiable at some point $x \in$ Unp $X \setminus X$ then its gradient is*

$$\nabla d_X(x) = \frac{x - \Pi_X(x)}{|x - \Pi_X(x)|}.$$

Further, d_X is continuously differentiable on int (Unp $X \setminus X$).

Proof It follows from the triangle inequality that d_X is 1-Lipschitz, i.e.,

$$|d_X(y) - d_X(x)| \leq |y - x|, \quad x, y \in \mathbb{R}^d.$$

Thus, $|\nabla d_X(x)| \le 1$ whenever d_X is differentiable at x. Note further that if $x \in \operatorname{Unp} X \setminus X$ and $u := \frac{x-\Pi_X(x)}{|x-\Pi_X(x)|}$ then

$$d_X(x - tu) = d_X(x) - t, \quad 0 \le t \le d_X(x).$$

Consequently, $\nabla d_X(x) = u$ whenever $x \in \operatorname{Unp} X \setminus X$ and d_X is differentiable at x. The second statement follows from Lemma 4.3 and the continuity of Π_X. □

Lemma 4.3 *Let $f : U \to \mathbb{R}$ be Lipschitz, $U \subset \mathbb{R}^d$ open, $g : U \to \mathbb{R}$ continuous and $1 \le i \le d$. If*

$$\frac{\partial}{\partial x_i} f(x) = g(x) \text{ whenever } f \text{ is differentiable at } x \in U,$$

then

$$\frac{\partial}{\partial x_i} f(x) = g(x) \text{ for all } x \in U.$$

Proof Let $x \in U$ and $r > 0$ be such that $B(x, 2r) \subset U$, and let e_i denote the ith vector of the canonical basis of $\mathbb{R}^d$. Since f is clearly absolutely continuous in the ith coordinate, we have

$$f(y + te_i) - f(y) = \int_0^t \frac{\partial f}{\partial x_i}(y + se_i)\, ds, \quad y \in B(x, r),\ |t| < r.$$

We know by Rademacher's theorem (Theorem 1.8) that f is differentiable almost everywhere on U. Hence, using the assumption and Fubini,

$$f(y + te_i) - f(y) = \int_0^t g(y + se_i)\, ds \text{ for almost all } y \in B(x, r) \text{ and all } |t| < r.$$

Since both f and g are continuous, we infer

$$f(x + te_i) - f(x) = \int_0^t g(x + se_i)\, ds, \quad |t| < r,$$

hence $\frac{\partial f}{\partial x_i}(x) = g(x)$. □

Remark 4.4 Applying the implicit function theorem, we obtain that the level sets $d_X^{-1}\{r\}$ are $(d-1)$-dimensional C^1-manifolds, $0 < r < \operatorname{reach} X$. Note also that $d_X^{-1}\{r\} = \partial X_r$ if $0 < r < \operatorname{reach} X$. A stronger result will be obtained later (Corollary 4.22).

Recall that the *tangent cone* of $X \subset \mathbb{R}^d$ at a point $a \in X$ is defined as the set of all vectors $u \in \mathbb{R}^d$ such that either $u = 0$ or there exists a sequence of points $a_i \in X \setminus \{a\}$ with $a_i \to a$ and $r_i > 0$ such that $r_i(a_i - a) \to u$, $i \to \infty$. For general sets X, $\mathrm{Tan}(X, a)$ is always a closed cone. We define further the *normal cone* of X at $a \in X$ as the polar cone to $\mathrm{Tan}(X, a)$ (see Definition 2.6), i.e.,

$$\mathrm{Nor}(X, a) := \mathrm{Tan}(X, a)^o = \{v : v \cdot u \leq 0 \text{ for any } u \in \mathrm{Tan}(X, a)\}.$$

Clearly, $\mathrm{Nor}(X, a)$ is always a closed convex cone. The following geometric property is crucial for sets with positive reach.

Lemma 4.5 *Let* reach $(X, a) =: r > 0$, $a \in \partial X$ *and* $n \in S^{d-1}$. *The following statements are equivalent.*

(i) $\Pi_X(a + tn) = a$ *for some* $t > 0$,
(ii) $\Pi_X(a + tn) = a$ *for all* $0 < t < r$,
(iii) $X \cap \mathring{B}(a + rn, r) = \emptyset$,
(iv) $n \in \mathrm{Nor}(X, a)$.

Proof We start by showing (i) $\Longrightarrow$ (ii). Let (i) be true and denote

$$t_0 := \sup\{t > 0 : \Pi_X(a + tn) = a\}.$$

Note that $t_0 > 0$ by assumption, and that clearly $\Pi_X(a+tn) = a$ for any $0 < t < t_0$. In order to show (ii), we have to verify that $t_0 \geq r$. Assume, for the contrary, that $t_0 < r$. Then $x_0 := a + t_0 n \in \mathrm{int}\,\mathrm{Unp}\,X$ (since reach $X > t_0$). Consider the differential equation

$$g'(s) = \nabla d_X \circ g(s), \quad g(0) = x_0.$$

By the Peano existence theorem, there exists $\delta > 0$ and a differentiable function $g : (-\delta, \delta) \to \mathbb{R}^d$ such that $g(0) = x_0$ and $g'(s) = \nabla d_X(g(s))$ for $|s| < \delta$. Since ∇d_X is always a unit vector (see Lemma 4.2), we have $|g'(s)| = 1$. Further,

$$(d_X \circ g)'(s) = \nabla d_X \circ g(s) \cdot g'(s) = g'(s) \cdot g'(s) = 1,$$

hence, for any $-\delta < s_1 < s_2 < \delta$,

$$\begin{aligned} s_2 - s_1 &= \int_{s_1}^{s_2} |g'(s)|\, ds = \int_{s_1}^{s_2} (d_X \circ g)'(s)\, ds \\ &= d_X(g(s_2)) - d_X(g(s_1)) \leq |g(s_2) - g(s_1)|. \end{aligned}$$

It follows that the image of g must be a straight segment of length 2δ, namely

$$(x_0 + (t_0 - \delta)n, x_0 + (t_0 + \delta)n).$$

By the definition of g, any point of this segment will have its metric projection onto X in a, which is a contradiction with the definition of t_0.

The implications (ii) $\Longrightarrow$ (iii) $\Longrightarrow$ (iv) are easy. We will show that (iv) $\Longrightarrow$ (i). Assume for the contrary that $n \in \operatorname{Nor}(X, a) \cap S^{d-1}$ but (i) is not true, and denote $x(t) := a + tn$, $t > 0$. Clearly $x(t) \notin X$ for sufficiently small $t > 0$ (since otherwise, n would be a tangent vector to X at x, contradicting our assumption). Then, $a(t) := \Pi_X(x(t)) \neq a$ for all sufficiently small $t > 0$ by our assumptions, nevertheless $a(t) \to a$, $t \to 0$, by the continuity of Π_X. We will show that the unit vectors $n(t) := \frac{x(t)-a(t)}{|x(t)-a(t)|}$ converge to n as $t \to 0$. We have

$$\limsup_{t\to 0}(a(t) - a) \cdot n \le 0$$

since $n \in \operatorname{Nor}(X, a)$. Thus

$$\begin{aligned}\liminf_{t\to 0} n(t) \cdot n &= \liminf_{t\to 0} \frac{(x(t) - a) \cdot n + (a - a(t)) \cdot n}{|x(t) - a(t)|} \\ &\ge \liminf_{t\to 0} \frac{t}{|x(t) - a(t)|} \ge 1,\end{aligned}$$

since $|x(t) - a(t)| \le |x(t) - a| = t$ for all t. Hence $n(t) \to n$, $t \to 0$. By the already proven implication (i) $\Longrightarrow$ (iii) applied to $a(t)$ and $n(t)$, the open balls $\mathring{B}(a(t) + rn(t), r)$ do not intersect X. But we have

$$\mathring{B}(a + rn, r) \subset \bigcup_{t>0} \mathring{B}(a(t) + rn(t), r),$$

hence $\mathring{B}(a + rn, r) \cap X = \emptyset$. We have thus shown (iii) which clearly implies (i), a contradiction. □

Note that for any $a, b \in \mathbb{R}^d$, $n \in S^{d-1}$ and $r > 0$,

$$b \in \mathring{B}(a + rn, r) \text{ if and only if } (b - a) \cdot n > \frac{|b - a|^2}{2r}. \tag{4.1}$$

Reformulating Lemma 4.5 (iii), we thus obtain the following.

Corollary 4.6 *If* reach $X > 0$, $a, b \in X$ *and* $v \in \operatorname{Nor}(X, a)$ *then*

$$(b - a) \cdot v \le \frac{|b - a|^2 |v|}{2\operatorname{reach} X}.$$

Recall that X_r denotes the r-parallel set to X (see (2.1)). In the following we show that Π_X is Lipschitz on X_r if $r <$ reach X.

Lemma 4.7 *If* reach $X > r$ *and* $x, y \in X_r$ *then*

$$|\Pi_X(y) - \Pi_X(x)| \le \frac{\operatorname{reach} X}{\operatorname{reach} X - r} |y - x|.$$

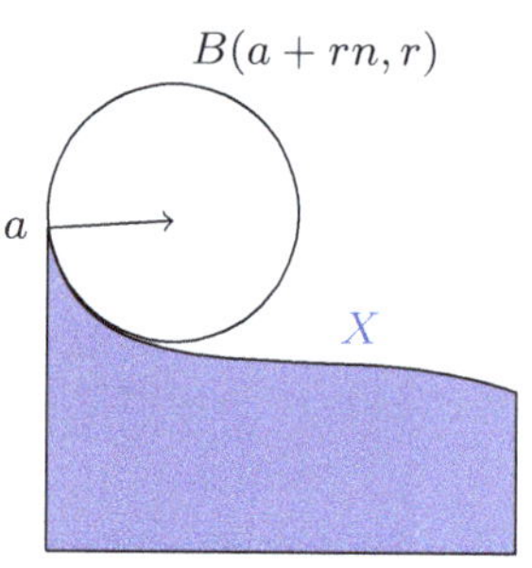

Fig. 4.2 X does not intersect the interior of $B(a+rn,n)$ if $r = \operatorname{reach} X$ and $n \in \operatorname{Nor}(X,a) \cap S^{d-1}$

Proof Denote $a := \Pi_X(x)$ and $b := \Pi_X(y)$. Using Corollary 4.6, we get

$$\begin{aligned}|y-x||b-a| &\geq (y-x)\cdot(b-a)\\ &= ((y-b)+(b-a)+(a-x))\cdot(b-a)\\ &\geq -\frac{|b-a|^2|y-b|}{2\operatorname{reach} X} + |b-a|^2 - \frac{|b-a|^2|x-a|}{2\operatorname{reach} X}\\ &\geq |b-a|^2\left(1-\frac{r}{\operatorname{reach} X}\right),\end{aligned}$$

which implies the assertion (Fig. 4.2). □

There is a connection of the reach of a set with a morphological property of closedness. A set $X \subset \mathbb{R}^d$ is said to be *closed with respect to a set* $B \subset \mathbb{R}^d$ if $X = (X \oplus (-B)) \ominus B$ (cf. [Mat75], where the Minkowski addition $\oplus$ and Minkowski subtraction $\ominus$ are defined). Note that X is closed with respect to B if and only if the complement $\mathbb{R}^d \setminus X$ can be written as a union of translates of $-B$.

Lemma 4.8 *If* reach $X \geq r$ *then* X *is closed with respect to the open ball* $\mathring{B}(0,r)$ *of radius* r.

Proof Assume that reach $X \geq r$. We will show that $\mathbb{R}^d \setminus X$ is the union of open balls of radius r disjoint with X. Let a point $x \in \mathbb{R}^d \setminus X$ be given. If $d_X(x) \geq r$ then clearly $\mathring{B}(x,r) \cap X = \emptyset$. If $0 < d_X(x) < r$ then $x \in \operatorname{Unp} X$ and, denoting $a := \Pi_X(x)$ and $n := \frac{x-a}{|x-a|}$, we have $\mathring{B}(a+rn,r) \cap X = \emptyset$ by Lemma 4.5 (iii), and we are done. □

Example 4.9 The following example shows that the reverse implication in Lemma 4.8 is not true in general. Let $0 < s < r$, let x, y be two points in $\mathbb{R}^2$ with $|x-y| = 2s$ and $X := \mathbb{R}^2 \setminus (\mathring{B}(x,r) \cup \mathring{B}(y,r))$. Then X is closed with respect to $\mathring{B}(r)$, nevertheless, reach $X = \sqrt{r^2-s^2} < r$.

Definition 4.10 Given a set $X \subset \mathbb{R}^d$ with reach $X > 0$, we define the *unit normal bundle* of X as

$$\operatorname{nor} X := \{(a,n) : a \in \partial X,\ n \in \operatorname{Nor}(X,x) \cap S^{d-1}\}.$$

Lemma 4.11 nor X *is a closed subset of* $\mathbb{R}^{2d}$.

Proof Let $(a, n) = \lim_{i\to\infty}(a_i, n_i)$, $(a_i, n_i) \in \operatorname{nor} X$ for all i. We have $\mathring{B}(a_i + rn_i, r) \cap X = \emptyset$ by Lemma 4.5 for $r = \operatorname{reach} X$ and all i, hence also $\mathring{B}(a + rn, r) \cap X = \emptyset$. Thus $(a, n) \in \operatorname{nor} X$, again by Lemma 4.5. □

The closedness of nor X has important consequences. We present two of them.

Corollary 4.12 *Let* reach $X > 0$ *and* $a \in X$.

(i) *If* $a \in \partial X$ *then* $\operatorname{Nor}(X, x) \neq \{0\}$.
(ii) *If* $u \in S^{d-1}$ *belongs to the topological interior of* $\operatorname{Tan}(X, a)$ *then the segment* $[a, a + \varepsilon u]$ *is included in* X *for some* $\varepsilon > 0$.

Proof

(i). If $a \in \partial X$ there exist points $x_i \in \mathbb{R}^d \setminus X$, $x_i \to a$. Then the metric projections $a_i := \Pi_X(x_i)$ also converge to a and $n_i := \frac{x_i - a_i}{|x_i - a_i|} \in \operatorname{Nor}(X, a_i)$ by Lemma 4.5. Passing to a subsequence, we can achieve that $n_i \to n \in S^{d-1}$. Then $(a, n) \in \operatorname{nor} X$ since nor X is closed.
(ii). Let $u \in \operatorname{int} \operatorname{Tan}(X, a) \cap S^{d-1}$ be given. Assume, for the contrary, that there exists a sequence $\varepsilon_i \to 0$ such that $x_i := a + \varepsilon_i u \notin X$ for all i. Then, as in the proof of (i), $a_i := \Pi_X(a + \varepsilon_i u) \to a$, $n_i := \frac{x_i - a_i}{|x_i - a_i|} \in \operatorname{Nor}(X, a_i)$ and we can assume that $n_i \to n \in S^{d-1} \cap \operatorname{Nor}(X, a)$. Corollary 4.6 implies that $(a - a_i) \cdot n_i \leq (2r)^{-1}|a_i - a|^2$ with $r := \operatorname{reach} X$, hence

$$u \cdot n_i = \frac{1}{\varepsilon_i}(x_i - a) \cdot n_i \geq \frac{1}{\varepsilon_i}(a_i - a) \cdot n_i \geq -\frac{|a_i - a|^2}{2r\varepsilon_i}.$$

Letting $i \to \infty$ we get $u \cdot n \geq 0$ (note that $|a_i - a| \leq 2\varepsilon_i$) which, however, contradits the facts that $n \in \operatorname{Nor}(X, a) = \operatorname{Tan}(X, a)^o$ and $u \in \operatorname{int} \operatorname{Tan}(X, a)$. □

An important property of sets with positive reach is the convexity of the tangent cones. Recall that C^o denotes the polar cone to a cone C (see Definition 2.6).

Lemma 4.13 *If* reach $X > 0$ *and* $a \in \partial X$ *then* $\operatorname{Tan}(X, a) = \operatorname{Nor}(X, a)^o$, *hence* $\operatorname{Tan}(X, a)$ *is convex.*

Proof First note that

$$\operatorname{Nor}(X, a)^o = \operatorname{Tan}(X, a)^{oo} \supset \operatorname{Tan}(X, a)$$

by (2.6). Hence, it is enough to show that

$$\operatorname{Nor}(X, a)^o \subset \operatorname{Tan}(X, a). \tag{4.2}$$

Let u be a unit vector which is not a tangent vector to X at a. Then, by the definition of tangent vectors, there exist $\varepsilon, \gamma > 0$ such that the cone

$$V := \{v : (v - a) \cdot u > |v - a| \cos \gamma\}$$

does not intersect $X \cap B(a, \varepsilon)$. We can assume that $\varepsilon \leq \text{reach}\, X$. As in the proof of Lemma 4.5, denote $x(t) := a + tu$, $a(t) := \Pi_X x(t)$, $n(t) := \frac{x(t)-a(t)}{|x(t)-a(t)|}$, $0 < t < \varepsilon$. An easy geometric observation is that the open cone V contains the open ball $\mathring{B}(x(t), t \sin \gamma)$. Clearly $|x(t) - a(t)| \leq t$, and since $a(t) \in X$, it must lie outside V and we have thus

$$t \sin \gamma \leq |x(t) - a(t)| \leq t, \quad 0 < t < \varepsilon.$$

Further, Corollary 4.6 implies

$$(a - a(t)) \cdot n(t) \leq \frac{|a - a(t)|^2}{2\text{reach}\, X}, \quad 0 < t < \varepsilon.$$

Using these two estimates, we obtain

$$\begin{aligned} u \cdot n(t) &= t^{-1}((x(t) - a(t)) - (a - a(t))) \cdot n(t) \\ &= t^{-1} (|x(t) - a(t)| - (a - a(t)) \cdot n(t)) \\ &\geq \sin \gamma - \frac{|a(t) - a|}{\text{reach}\, X}. \end{aligned}$$

In view of the compactness of the unit sphere, we can find a sequence $t_i \to 0_+$ such that $n(t_i) \to n \in S^{d-1}$. Since also $a(t_i) \to a$, we have $(a, n) \in \text{nor}\, X$ by Lemma 4.11, hence, $n \in \text{Nor}(X, x)$. As

$$u \cdot n \geq \liminf_{t \to 0} u \cdot n(t) > 0,$$

we infer that $u \notin \text{Nor}(X, x)^o$, and the proof (4.2) is finished. □

We conclude this subsection with an equivalent condition for reach $X \geq r$.

Proposition 4.14 *Given a closed set $X \subset \mathbb{R}^d$ and $r > 0$, the following conditions are equivalent:*

(a) reach $X \geq r$,
(b) *for any $a, b \in X$, $d_{\text{Tan}(X,a)}(b - a) \leq \frac{|b-a|^2}{2r}$.*

Remark 4.15 Note that condition (b) can be reformulated as follows: If $a, b \in X$ with $0 < |b - a| < 2r$ then there exists a tangent vector $0 \neq u$ to X at a with $\angle(b - a, u) \leq \arcsin \frac{b-a}{2r}$.

Proof (a) $\Longrightarrow$ (b). Assume that reach $X \geq r$ and $a, b \in X$. If $a = b$ or $|b-a| \geq 2r$ then condition (b) is clearly satisfied with $0 \in \mathrm{Tan}(X, a)$. Assume thus that $0 < |b-a| < 2r$ and denote $u_0 := \frac{b-a}{|b-a|}$, $\gamma := \arcsin \frac{|b-a|}{2r}$ and

$$C := \{u : u \cdot u_0 \geq |u| \cos \gamma\}.$$

We have to show that $\mathrm{Tan}(X, a)$ has nontrivial intersection with C. Assume, for the contrary, that the intersection is trivial. Since both are closed convex cones, there must be a hyperplane strictly separating them, i.e., there exists a unit vector w such that $u \cdot w < 0$ if $u \in \mathrm{Tan}(X, a)$ and $v \cdot w > 0$ if $v \in C$. Note that clearly $w \in \mathrm{Nor}(X, a) = \mathrm{Tan}(X, a)^o$. Consider the vector $w_0 := u_0 - (\sin \gamma)w$. Clearly, the angle formed by u_0 and w_0 is less or equal to γ, hence, $w_0 \in C$. Further, we have

$$w \cdot \frac{b-a}{|b-a|} = w \cdot u_0 = w \cdot (u_0 - w_0) + w \cdot w_0 > w \cdot (u_0 - w_0) = \sin \gamma = \frac{|b-a|}{2r}.$$

But this contradicts Corollary 4.6.

(b) $\Longrightarrow$ (a). Assume that (b) holds but reach $X < r$. Then there exists a point x with $s := d_X(x) < r$ which has at least two nearest neighbours in X, i.e., there are two different point $a, b \in X \cap \partial B(x, s)$. Since X does not intersect $\mathring{B}(x, s)$, any tangent vector to X at a must form an angle with $b - a$ of size at least $\arcsin \frac{|b-a|}{2s}$, which contradicts property (b). □

4.2 Reach of Intersection

Example 4.16 Let X be the x-axis and Y the graph of the function $f(x) = x^4 \sin \frac{1}{x}$ if $x \in (-1, 0) \cup (0, 1)$ and $f(0) = 0$. Then reach X, reach $Y > 0$, but reach $(X \cap Y, 0) = 0$. Hence, the property to have positive reach is not preserved by intersections.

Under additional assumptions, the intersection preserves positive reach, as the following proposition shows.

Proposition 4.17 *Let* reach X, reach $Y \geq r > 0$. *Given* $a \in \partial(X \cap Y)$, *denote*

$$\eta(a) = \inf \left\{ \frac{|u+v|}{|u|+|v|} : u \in \mathrm{Nor}(X, a), v \in \mathrm{Nor}(Y, a), |u| + |v| > 0 \right\}.$$

Set $\eta = \inf_{a \in \partial(X \cap Y)} \eta(a)$, *and assume that* $\eta > 0$. *Then for all* $a \in \partial(X \cap Y)$,

(i) $\mathrm{Tan}(X \cap Y, a) = \mathrm{Tan}(X, a) \cap \mathrm{Tan}(Y, a)$,
(ii) $\mathrm{Nor}(X \cap Y, a) = \mathrm{Nor}(X, a) + \mathrm{Nor}(Y, a)$,
(iii) reach $(X \cap Y) \geq r\eta$.

Proof First, note that (i) and (ii) are equivalent by (2.7). We will show (ii).

Assume that reach X, reach $Y \geq r$ and let $a \in X \cap Y$ be given. If $a \notin \partial X \cap \partial Y$ then (ii) is obvious. If $a \in \partial X \cap \partial Y$ and $m \in \mathrm{Nor}(X, a)$, $n \in \mathrm{Nor}(Y, a)$ are unit vectors, then $\mathring{B}(a+rm, r) \cap X = \mathring{B}(a+rn, r) \cap Y = \emptyset$ by Lemma 4.5 (iii). Hence,

$$(\mathring{B}(a + rm, r) \cup \mathring{B}(a + rn, r)) \cap (X \cap Y) = \emptyset.$$

This implies that both m, n belong to $\mathrm{Nor}(X \cup Y, a)$, and if m, n are linearly independent, then for any point z from the line segment $[rm, rn]$, the open ball $\mathring{B}(z, |z - a|)$ is contained in $\mathring{B}(a + rm, r) \cup \mathring{B}(a + rn, r)$, beeing thus disjoint with $X \cap Y$, which implies that $z - a \in \mathrm{Nor}(X \cap Y, a)$ (indeed, $\mathrm{Tan}(X \cap Y, a) \subset \mathrm{Tan}(\mathbb{R}^d \setminus \mathring{B}(z, |z-a|), a) = \{u : u \cdot (z-a) \leq 0\}$). We have thus shown the inclusion

$$\mathrm{Nor}(X, a) + \mathrm{Nor}(Y, a) \subset \mathrm{Nor}(X \cap Y, a).$$

For the opposite inclusion, take a vector $u \notin \mathrm{Nor}(X, a) + \mathrm{Nor}(Y, a)$, let H be a hyperplane separating u and the convex cone $\mathrm{Nor}(X, x) + \mathrm{Nor}(Y, x)$, and let v be a vector perpendicular to H and forming an acute angle with u. Then $u \in \mathrm{int}\,\mathrm{Tan}(X, a) \cap \mathrm{int}\,\mathrm{Tan}(Y, a)$, hence, $a + \varepsilon v \in X \cap Y$ for sufficiently small ε by Lemma 4.12. Hence $v \in \mathrm{Tan}(X \cap Y, x)$ and $u \notin \mathrm{Nor}(X \cap Y, x)$. Thus (ii) is verified (Fig. 4.3).

To show (iii), we will use Proposition 4.14. Let $a, b \in X \cap Y$ be given. We show that

$$(b - a) \cdot w \leq \frac{|b - a|^2 |w|}{2\eta r}, \quad w \in \mathrm{Nor}(X \cap Y, a). \tag{4.3}$$

If $w = 0$ the inequality is obvious. If $w \neq 0$ we can represent $w = u + v$ with some $u \in \mathrm{Nor}(X, a)$, $v \in \mathrm{Nor}(Y, a)$ by (ii), at least one of u, v being nonzero. We have

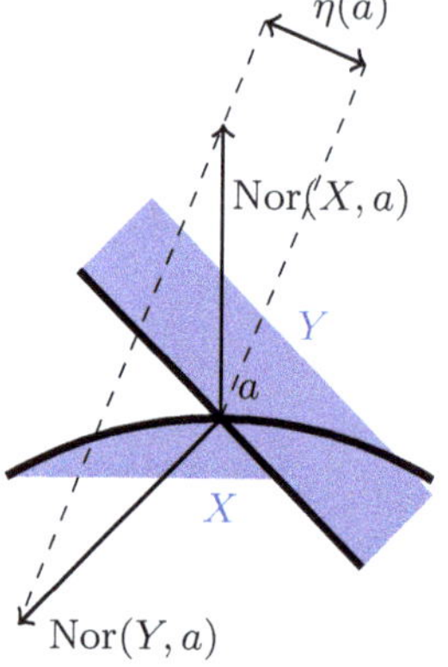

Fig. 4.3 Two sets X, Y with positive reach and the value of $\eta(a)$ at a point $a \in \partial X \cap \partial Y$

$(b-a)\cdot u \le \frac{|b-a|^2|u|}{2r}$ and $(b-a)\cdot v \le \frac{|b-a|^2|v|}{2r}$ by Corollary 4.6, hence

$$(b-a)\cdot(u+v) \le \frac{|b-a|^2(|u|+|v|)}{2r} \le \frac{|b-a|^2|u+v|}{2\eta r},$$

proving (4.3).

We know already from (i) and (ii) that $\mathrm{Tan}(X\cap Y, a)$ is a convex cone and $\mathrm{Nor}(X\cap Y, a) = \mathrm{Tan}(X\cap Y, a)^o$. Denote $u = \Pi_{\mathrm{Tan}(X\cap Y,a)}(b-a)$ and $w = b-a-u$. Since $\mathrm{Tan}(X\cap Y, a)$ is a cone we have $u\cdot w = 0$ and, hence, $|w| = |b-a|\cos\theta$, where θ is the angle formed by w and $b-a$. Also $w\in \mathrm{Nor}(X\cap Y, a)$, and we infer from (4.3) that $\cos\theta \le |b-a|/(2\eta r)$, hence $|v| \le |b-a|^2/(2\eta r)$. Thus we have shown that

$$d_{\mathrm{Tan}(X\cap Y,a)}(b-a) \le \frac{|b-a|^2}{2\eta r},$$

which implies that reach $(X\cap Y)\ge \eta r$ by Proposition 4.14. □

Definition 4.18 We say that two sets X, Y with positive reach *touch* if there exist $(a,n)\in \operatorname{nor} X$ such that $(a,-n)\in \operatorname{nor} Y$.

Note that X, Y do not touch whenever the number $\eta > 0$ from Proposition 4.17 is positive.

We will need later an extension of Proposition 4.17 for more than two sets:

Theorem 4.19 *Let $X^1,\dots,X^k$ be sets with* reach $X^i \ge r > 0$, $i = 1,\dots,k$. *Let η be the infimum of all numbers*

$$\frac{|v_1+\cdots+v_k|}{|v_1|+\cdots+|v_k|}$$

such that $v_i\in \mathrm{Nor}(X^i, a)$, $i = 1,\dots,k$, $a\in \bigcap_{i=1}^k X^i$ and $|v_1|+\cdots+|v_k| > 0$. Assume that $\eta > 0$. Then for all $a\in\partial(X^1\cap\cdots\cap X^k)$,

(i) $\mathrm{Tan}(X^1\cap\cdots\cap X^k, a) = \mathrm{Tan}(X^1, a)\cap\cdots\cap \mathrm{Tan}(X^k, a)$,
(ii) $\mathrm{Nor}(X^1\cap\cdots\cap X^k, a) = \mathrm{Nor}(X^1, a)+\cdots+\mathrm{Nor}(X^k, a)$,
(iii) reach $\left(\bigcap_{i=1}^k X^i\right) \ge r\eta^{k-1}$.

Proof For $k = 2$, this is just Proposition 4.17. For general k it follows by induction. Since the proof of (i) and (ii) is easy, we will show only (iii).

Suppose that (iii) is valid for at most $k-1$ sets and let $X^1,\dots,X^k$ be sets with reach greater or equal to r fulfilling the assumptions. The sets $X^1,\dots,X^{k-1}$ also satisfy the assumptions with some $\eta'\ge\eta$ and, hence, reach $(\bigcap_{i=1}^{k-1} X^i) \ge r\eta^{k-2}$ by induction assumption. Further, let $u\in \mathrm{Nor}(\bigcap_{i=1}^{k-1} X^i, a)$ and $v\in \mathrm{Nor}(X^k, a)$, $|u|+|v| > 0$. By (ii), we can write $u = u_1+\cdots+u_{k-1}$ for some $u_i\in \mathrm{Nor}(X^i, a)$. Then we have using the assumption and the triangle inequality

$$|u+v| = |u_1+\cdots+u_{k-1}+v| \ge \eta(|u_1|+\cdots+|u_{k-1}|+|v|) \ge \eta(|u|+|v|).$$

Hence, the two sets $\bigcap_{i=1}^{k-1} X^i$ and X^k satisfy the conditions of Proposition 4.17 with $\tilde{r} = r\eta^{k-2} \leq r$ and we infer reach $(\bigcap_{i=1}^{k} X^i) \geq \tilde{r}\eta = r\eta^{k-1}$. □

We close this section with a result about sections with small balls.

Proposition 4.20 *If* reach $X \geq r > 0$ *then for all* $z \in \mathbb{R}^d$, $X \cap B(z, r)$ *is either empty or* reach $(X \cap B(z, r)) \geq r$.

Proof Assume that $X \cap B(z, r) \neq \emptyset$. We will prove that reach $(X \cap B(z, r)) \geq r$. Let $a, b \in X \cap B(z, r)$. Due to Proposition 4.14, we have to show that

$$d_{\mathrm{Tan}(X \cap B(z,r))}(b - a) \leq \frac{|b - a|^2}{2r} =: \rho.$$

Obviously, $|b - a| \leq 2r$ and $B(z, r)$ contains the spindle-shaped set $L_{a,b}$ given as the intersection of all closed balls of radius r containing a and b. One easily verifies that $\mathrm{Tan}(L_{a,b}, a)$ contains the ball $B(b - a, \rho)$. Since reach $X \geq r$, we have by Proposition 4.14

$$d_{\mathrm{Tan}(X,a)}(b - a) \leq \rho.$$

Thus there is a vector $v \in \mathrm{Tan}(X, a)$ with $|v - (b - a)| \leq \rho$.Then, also $v \in B(b - a, \rho) \subset \mathrm{Tan}(L_{a,b}, a) \subset \mathrm{Tan}(B(z, r), a)$. Since X and $B(z, r)$ do not touch, we infer

$$v \in \mathrm{Tan}(X, a) \cap \mathrm{Tan}(B(z, r), a) = \mathrm{Tan}(X \cap B(z, r), a)$$

by Proposition 4.17(i), and we are done. □

4.3 Unit Normal Bundle

Recall that the unit normal bundle of a set X with positive reach is defined as

$$\mathrm{nor}\, X = \{(a, n) : a \in \partial X, n \in S^{d-1} \cap \mathrm{Nor}(X, a)\}.$$

By means of the following mappings we parametrize the "collars" $X_r \setminus X$ and boundaries ∂X_r for $0 < r <$ reach X. We define

$$f : (a, n, t) \mapsto a + tn, \quad (a, n, t) \in \mathrm{nor}\, X \times (0, \mathrm{reach}\, X),$$

and the sections

$$f^{(t)} : (a, n) \mapsto f(a, n, t), \quad (a, n) \in \mathrm{nor}\, X, \quad 0 < t < \mathrm{reach}\, X.$$

Lemma 4.21 *For each* $0 < r < \operatorname{reach} X$, $f^{(r)}$ *is a bi-Lipschitz homeomorphism from* $\operatorname{nor} X$ *onto* ∂X_r *and the restriction of* f *to* $\operatorname{nor} X \times (0, r]$ *is a locally bi-Lipschitz homeomorphism from* $\operatorname{nor} X \times (0, r]$ *onto* $X_r \setminus X$.

Proof The mapping f is clearly Lipschitz on $\operatorname{nor} X \times (0, r]$. Consider now the inverse mapping

$$f^{-1} : z \mapsto \left(\Pi_X(z), \frac{z - \Pi_X(z)}{|z - \Pi_X(z)|}, d_X(z) \right), \quad z \in X_r \setminus X.$$

Its first and third component are Lipschitz on $X_r \setminus X$ by Lemma 4.7. It is easy to verify that if g is a Lipschitz vector function on $A \subset \mathbb{R}^d$ and there exist $c, C > 0$ such that $c \le |g(x)| \le C$, $x \in A$, then the normalized vector function $g/|g|$ is Lipschitz on A as well. Thus, the vector function

$$z \mapsto \frac{z - \Pi_X(z)}{|z - \Pi_X(z)|}$$

is Lipschitz on $X_r \setminus X_s$ for any $0 < s < r < \operatorname{reach} X$. Consequently, f^{-1} is Lipschitz on $\operatorname{nor} X \times (s, r]$ for any $0 < s < r$, hence, locally Lipschitz on $\operatorname{nor} X \times (0, r)$. The bi-Lipschitz property of $f^{(r)}$ follows immediately (Fig. 4.4). □

Corollary 4.22 *If* $0 < r < \operatorname{reach} X$ *then* X_r *is a closed* $C^{1,1}$*-domain (see Definition* 3.8*). Further, the unit normal bundle* $\operatorname{nor} X$ *is a* $(d-1)$*-dimensional Lipschitz submanifold of* $\mathbb{R}^{2d}$ *(in particular, it is locally* $(d-1)$*-rectifiable).*

Proof The Lipschitz property of the Gauss map $z \mapsto \frac{z - \Pi_X(z)}{|z - \Pi_X(z)|}$ on ∂X_r follows from Lemma 4.21. Also, since ∂X_r is C^1-submanifold, $\operatorname{nor} X$ as its bi-Lipschitz image is a Lipschitz submanifold. □

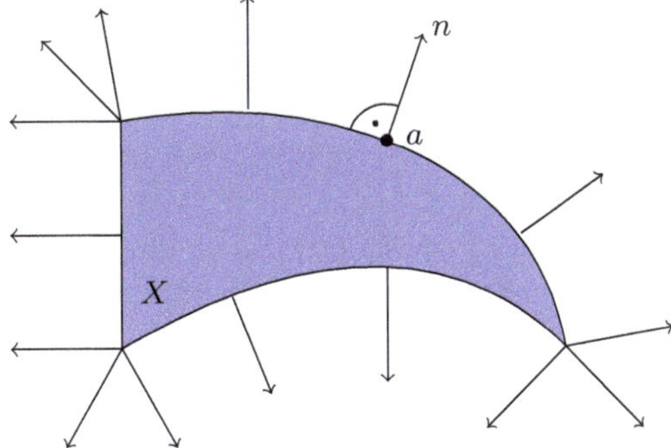

Fig. 4.4 A set X with positive reach and its unit normal bundle

4.4 Principal Curvatures and Directions

Let X be a set of positive reach in $\mathbb{R}^d$ and let $(x, n) \in \operatorname{nor} X$ be given. For any $0 < r < \operatorname{reach} X$, $x + rn \in \partial X_r$ and the Gauss map ν_r on X_r (cf. Sect. 3.1) satisfies $\nu_r(x+rn) = n$. Thus, if $0 < t < t+s < \operatorname{reach} X$ then the mapping $y \mapsto y + s\nu_t(y)$ is a bi-Lipschitz homeomorphism between ∂X_t and ∂X_{t+s} (cf. Lemma 4.21), and its inverse is $z \mapsto z - s\nu_{t+s}(z)$. Consequently, ν_t is differentiable at y if and only if ν_{t+s} is differentiable at $y + sn$. Further, note that by Rademacher's theorem, for each $0 < t < \operatorname{reach} X$, since ν_t is Lipschitz on the (C^1-submanifold) ∂X_t, it is differentiable $\mathcal{H}^{d-1}$-almost everywhere on ∂X_t. It follows that for $\mathcal{H}^{d-1}$-almost all points $(x, n) \in \operatorname{nor} X$, ν_t is differentiable at $x + tn$, for all $0 < t < \operatorname{reach} X$. We call points $(x, n) \in \operatorname{nor} X$ with this property *regular*.

Proposition 4.23 *If* $\operatorname{reach} X > 0$ *and* $(x, n) \in \operatorname{nor} X$ *is regular then* $\operatorname{Tan}(\operatorname{nor} X, (x, n))$ *is a* $(d-1)$*-dimensional subspace and there exist vectors* $b_1(x, n), \dots, b_{d-1}(x, n)$ *in* $\mathbb{R}^d$ *and numbers* $\kappa_1(x, n), \dots, \kappa_{d-1}(x, n) \in [-\operatorname{reach} X, \infty]$ *such that* $b_1(x, n), \dots, b_{d-1}(x, n), n$ *form a positively oriented orthonormal basis of* $\mathbb{R}^d$ *and the vectors*

$$\left(\frac{1}{\sqrt{1+\kappa_i^2(x,n)}} b_i(x,n), \frac{\kappa_i(x,n)}{\sqrt{1+\kappa_i^2(x,n)}} b_i(x,n) \right), \quad i = 1, \dots, d-1,$$

form an orthonormal basis of $\operatorname{Tan}(\operatorname{nor} X, (x, n))$. *(We set* $\frac{1}{\sqrt{1+\infty^2}} = 0$ *and* $\frac{\infty}{\sqrt{1+\infty^2}} = 1$.*)*

Proof Fix a regular point $(x, n) \in \operatorname{nor} X$ and $0 < r < \operatorname{reach} X$. Recall that $\nu_r(y) = n$ at $y = x + rn$, the Weingarten mapping $-D\nu_r(y)$ exists and, consequently, there exist principal directions $b_i^r(y) \in \operatorname{Tan}(\partial X_r, y)$ and principal values (curvatures) $\kappa_i^r(y)$; we may assume that the vectors $b_1^r(y), \dots, b_{d-1}^r(y), n$ form a positively oriented orthonormal basis of $\mathbb{R}^d$. We will show that

$$-\frac{1}{\operatorname{reach} X - r} \le \kappa_i^r(y) \le \frac{1}{r}, \quad i = 1, \dots, d-1. \tag{4.4}$$

This follows from the fact that all directional derivatives of ν_r at y lie within the given bounds, which is a consequence of

$$-\frac{1}{\operatorname{reach} X - r} \le \frac{\nu_r(z) - \nu_r(y)}{|z-y|} \cdot \frac{z-y}{|z-y|} \le \frac{1}{r}, \quad y, z \in \partial X_r.$$

Here the left-hand side inequality follows from Corollary 4.6 since reach $X_r \geq$ reach $X - r$. For the right-hand side inequality, note that

$$\frac{\nu_r(z) - \nu_r(y)}{|z - y|} \cdot \frac{z - y}{|z - y|} = \frac{1}{s} \frac{((z - \Pi_X z) - (y - \Pi_X y)) \cdot (z - y)}{|z - y|^2}$$
$$= \frac{1}{s}\left(1 - \frac{(\Pi_X z - \Pi_X y) \cdot (z - y)}{|z - y|^2}\right) \leq \frac{1}{s},$$

since we have shown in the proof of Lemma 4.7 that $(\Pi_X z - \Pi_X y) \cdot (z - y) \geq 0$. Thus, (4.4) is proved.

The vectors $b_i^r(y)$, $i = 1, \dots, d-1$, form a basis of $\operatorname{Tan}(\partial X_r, y)$. The inverse mapping to $f^{(r)}$ can be written in the form

$$(f^{(r)})^{-1}(y) = (y - r\nu_r(y), \nu_r(y)),$$

hence, it is differentiable as well, with differential

$$D(f^{(r)})^{-1}(y) = (I - rD\nu_r(y), D\nu_r(y)),$$

mapping $\operatorname{Tan}(\partial X_r, y)$ onto $\operatorname{Tan}(\operatorname{nor} X, (x, n))$. Thus, the vectors

$$D(f^{(r)})^{-1}(y)b_i^r(y) = ((1 - \kappa_i^{(r)} r)b_i^r, \kappa_i^{(r)} b_i^r), \quad i = 1, \dots, d-1, \tag{4.5}$$

form a basis of $\operatorname{Tan}(\operatorname{nor} X, (x, n))$. Setting $b_i(x, n) := b_i^r(y)$ and

$$\kappa_i(x, n) := \begin{cases} \frac{\kappa_i^{(r)}(y)}{1 - \kappa_i^{(r)}(y)r} & \text{if } \kappa_i^{(r)} r < 1, \\ \infty & \text{if } \kappa_i^{(r)} = 1, \end{cases} \qquad i = 1, \dots, d-1,$$

(recall that $y = x + rn$), we get the assertion. □

Lemma 4.24 *The values $\kappa_i(x, n)$ from Proposition 4.23 are uniquely determined at any regular point $(x, n) \in \operatorname{nor} X$, up to the order. Furthermore, for any $1 \leq i \leq d-1$, the subspace*

$$\operatorname{Lin}\{b_j(x, n) : \kappa_j(x, n) = \kappa_i(x, n)\}$$

is uniquely determined.

Proof Throughout the proof, we shall omit the argument (x, n) at κ_i and b_i. Assume that

$$\left(\frac{1}{\sqrt{1 + \kappa_i^2}} b_i, \frac{\kappa_i}{\sqrt{1 + \kappa_i^2}} b_i\right), \quad i = 1, \dots, d-1,$$

and

$$\left(\frac{1}{\sqrt{1+(\kappa_i')^2}}b_i', \frac{\kappa_i'}{\sqrt{1+(\kappa_i')^2}}b_i'\right), \quad i = 1, \dots, d-1,$$

are two orthonormal bases of Tan(nor X, (x, n)), where $\{b_i\}$, $\{b_i'\}$, are two orthonormal bases of $n^\perp$. Then there exist coefficients c_{ij} such that

$$\frac{1}{\sqrt{1+(\kappa_i')^2}}b_i' = \sum_j c_{ij}\frac{1}{\sqrt{1+\kappa_j^2}}b_j, \tag{4.6}$$

$$\frac{\kappa_i'}{\sqrt{1+(\kappa_i')^2}}b_i' = \sum_j c_{ij}\frac{\kappa_j}{\sqrt{1+\kappa_j^2}}b_j. \tag{4.7}$$

Fix some $1 \le i \le d-1$ and assume first that $\kappa_i' < \infty$. Multiplying (4.6) with κ_i', we get

$$\frac{\kappa_i'}{\sqrt{1+(\kappa_i')^2}}b_i' = \sum_j c_{ij}\frac{\kappa_i'}{\sqrt{1+\kappa_j^2}}b_j \tag{4.8}$$

and, comparing (4.7) and (4.8), we obtain that

$$c_{ij}\left(\frac{\kappa_j}{\sqrt{1+\kappa_j^2}} - \frac{\kappa_i'}{\sqrt{1+\kappa_j^2}}\right) = 0$$

for all j. Consequently, we have $\kappa_j < \infty$ and $c_{ij}\kappa_i' = c_{ij}\kappa_j$ for all j, hence, the alternative

$$c_{ij} = 0 \text{ or } \kappa_i' = \kappa_j \tag{4.9}$$

holds for any j.

Assume now that $\kappa_i' = \infty$. Then we have zero on the left hand side of (4.6) which implies that $c_{ij}/\sqrt{1+\kappa_j^2} = 0$, hence $c_{ij} = 0$ or $\kappa_j = \infty$, for all j. Thus (4.9) holds again for all j. It follows from (4.9) that the sets of numbers $\{\kappa_i : 1 \le i \le d-1\}$ and $\{\kappa_i' : 1 \le i \le d-1\}$ coincide and that any b_i' is a linear combination of those b_j belonging to the same κ_i. From this property, the assertion follows. □

Definition 4.25 The numbers $\kappa_1(x,n),\ldots,\kappa_{d-1}(x,n) \in [-\operatorname{reach} X,\infty]$ are called *(generalized) principal curvatures* of X at (x,n), and the unit vectors $b_1(x,n),\ldots,b_{d-1}(x,n)$ are the associated *principal directions*. The function

$$s_k(X;x,n) = \sum_{1\le i_1<\cdots<i_k\le d-1} \frac{\prod_{j=1}^k \kappa_{i_j}(x,n)}{\prod_{i=1}^{d-1}\sqrt{1+\kappa_i^2(x,n)}}$$

defined at regular points of $\operatorname{nor} X$ will be called the *(generalized) symmetric function of principal curvatures of order k of X at (x,n)*. (The summands should be understood as products of $d-1$ fractions of the type $\frac{1}{\sqrt{1+\kappa_i^2}}$ or $\frac{\kappa_i}{\sqrt{1+\kappa_i^2}}$ which are defined even if $\kappa_i = \infty$ by the convention from Proposition 4.23.)

As a corollary, we obtain a relation between the principal curvatures and directions of X and of the parallel sets.

Corollary 4.26 *Let (x,n) be a regular point of* $\operatorname{nor} X$ *and* $0 < r < \operatorname{reach} X$. *Then, the principal curvatures $\kappa_i^{(r)}$ of X_r at $x+rn$ and the (generalized) principal curvatures $\kappa_i(x,n)$ of X at (x,n) are related by*

$$\kappa_i(x,n) = \begin{cases} \frac{\kappa_i^{(r)}(x+rn)}{1-r\kappa_i^{(r)}(x+rn)} & \text{if } \kappa_i^{(r)}(x+rn) < \frac{1}{r}, \\ \infty & \text{if } \kappa_i^{(r)}(x+rn) = \frac{1}{r}, \end{cases} \qquad i=1,\ldots,d-1,$$

after an eventual change of order. Consequently,

$$\lim_{r\to 0_+} \kappa_i^{(r)}(x+rn) = \kappa_i(x,n), \quad i=1,\ldots,d-1.$$

The (generalized) principal directions $b_i(x,n)$, $i=1,\ldots,d-1$ are the principal directions $b_i^r(x+rn)$ of ∂X_r at $x+rn$, for all $r <$ $\operatorname{reach} X$.

Remark 4.27 Note that, by construction, the principal curvatures are locally determined in the following sense: If X,Y are two sets of positive reach such that $X\cap U = Y\cap U$ for some nonempty open set U, then X and Y have the same set of principal curvatures at $\mathcal{H}^{d-1}$-almost all points of

$$\operatorname{nor} X \cap (U\times S^{d-1}) = \operatorname{nor} Y \cap (U\times S^{d-1}).$$

4.5 Curvature Measures and Steiner Formula

Recall the definition of the function $f : \operatorname{nor} X \times (0,\operatorname{reach} X) \to \{y : 0 < d_X(y) < \operatorname{reach} X\}$ from Sect. 4.3. Applying the Area formula (Theorem 1.14) for the mapping $g = f^{-1}$, we get for any $0 < r < \operatorname{reach} X$ and for any nonnegative measurable function h on $X_r \setminus X$ with bounded support

$$\int_{X_r \setminus X} J_d g(y) h(y)\, \mathcal{L}^d(\mathrm{d}y) = \int_{\operatorname{nor} X} \int_0^r h(x+tn)\, \mathrm{d}t\, \mathcal{H}^{d-1}(\mathrm{d}(x,n)). \tag{4.10}$$

This will turn to be the Steiner formula. It remains to compute the Jacobian $J_d g$.

From the expression

$$g(y) = (y - t\nu_t(y), \nu_t(y), t), \quad y \in \partial X_t,$$

we get (with the principal directions $b_i = b_i^t = b_i(y)$ and principal curvatures $\kappa_i^t = \kappa_i^t(y)$ at $y = x + tn$)

$$Dg(y)b_i = (b_i - t\kappa_i^t b_i, \kappa_i^t b_i, 0), \quad i = 1, \dots, d-1,$$

and

$$Dg(y)n = (0, 0, 1).$$

Thus, with respect to the orthonormal basis $\{b_1, \dots, b_{d-1}, n\}$, $Dg(y)^\top Dg(y)$ has a diagonal matrix with diagonal elements

$$(1 - t\kappa_1^t)^2 + (\kappa_1^t)^2, \dots, (1 - t\kappa_{d-1}^t)^2 + (\kappa_{d-1}^t)^2, 1.$$

Using the relation between κ_i^t and the principal directions $\kappa_i = \kappa_i(x, n)$ from Corollary 4.26, we get

$$(1 - t\kappa_i^t)^2 + (\kappa_i^t)^2 = \frac{1 + \kappa_i^2}{(1 + t\kappa_i)^2},$$

hence, since $1 + t\kappa_i > 0$, we get

$$J_d g(y) = \prod_{i=1}^{d-1} \frac{\sqrt{1 + \kappa_i^2}}{1 + t\kappa_i}.$$

Notice that the Jacobian is positive. Using Definition 4.25, we can write

$$(J_d g(y))^{-1} = \sum_{k=0}^{d-1} s_k(X, x, n) t^k. \tag{4.11}$$

Hence, assuming that X is bounded, and choosing the function $h = 1/J_d g$, we infer from (4.10) (after renumeration)

$$\mathcal{L}^d(X_r \setminus X) = \sum_{k=0}^{d-1} \frac{r^{d-k}}{d-k} \int_{\operatorname{nor} X} s_{d-1-k}(X; x, n)\, \mathcal{H}^{d-1}(\mathrm{d}(x, n)). \tag{4.12}$$

This is the (global) Steiner formula for X. To obtain the local version, denote

$$\widetilde{\Pi}_X : z \mapsto \left(\Pi_X(z), \frac{z - \Pi_X(z)}{|z - \Pi_X(z)|} \right),$$

from $\operatorname{Unp} X$ to $\operatorname{nor} X$, which is in fact $g = f^{-1}$ without the last coordinate. Applying now (4.10) with $h(y) = \mathbf{1}_E \circ \widetilde{\Pi}_X(y)/J_d g(y)$ with a bounded Borel subset $E \subset \mathbb{R}^d \times \mathbb{R}^d$, we get

$$\begin{aligned} &\mathcal{L}^d((X_r \setminus X) \cap (\widetilde{\Pi}_X)^{-1}(E)) \\ &\quad = \sum_{k=0}^{d-1} \frac{r^{d-k}}{d-k} \int_{E \cap \operatorname{nor} X} s_{d-1-k}(X; x, n)\, \mathcal{H}^{d-1}(\mathrm{d}(x, n)). \end{aligned} \tag{4.13}$$

(Here we do not need the assumption X being bounded.)

Definition 4.28 The *curvature-direction measure of order* $k \in \{0, \dots, d-1\}$ of X is the signed Radon measure $\widetilde{C}_k(X, \cdot)$ on $\mathbb{R}^d \times \mathbb{R}^d$ given by

$$\widetilde{C}_k(X, E) = \frac{1}{(d-k)\omega_{d-k}} \int_{E \cap \operatorname{nor} X} s_{d-1-k}(X; x, n) \mathcal{H}^{d-1}(\mathrm{d}(x, n))$$

for any bounded Borel set $E \subset \mathbb{R}^d \times \mathbb{R}^d$.

Remark 4.29

1. Since the Jacobian of a Lipschitz mapping is always locally integrable (cf. Theorem 1.14), we infer from (4.11) that the functions $s_k(X; \cdot)$ are locally $\mathcal{H}^{d-1}$-integrable on $\operatorname{nor} X$, $k = 0, \dots, d-1$. (In particular, they are $\mathcal{H}^{d-1}$-measurable.)
2. Note that for convex bodies, the curvature-direction measures are the support measures from convex geometry, see Chap. 2.

Using Definition 4.28, formula (4.13) can be written in the following form.

Theorem 4.30 (Refined Steiner formula for sets with positive reach) *Assume that* $0 < r < \operatorname{reach} X$ *and let* E *be a bounded Borel subset of* $\mathbb{R}^d \times \mathbb{R}^d$. *Then*

$$\mathcal{L}^d\big((X_r \setminus X) \cap \widetilde{\Pi}_X^{-1}(E)\big) = \sum_{k=0}^{d-1} \omega_{d-k} r^{d-k} \widetilde{C}_k(X, E).$$

The "classical" curvature measures are projections of the curvature-direction measures $\widetilde{C}_k$ to the first coordinate vector.

Definition 4.31 Given a set X with positive reach, a bounded Borel set $B \subset \mathbb{R}^d$ and $k = 0, \ldots, d-1$, we define

$$C_k(X, B) := \widetilde{C}_k(X, B \times S^{d-1}).$$

$C_k(X, \cdot)$ is a signed Radon measure on $\mathbb{R}^d$ and will be called *curvature measure of order* k *of* X. We complete the definition by setting

$$C_d(X, B) = \mathcal{L}^d(X \cap B).$$

If the boundary ∂X is moreover bounded we define the total curvature measures

$$\mathbf{C}_k(X) = C_k(X, \mathbb{R}^d), \quad k = 0, \ldots, d,$$

and call them (as in the setting of differential geometry) *Lipschitz-Killing curvatures of* X.

Note that $\mathbf{C}_k$ corresponds to the k-th intrinsic volume V_k in convex geometry, see Chap. 2.

Remark 4.32 If X is a compact C^2-domain then $C_k(X, \cdot)$ agrees with the curvature measure given by (3.1). Indeed, $\kappa_i(x) = \kappa_i(x, \nu_X(x))$ are the "classical" principal curvature in this case (see Corollary 4.26),

$$(x, n) \mapsto \left(\prod_{i=1}^{d-1} \sqrt{1 + \kappa_i(x, n)^2} \right)^{-1}$$

is the Jacobian of the projection $(x, n) \mapsto x$ from $\operatorname{nor} X$ onto ∂X, and we get from Definition 4.28 equation (3.1) applying the Area formula (Theorem 1.14).

An immediate consequence of Theorem 4.30 is the following result.

Corollary 4.33 (Steiner formula for sets with positive reach) *If* $0 < r < \operatorname{reach} X$ *and* B *is a bounded Borel subset of* $\mathbb{R}^d$ *then*

$$\mathcal{L}^d(X_r \cap \Pi_X^{-1}(B)) = \sum_{k=0}^{d} \omega_{d-k} r^{d-k} C_k(X, B).$$

If X *is moreover bounded then*

$$\mathcal{L}^d(X_r) = \sum_{k=0}^{d} \omega_{d-k} r^{d-k} \mathbf{C}_k(X).$$

Directly from the (local) Steiner formula, we can derive an analogous polynomial expression for the curvature(-direction) measures.

Proposition 4.34 *If* $0 < r <$ reach X, $k \in \{0, 1, \dots, d-1\}$ *and* $E \subset \mathbb{R}^{2d}$ *is a bounded Borel set then*

$$\widetilde{C}_k(X_r, E) = \sum_{j=0}^{k-1} \frac{\omega_{d-j}}{\omega_{d-k}} \binom{d-j}{d-k} r^{k-j} \widetilde{C}_j(X, E).$$

Proof From Theorem 4.30 we get for $0 < s <$ reach $X - r$

$$\begin{aligned}\mathcal{L}^d\left((X_{r+s} \setminus X) \cap \widetilde{\Pi}_X^{-1}(E)\right) &= \sum_{j=0}^{d-1} \omega_{d-j}(r+s)^{d-j} \widetilde{C}_j(X, E) \\ &= \sum_{j=0}^{d-1} \omega_{d-j} \left(\sum_{i=0}^{d-j} \binom{d-j}{i} r^i s^{d-j-i} \right) \widetilde{C}_j(X, E).\end{aligned}$$

On the other hand, since reach $X_r \geq$ reach $X - r > s$ (indeed, if $d_X(y) = r + s$ and $x = \Pi_X(y)$ then $\Pi_{X_r}(y) = \frac{sx+ry}{r+s}$) and $X_{r+s} = (X_r)_s$, we have also by the Steiner formula

$$\mathcal{L}^d\left((X_{r+s} \setminus X) \cap \widetilde{\Pi}_X^{-1}(E)\right) = \sum_{k=0}^{d-1} \omega_{d-k} s^{d-k} \widetilde{C}_k(X_r, E).$$

Comparing the coefficients at s_{d-k} in the above two expressions, we obtain the desired formula. □

As a corollary we infer the vague convergence of curvature-direction measures of X_r to those of X. A much stronger continuity result will be shown in Chap. 7 (Theorem 7.3).

Corollary 4.35 *If* reach $X > 0$ *and* $k \in \{0, 1, \dots, d-1\}$ *then*

$$\widetilde{C}_k(X_r, \cdot) \xrightarrow{v} \widetilde{C}_k(X, \cdot), \quad r \to 0+.$$

Consequently, also $C_k(X_r, \cdot) \xrightarrow{v} C_k(X, \cdot)$, $r \to 0+$. *If* X *is bounded then also* $\mathbf{C}_k(X_r) \to \mathbf{C}_k(X)$, $r \to 0+$.

Let F be an affine j-subspace of $\mathbb{R}^d$ $(0 < j < d)$ and let $Y \subset F$ have positive reach. We can consider the curvature (curvature-direction) measures of Y to be defined in $\mathbb{R}^d$ as above, or to be defined relatively in F. (In fact, the curvature-direction measures are determined through the refined Steiner formula, cf. Theorem 4.30, using the j-volume in F.) We will use the upper index (F) for the curvature(-direction) measures defined relatively in F. It turns out that the curvature measures are the same, whatever embedding space we choose, and the curvature-direction measures (which live on different spaces) are simply related, see the following lemma. The symbol μ^{var} denotes the variation of a (signed) measure μ.

Lemma 4.36 *Let $Y \subset F \subset \mathbb{R}^d$ have positive reach, where $F \in \mathcal{A}^d_j$. Then,*

(i) $C_k^{(F)}(Y, B) = C_k(Y, B)$, $B \subset F$ *Borel*, $k = 0, \dots, j$,
(ii) $\widetilde{C}_k^{(F)}(Y, E) = \widetilde{C}_k(Y, \xi_L^{-1}(E))$, $E \subset F \times S^{j-1}$ *Borel*, $k = 0, \dots, j-1$,
(iii) $(\widetilde{C}_k^{(F)})^{\mathrm{var}}(Y, E) = \widetilde{C}_k^{\mathrm{var}}(Y, \xi_L^{-1}(E))$, $E \subset F \times S^{j-1}$ *Borel*, $k = 0, \dots, j-1$,

where $L \in G(d, j)$ is the j-subspace parallel to F and

$$\xi_L : F \times (S^{d-1} \setminus L^\perp) \to F \times S^{j-1}, \quad (x, n) \mapsto \left(x, \frac{p_L n}{|p_L n|}\right).$$

Proof Let $B \subset F$ be a bounded Borel set. Since $Y \subset F$, we can decompose the parallel set (in $\mathbb{R}^d$) as follows:

$$Y_r \cap \Pi_Y^{-1}(B) = \{y + z : y \in F \cap \Pi_Y^{-1}(B),\ z \in L^\perp,\ \mathrm{dist}\,(y, Y)^2 + |z|^2 \le r^2\},$$

and using subsequently the Fubini theorem and the Coarea theorem with $z \mapsto |z|$, we get

$$\begin{aligned}\mathcal{L}^d(Y_r \cap \Pi_Y^{-1}(B)) &= \int_{L^\perp \cap B(0,r)} \mathcal{L}^j(Y_{\sqrt{r^2-|z|^2}} \cap \Pi_Y^{-1}(B))\, \mathcal{L}^{d-j}(\mathrm{d}z)\\ &= (d-j)\omega_{d-j} \int_0^r s^{d-j-1} \mathcal{L}^j(Y_{\sqrt{r^2-s^2}} \cap \Pi_Y^{-1}(B))\, \mathrm{d}s.\end{aligned}$$

Applying the Steiner formula (Corollary 4.33) relatively in F, we obtain

$$\mathcal{L}^j(Y_{\sqrt{r^2-s^2}} \cap \Pi_Y^{-1}(B)) = \sum_{k=0}^{j} \omega_{j-k}(r^2 - s^2)^{\frac{j-k}{2}} C_k^{(F)}(Y, \Pi_Y^{-1}(B)).$$

A routine calculation (use substitution $s^2 = r^2 t$) yields

$$(d-j)\omega_{d-j}\omega_{j-k} \int_0^r s^{d-j-1}(r^2 - s^2)^{\frac{j-k}{2}}\, \mathrm{d}s = \omega_{d-k} r^{d-k},$$

hence,

$$\mathcal{L}^d(Y_r \cap \Pi_Y^{-1}(B)) = \sum_{k=0}^{j} \omega_{d-k} r^{d-k} C_k^{(F)}(Y, \Pi_Y^{-1}(B)).$$

Comparing the polynomial coefficints with those of the Steiner formula (Corollary 4.33, now in $\mathbb{R}^d$), we arrive at (i).

In order to show (ii), we use a similar decomposition with a bounded Borel set $E \subset Y \times S^{j-1}$:

$$\begin{aligned}&(Y_r \setminus Y) \cap \widetilde{\Pi}_Y^{-1}((\xi_L)^{-1}(E))\\&= \{y + z : y \in F \cap (Y_r \setminus Y) \cap \widetilde{\Pi}_Y^{-1}(E),\ z \in L^\perp,\ \operatorname{dist}(y, Y)^2 + |z|^2 \le r^2\},\end{aligned}$$

and we proceed exactly in the same way as when proving (i).

Finally, in order to prove (iii), we use the description of the relative (in F) unit normal bundle of Y

$$\operatorname{nor}^{(F)} Y = \xi_L \left(\operatorname{nor} Y \cap (Y \times (S^{d-1} \setminus L^\perp))\right).$$

Let $(y, u) \in \operatorname{nor}^{(F)} Y$ be a regular point with principal curvatures $\kappa_1(y, u), \ldots, \kappa_{j-1}(y, u)$. Any point $(y, n) \in \operatorname{nor} Y \cap \xi_L^{-1}(y, u)$ can be written in the form $n = (\cos t)u + (\sin t)v$ with some $v \in L^\perp \cap S^{d-1}$ and $t \in [0, \frac{\pi}{2})$, and the principal curvatures at such a point (y, n) are $\kappa_i(y, n) = (\cos t)\kappa_i(y, u)$, $i = 1, \ldots, j-1$, and $\kappa_i(y, n) = \infty$, $i = j, \ldots, d-1$. Thus the symmetric functions of principal curvatures do not change signs on the fibres $\xi_L^{-1}(y, v)$ and, hence, (iii) follows from (ii). □

4.6 Normal Cycle

Proposition 4.23 provided an orthonormal basis of the tangent space to $\operatorname{nor} X$ at regular points $(x, n) \in \operatorname{nor} X$. Thus,

$$a_X(x, n) := \bigwedge_{i=1}^{d-1} \left(\frac{1}{\sqrt{1 + \kappa_i^2(x, n)}} b_i(x, n), \frac{\kappa_i(x, n)}{\sqrt{1 + \kappa_i^2(x, n)}} b_i(x, n) \right)$$

is a unit simple $(d-1)$-vectorfield orienting the locally $(d-1)$-rectifiable set $\operatorname{nor} X$.

Recall the definition of the bi-Lipschitz mapping $f^{(r)} : \operatorname{nor} X \to \partial X_r$ from Sect. 4.4 ($0 < r < \operatorname{reach} X$), the notation for the Gauss mapping ν_r of ∂X_r, and the principal directions $b_1^r(y), \ldots, b_{d-1}^r(y)$ of X_r at a point $y \in \partial X_r$.

Lemma 4.37 *We have for $\mathcal{H}^{d-1}$-almost all $(x, n) \in \operatorname{nor} X$*

$$a_X(x, n) = \left(\bigwedge\nolimits_{d-1} D(f^{(r)})^{-1}(x + rn)\right) (b_1^r(x + rn) \wedge \cdots \wedge b_{d-1}^r(x + rn)).$$

Further, the mapping $(x, n) \mapsto a_X(x, n)$ is $\mathcal{H}^{d-1}$-measurable on $\operatorname{nor} X$.

Proof The formula was, in fact, shown in the proof of Proposition 4.23. The measurability of a_X follows then from the Area formula for currents (Theorem 1.49) applied with $T = (\mathcal{H}^{d-1} \llcorner \partial X_r) \wedge (b_1^r \wedge \cdots \wedge b_{d-1}^r)$ and the mapping $(f^{(r)})^{-1}$. □

Integration over oriented Hausdorff rectifiable sets can be treated by means of rectifiable currents (see Definition 1.43).

Definition 4.38 If reach $X > 0$, we define the current $N_X \in \mathcal{D}_{d-1}(\mathbb{R}^{2d})$ by

$$N_X = (\mathcal{H}^{d-1} \llcorner \operatorname{nor} X) \wedge a_X$$

and call it *normal cycle associated with* X.

We state several important properties of the normal cycle. We will use the following notion.

Definition 4.39 The *contact 1-form* is the differential form $\alpha \in \mathcal{D}^1(\mathbb{R}^{2d})$ given by

$$\langle (u, v), \alpha(x, n) \rangle = u \cdot n, \quad u, v, x, n \in \mathbb{R}^d.$$

A current $T \in \mathcal{D}_k(\mathbb{R}^{2d})$ $(k \geq 1)$ is called *Legendrian* if $T \llcorner \alpha = 0$.

Proposition 4.40 *The current N_X has the following properties for any set X with positive reach.*

(i) *N_X is a locally integral current.*
(ii) $\operatorname{spt} N_X \subset \partial X \times S^{d-1}$.
(iii) $\partial N_X = 0$ *(N_X is a cycle).*
(iv) *N_X is Legendrian.*

Proof The first two assertions follow immediately from the definition. To prove (iii), recall (Lemma 4.21) that there exists a bi-Lipschitz homeomorphism $f^{(r)} : \operatorname{nor} X \to \partial X_r$ whenever $0 < r <$ reach X. Then, the push-forward $T_r := (f^{(r)})_\# N_X$ is a $(d-1)$-current given by integration over the oriented boundary ∂X_r which is a C_1-manifold without boundary (see Corollary 4.22), hence, $\partial T_r = 0$ by Theorem 1.41. Since $f^{(r)}$ is bi-Lipschitz, we have also $N_X = (f^{(r)})_\#^{-1} T_r$ and $\partial N_X = (f^{(r)})_\#^{-1} \partial T_r = 0$ due to (1.29).

(iv) follows from the fact that

$$(N_X \llcorner \alpha)(\phi) = \int_{\operatorname{nor} X} \langle a_X, \alpha \wedge \phi \rangle \, d\mathcal{H}^{d-1}$$

and $\langle a_X, \alpha \wedge \phi \rangle = 0$ almost everywhere on nor X since all the principal directions $b_i(x, n)$ are perpendicular to n. □

In order to relate the normal cycle to the curvature measures, we need the following algebraic notion. Recall that $\pi_0(x, y) = x$, $\pi_1(x, y) = y$.

Definition 4.41 The *Lipschitz-Killing curvature form of order* $k \in \{0, \dots, d-1\}$ is the $(d-1)$-form φ_k on $\mathbb{R}^{2d}$ given by

$$\begin{aligned}&\langle a_1 \wedge \cdots \wedge a_{d-1}, \varphi_k(x,n)\rangle\\&= \frac{1}{(d-k)\omega_{d-k}} \sum_{\substack{j_1+\cdots+j_{d-1}=d-k-1\\ j_i\in\{0,1\}}} \langle \pi_{j_1}(a_1) \wedge \cdots \wedge \pi_{j_{d-1}}(a_{d-1}) \wedge n, \Omega_d\rangle,\end{aligned}$$

$a_1, \dots, a_{d-1} \in \mathbb{R}^{2d}$, $(x, n) \in \mathbb{R}^{2d}$. Since $\varphi_k(x, n)$ does not depend on the first vector coordinate x, we write usually only $\varphi_k(n)$. In particular,

$$\varphi_0 = (d\omega_d)^{-1}(\pi_1)^{\#}(\mathrm{id} \lrcorner \Omega_d).$$

Note that, applying φ_k to the vectorfield a_X defined above, we get

$$\langle a_X(x,n), \varphi_k(n)\rangle = \frac{1}{(d-k)\omega_{d-k}} s_{d-1-k}(X; x, n) \tag{4.14}$$

at any regular point $(x, n) \in \operatorname{nor} X$. It follows from Definitions 4.28 and 4.38 that

Theorem 4.42 *If* $\operatorname{reach} X > 0$, $0 \le k \le d-1$ *and* E *is a bounded Borel subset of* $\mathbb{R}^{2d}$ *then*

$$\widetilde{C}_k(X, E) = (N_X \llcorner \mathbf{1}_E)(\varphi_k).$$

Remark 4.43 The orientation of a_X is chosen so that, for $\mathcal{H}^{d-1}$-almost all $(x, n) \in \operatorname{nor} X$,

$$\langle a_X(x,n), \varphi_{k(x,n)}(n)\rangle > 0, \tag{4.15}$$

where

$$k(x,n) = \max\{k \le d-1 : \langle a_X(x,n), \varphi_{k(x,n)}(n)\rangle \ne 0\}.$$

(In other words, $k(x, n)$ is the number of nonzero principal curvatures at (x, n).)

Lemma 4.44 *We have for* $\mathcal{H}^{d-1}$*-almost all* $(x, n) \in \operatorname{nor} X$,

$$\frac{\bigwedge_{d-1}(\pi_0 + t\pi_1)a_X(x,n)}{|\bigwedge_{d-1}(\pi_0 + t\pi_1)a_X(x,n)|} = \star n, \quad 0 < t < \operatorname{reach} X$$

(for the notion of the Hodge star operator $\star$*, see Definition* 1.29*). Consequently,*

$$\big\langle \big(\textstyle\bigwedge_{d-1}(\pi_0 + t\pi_1)a_X(x,n)\big) \wedge n, \Omega_d\big\rangle > 0, \quad 0 < t < \operatorname{reach} X.$$

Proof In fact, both sides of the first equation are equal to

$$b_1(x,n) \wedge \cdots \wedge b_{d-1}(x,n)$$

since $(b_1(x,n),\ldots,b_{d-1}(x,n),n)$ is supposed to be a positively oriented orthonormal basis. Note further that

$$|\bigwedge_{d-1}(\pi_0 + t\pi_1)a_X(x,n)| = J_{d-1}f^{(t)}(x,n)$$

with the mapping $f^{(t)}$ from Lemma 4.21 which is bi-Lipschitz and, hence, its Jacobian is nonzero. □

Corollary 4.45 *Let* $X, Y \subset \mathbb{R}^d$ *be two sets with positive reach, assume that* (x,n) *is regular in both* nor X *and* nor Y *and that*

$$\operatorname{Tan}(\operatorname{nor} X, (x,n)) = \operatorname{Tan}(\operatorname{nor} Y, (x,n)).$$

Then $a_X(x,n) = a_Y(x,n)$.

Proof Let T denote the common tangent space of nor X and nor Y at (x,n) and denote $L_t : n^\perp \to T$ the inverse to the linear mapping $\pi_0 + t\pi_1$ restricted to T. Then we have, using Lemma 4.44,

$$a_X(x,n) = a_Y(x,n) = \frac{(\bigwedge_{d-1} L_t)(\star n)}{|(\bigwedge_{d-1} L_t)(\star n)|}.$$ □

4.7 Basic Properties of Curvature Measures

Given a Euclidean motion $g \in \mathcal{G}_d$, we denote the associated mapping

$$\tilde{g}(x,n) = (g(x), O(n)), \quad (x,n) \in \mathbb{R}^{2d},$$

where $g(x) = O(x) + z$ is the decomposition of g into the shift by $z \in \mathbb{R}^d$ and orientation preserving orthogonal mapping $O \in \mathrm{SO}(d)$.

Theorem 4.46 (Motion covariance) *If* $X \subset \mathbb{R}^d$ *has positive reach and* $g \in \mathcal{G}_d$ *then*

$$N_{gX} = \tilde{g}_\# N_X. \tag{4.16}$$

Consequently, for any $0 \le k \le d-1$,

$$\widetilde{C}_k(gX, \tilde{g}E) = \widetilde{C}_k(X, E) \text{ and } C_k(gX, gB) = C_k(X, B) \tag{4.17}$$

whenever $E \subset \mathbb{R}^{2d}$, $B \subset \mathbb{R}^d$ *are bounded Borel sets.*

Proof From the definition of the unit normal bundle we find that

$$\operatorname{nor} gX = \tilde{g}(\operatorname{nor} X).$$

It is well-known in classical differential geometry that the principal curvatures are motion invariant and the principal directions are motion covariant. We infer from Corollary 4.26 that the same is true for sets with positive reach:

$$\kappa_i(gX; \tilde{g}(x,n)) = \kappa_i(X; x, n), \quad b_i(gX; \tilde{g}(x,n)) = Ob_i(X; x, n),$$

$i = 1, \dots, d-1$ (we use here an extended notation for principal curvatures and directions refering to the set the are associated with). Consequently, since $D\tilde{g}(x,n)(u,v) = (O(u), O(v))$, we can write

$$a_{gX}(\tilde{g}(x,n)) = \left(\textstyle\bigwedge_{d-1} D\tilde{g}(x,n)\right) a_X(x,n).$$

Applying the Area formula for currents (Theorem 1.49), we obtain (4.16) (note that $\tilde{g}$ is a bijection). Formulas in (4.17) follows directly using Theorem 4.42. □

The next result is the homogeneity of curvature measures. Note that reach $tX > 0$ whenever reach $X > 0$ and $t > 0$.

Proposition 4.47 (Homogeneity) *If* reach $X > 0$, $t > 0$ *and* $B \subset \mathbb{R}^d$ *is a bounded Borel set then*

$$C_k(tX, tB) = t^k C_k(X, B), \quad i = 0, \dots, d.$$

Proof Applying the Steiner formula (Corollary 4.33) to tX and tB we get for $0 < tr <$ reach X

$$\mathcal{L}^d\left((tX)_{tr} \cap \Pi_{tX}^{-1}(tE)\right) = \sum_{k=0}^{d} \omega_{d-k} r^{d-k} t^{d-k} C_k(tX, tE).$$

Since the Lebesgue measure is homogeneous of order d, the left-hand side equals

$$t^d \mathcal{L}^d(X_r \cap \Pi_X^{-1}(B)) = t^d \sum_{k=0}^{d} \omega_{d-k} r^{d-k} C_k(X, B),$$

where we applied again the Steiner formula. Comparing the coefficients in the polynomials, we get the assertion. □

Remark 4.48 Proposition 4.47 can be shown also directly from the integral representation of curvature measures in Definition 4.28, using the fact that the principal curvatures of tXat a point (tx, n) are those of X at (x, n) divided by t.

An important property is that both N_X and $C_k(X, \cdot)$ depend additively on X.

Theorem 4.49 (Additivity) *Let $X, Y \subset \mathbb{R}^d$ be such that*

$$r_0 := \min\{\text{reach}\, X, \text{reach}\, Y, \text{reach}\,(X \cup Y)\} > 0.$$

Then we have:

(i) reach $(X \cap Y) \geq r_0$;
(ii) $\mathbf{1}_{\text{nor}(X \cup Y)}(x, n) = \mathbf{1}_{\text{nor}(X)}(x, n) + \mathbf{1}_{\text{nor}(Y)}(x, n) - \mathbf{1}_{\text{nor}(X \cap Y)}(x, n)$, $(x, n) \in \mathbb{R}^d \times S^{d-1}$;
(iii) $N_{X \cup Y} = N_X + N_Y - N_{X \cap Y}$;
(iv) $\widetilde{C}(X \cup Y, \cdot) = \widetilde{C}(X, \cdot) + \widetilde{C}(Y, \cdot) - \widetilde{C}(X \cap Y, \cdot)$.

Proof Let a point z satisfy $d_X(z) < r_0$ and $d_Y(z) < r_0$. If $d_X(z) = d_Y(z)$ then necessarily $\Pi_X(z) = \Pi_Y(z)$ (since otherwise, we would get a contradiction with the unique footpoint property for $X \cup Y$). If, on the other hand, say $d_X(z) < d_Y(z)$ then $\Pi_X(z) = \Pi_{X \cup Y}(z)$ and $\Pi_Y(z) = \Pi_{X \cap Y}(z)$. The last equality can be seen as follows: by continuity of the distance, there must be a point w in the segment $[z, \Pi_Y(z)]$ having the same distance from X and Y, and by the same argument as above, $\Pi_X(w) = \Pi_Y(w)$, but this is only possible if $w = \Pi_Y(z) \in X$. The opposite inequality $d_X(z) > d_Y(z)$ can be treated analogously. In all cases, it follows that $z \in \text{Unp}\,(X \cap Y)$, hence, reach $(X \cap Y) \geq r_0$. Further, if follows from the above observation that (ii) holds whenever $(x, n) = (\Pi_X(z), \frac{z - \Pi_X(z)}{|z - \Pi_X(z)|})$ for $X = X$ or $X = Y$ and $z \notin X$. If, on the other hand, (x, n) is not of this type then clearly it does not belong to any of the four unit normal bundles in appearing (ii). This verifies (ii).

For (iii), we have to show that for $\mathcal{H}^{d-1}$-almost all (x, n),

$$\mathbf{1}_{\text{nor}(X \cup Y)}(x, n) a_{X \cup Y}(x, n) = \mathbf{1}_{\text{nor}(X)}(x, n) a_X(x, n) + \mathbf{1}_{\text{nor}(Y)}(x, n) a_Y(x, n)$$
$$- \mathbf{1}_{\text{nor}(X \cap Y)}(x, n) a_{X \cap Y}(x, n).$$

Due to Proposition 1.19, we can assume that Tan(nor X, (x, n)) is the same subspace whenever $(x, n) \in$ nor X and X is X, Y, $X \cup Y$ or $X \cap Y$. Then the last equality follows from (ii) and Corollary 4.45.

(iv) follows from (iii) by definition. □

The next property says that the curvature measures are locally determined.

Proposition 4.50 (Locality of curvature measures) *If X, Y are two sets of positive reach such that $U \cap X = U \cap Y$ for some bounded open set $U \subset \mathbb{R}^d$ then*

$$N_X \llcorner \mathbf{1}_{U \times S^{d-1}} = N_Y \llcorner \mathbf{1}_{U \times S^{d-1}}.$$

In particular, $\widetilde{C}_k(X, E) = \widetilde{C}_k(Y, E)$ *for any Borel set* $E \subset U \times S^{d-1}$ *and* $k = 0, \dots, d-1$.

Proof Follows from the locality of principal curvatures (see Remark 4.29) and Definition 4.28. □

In Chap. 8 we will show that the above properties characterize the curvature measures as some basic Euclidean invariants. Note that they extend the well-known relationships from convex geometry (cf. Chap. 2).

4.8 Gauss-Bonnet Formula

The classical Gauss-Bonnet formula says that for a C^2-smooth compact manifold M without boundary,

$$\mathbf{C}_0(M) = \chi(M),$$

where χ is the Euler-Poincaré characteristic (cf. (3.4)). We will show that the same is true for sets with positive reach.

Assume first that X is a $C^{1,1}$ domain. Then, Corollary 3.10 says that for $\mathcal{H}^{d-1}$-almost all $v \in S^{d-1}$,

$$\chi(X) = \sum_{x \in \nu_X^{-1}\{v\}} (-1)^{\lambda_X(x)}, \tag{4.18}$$

where ν_X is the Gauss map of X and $\lambda_X(x)$ is the number of negative principal curvatures at $x \in \partial X$. We will show that this already implies the Gauss-Bonnet formula for $C^{1,1}$-domains.

Proposition 4.51 *If* $X \subset \mathbb{R}^d$ *is a compact* $C^{1,1}$ *domain then*

$$\mathbf{C}_0(X) = \chi(X).$$

Proof The total curvature of order zero is

$$\mathbf{C}_0(X) = (d\omega_d)^{-1} \int_{\partial X} \prod_{i=1}^{d-1} \kappa_i(x)\, \mathcal{H}^{d-1}(dx)$$

(cf. Sect. 3.1). The Jacobian of the Gauss map $\nu_X : x \mapsto \nu_X(x)$ on ∂X is

$$J_{d-1}\nu_X(x) = \left| \prod_{i=1}^{d-1} \kappa_i(x) \right|$$

whenever ν_X is differentiable at $x \in \partial X$. Thus, using the Area formula (Theorem 1.21), we get

$$\begin{aligned}\mathbf{C}_0(X) &= (d\omega_d)^{-1} \int_{\partial X} (-1)^{\lambda_X(x)} J_{d-1}\nu_X(x)\,\mathcal{H}^{d-1}(dx)\\ &= (d\omega_d)^{-1} \int_{S^{d-1}} \sum_{x\in\nu_X^{-1}\{v\}} (-1)^{\lambda_X(x)}\,\mathcal{H}^{d-1}(dv),\end{aligned}$$

and the last expression equals $\chi(X)$ by (4.18). □

We further approximate a general set X of positive reach by its parallel neighbourhoods which are already $C^{1,1}$ domains.

Lemma 4.52 *If* $0 < r <$ reach X *then the set* X *and its parallel neighbourhood* X_r *are homotopy equivalent.*

Proof Consider the mapping

$$h : X_r \times [0,1] \to X_r, \quad (z,t) \mapsto t\Pi_X(z) + (1-t)z.$$

It is an easy exercise to verify that h is a homotopy with $h(z,0) = z$ for all z and $h(X_r \times \{1\}) = X$. □

The last three statements, together with the continuity result (see Corollary 4.35)

$$\lim_{r\to 0+} \mathbf{C}_0(X_r) = \mathbf{C}_0(X)$$

imply already the Gauss-Bonnet formula.

Theorem 4.53 (Gauss-Bonnet formula for sets with positive reach) *If* $X \subset \mathbb{R}^d$ *is compact and* reach $X > 0$ *then* $\mathbf{C}_0(X) = \chi(X)$.

Finally, we show that (4.18) holds for compact sets with positive reach as well.

Theorem 4.54 *If* $X \subset \mathbb{R}^d$ *is compact and* reach $X > 0$ *then for* $\mathcal{H}^{d-1}$*-almost all* $n \in S^{d-1}$,

$$\chi(X) = \sum_{x:\,(x,n)\in \operatorname{nor} X} (-1)^{\lambda_X(x,n)},$$

where $\lambda_X(x,n)$ *is the number of negative principal curvatures of* X *at a regular point* $(x,n) \in$ nor X.

Remark 4.55 Since regular points in nor X have full $(d-1)$-dimensional measure, the right hand side in the above formula is well defined for $\mathcal{H}^{d-1}$-almost all $n \in S^{d-1}$.

Proof Let X be compact and $0 < r < \operatorname{reach} X$. Corollary 4.26 implies that the principal curvatures of X at a regular point (x, n) and of the parallel set X_r at $x + rn$ have the same signs. Thus, $\lambda_X(x, n) = \lambda_{X_r}(x + rn)$ for $\mathcal{H}^{d-1}$-almost all $(x, n) \in \operatorname{nor} X$ and the assertion follows from (4.18) by approximating X with the compact $C^{1,1}$-domain X_r. □

4.9 Bibliographical Notes

1. Sets with positive reach have been introduced in the fundamental paper by Federer in 1959 [Fed59] where also curvature measures, Steiner formula and Principal kinematic formula were proved. Representation of curvature measures as integrals of local curvatures, as well as the current representation by means of normal cycles, were proved in [Zäh86b].
2. *Sets with positive reach in Riemannian manifolds* have been considered by Kleinjohann [Kle81] and Bangert [Ban82] who showed that the property $\operatorname{reach}(X, x) > 0$ for all $x \in X$ is independent of the Riemannian structure in a smooth connected manifold. In particular, a compact set has (locally) positive reach if and only if it is the sublevel set of a proper, bounded from below and semiconvex function at a weakly regular value.
3. Further characterizations and properties of sets with positive reach in Riemannian manifolds were obtained by Lytchak [Lyt04, Lyt05]. In particular, a compact set X has positive reach if and only if there exist positive constants r, K such that any two points in $x, y \in X$ of distance $\operatorname{dist}(x, y) < r$ can be joined in X by a $C^{1,1}$-path of length less than $K \operatorname{dist}(x, y)$ and $C^{1,1}$-norm less than K (see [Lyt05, Theorems 1.2, 1.3]).
4. *Reach of graphs and subgraphs.* Federer [Fed59, §4.20] stated that the graph of a Lipschitz function $f : \mathbb{R}^d \to \mathbb{R}$ has positive reach if and only if f has a Lipschitz derivative. Fu [Fu85, Theorem 2.3] showed that if f is locally Lipschitz then the subgraph of f has positive reach if and only if f is semiconcave (which means that f can be written in the form $f(x) = C|x|^2 - g(x)$ for some convex function g and constant $C \geq 0$).
5. Let $X \subset \mathbb{R}^d$ have positive reach and topological dimension $0 < k \leq d$, and denote $X^{(j)} := \{x \in X : \dim \operatorname{Nor}(X, x) \geq d - j\}$, $0 \leq j \leq d$. Federer [Fed59] showed that $X \setminus X^{(k-1)}$ is open if $k = d$ and is a $C^{1,1}$ k-dimensional submanifold of $\mathbb{R}^d$ if $k < d$. Further, he showed that $X^{(k-1)}$ is countably $\mathcal{H}^{k-1}$-rectifiable. The last statement was strengthened in [RZ17] where it was shown that $X^{(k-1)}$ can be covered by locally finitely many DC surfaces of dimension $k - 1$.
6. *Lower dimensional sets of positive reach.* If $X \subset \mathbb{R}^d$ is a topological manifold of dimension $0 < k < d$ and $\operatorname{reach} X > 0$ then X is already a k-dimensional $C^{1,1}$ submanifold of $\mathbb{R}^d$ (i.e., it can be locally represented as the graph of a C^1 function with Lipschitz differential). This was stated already by Federer [Fed59, Remark 4.20] and proved essentially by Lytchak [Lyt05, Proposition 1.4]. Note

that, however, the mapping $a \mapsto \operatorname{Tan}(X, a)$ need not be globally Lipschitz on X (see [RZ17, Example 7.13]).

7. A full characterization of planar sets with positive reach and of one-dimensional sets of positive reach was obtained in [RZ17]. The topological structure of a general set in $X \subset \mathbb{R}^d$ of positive reach can be rather complicated; though, of course, X itself is locally contractible, its boundary need not be and its complement can have infinitely many components (consider the set $\{(x, y) : 0 \le x \le 1, 0 \le y \le d_K(x)^2\}$ with an uncountable totally disconnected set $K \subset [0, 1]$).
8. The *Steiner formula* goes back to Jacob Steiner who proved it for polytopes and smooth convex bodies in dimensions 2 and 3. The related tube formula was obtained by Weyl [Wey39]. Its proof for sets of positive reach (in particular, for general convex bodies) is due to Federer [Fed59].
9. Hug, Last and Weil [HLW04] proved the following Steiner-type formula valid for any closed set $X \subset \mathbb{R}^d$ and bounded measurable function $f : \mathbb{R}^d \to \mathbb{R}$ with compact support:

$$\begin{aligned}&\int_{\mathbb{R}^d \setminus X} f(y)\, \mathcal{L}^d(\mathrm{d}y)\\ &= \sum_{k=0}^{d-1} \mathcal{O}_{d-k} \int_{N^+(X)} \int_0^{\delta(X,x,n)} t^{d-1-k} f(x+tn)\, \mathrm{d}t\, \mu_k(X, \mathrm{d}(x, n)),\end{aligned}$$

where

$$N^+(X) = \left\{ \left(\Pi_X(y), \frac{y - \Pi_X(y)}{|y - \Pi_X(y)|} \right) : y \in \operatorname{Unp} X \setminus X \right\},$$

$$\delta(X, x, n) = \inf\{\delta > 0 : x + \delta n \notin \operatorname{Unp} X\}, \quad (x, n) \in N^+(X),$$

and $\mu_k(X, \cdot)$ are certain signed Borel measures on $N^+(X)$ for which the above integrals converge. If reach $X > 0$ then $\mu(X, \cdot) = \widetilde{C}_k(X, \cdot)$.

10. The *normal cycle* has been considered by Sulanke and Wintgen [SW72] for smooth submanifolds. For sets with positive reach, the notion was introduced in [Zäh86b]. Closed integral currents with the Legendrian property have been studied in more general context by Fu, see e.g. [Fu89a, Fu94].
11. *Stochastic processes* of sets with positive reach and their curvature measures were considered in [Zäh86a].

Chapter 5
Unions of Sets with Positive Reach

5.1 Topological Index Functions and Additive Extension of Normal Cycles

Though the *family* $\mathcal{PR}$ *of sets with positive reach* covers both the classical geometric sets, i.e., convex sets and $C^{1,1}$ smooth submanifolds it does not include their natural singular extensions: the polyconvex sets and the piecewise smooth submanifolds, respectively. Therefore the natural question arises whether we can consider unions of sets with positive reach under the above aspects. To this aim we recall the additivity properties of the unit normal bundles, the normal cycles and the curvature-direction measures of $\mathcal{PR}$-sets from Theorem 4.49. If $X, Y \subset \mathbb{R}^d$ are such that $r_0 := \min\{\text{reach}\, X, \text{reach}\, Y, \text{reach}\,(X \cup Y)\} > 0$. then we have:

(i) $\text{reach}\,(X \cap Y) \geq r_0$;
(ii) $\mathbf{1}_{\text{nor}(X\cup Y)}(x, n) = \mathbf{1}_{\text{nor}\, X}(x, n) + \mathbf{1}_{\text{nor}\, Y}(x, n) - \mathbf{1}_{\text{nor}(X\cap Y)}(x, n)$, $(x, n) \in \mathbb{R}^d \times S^{d-1}$;
(iii) $N_{X\cup Y} = N_X + N_Y - N_{X\cap Y}$;
(iv) $\widetilde{C}(X \cup Y, \cdot) = \widetilde{C}(X, \cdot) + \widetilde{C}(Y, \cdot) - \widetilde{C}(X \cap Y, \cdot)$.

A first aim of this section is to extend the notion of normal cycle and of the curvature-direction measures in such a way that formulas (iii) and (iv) remain valid if we merely suppose that $\text{reach}\, X > 0$, $\text{reach}\, Y > 0$ and $\text{reach}\,(X \cap Y) > 0$.

Definition 5.1 Let $\mathcal{U}_{\mathcal{PR}}$ denote the family of all subsets of $\mathbb{R}^d$ which can be represented as a locally finite union of sets with positive reach whose arbitrary finite intersections also have positive reach. If $X \in \mathcal{U}_{\mathcal{PR}}$, we call

$$X = \bigcup_{i=1}^{\infty} X^i \tag{5.1}$$

J. Rataj, M. Zähle, *Curvature Measures of Singular Sets*, Springer Monographs in Mathematics, https://doi.org/10.1007/978-3-030-18183-3_5

a $\mathcal{U}_{\mathcal{PR}}$*-representation of* X, provided that

1. $\{i \in \mathbb{N} : X^i \cap K \neq \emptyset\}$ is finite for any $K \subset \mathbb{R}^d$ compact,
2. reach $\bigcap_{i \in I} X^i > 0$ for any finite set $I \subset \mathbb{N}$.

Seeking for an additive extension of the normal cycle to $\mathcal{U}_{\mathcal{PR}}$, we could take a $\mathcal{U}_{\mathcal{PR}}$-representation (5.1) and apply the *inclusion-exclusion principle*:

$$N_X = \sum_{I \in \mathcal{P}_*(\mathbb{N})} (-1)^{\mathrm{card}(I)-1} N_{\bigcap_{i \in I} X^i}, \tag{5.2}$$

where $\mathcal{P}_*(\mathbb{N})$ denotes the family of nonempty finite sets of natural numbers. This formula could be used as a definition of the normal cycle of the set X if the right hand side is independent of its $\mathcal{U}_{\mathcal{PR}}$-representation. (Note that the sum is finite when applied to a differential form with compact support.) In order to make this rigorous we interpret formula (ii) in terms of a local topological index function via Euler numbers admitting the corresponding generalizations (Fig. 5.1).

Definition 5.2 For any $X \subset \mathbb{R}^d$, $x \in \mathbb{R}^d$ and $n \in S^{d-1}$, the *index* of X at x in direction n is defined by

$$i_X(x,n) = \mathbf{1}_X(x)\Big(1 - \lim_{\varepsilon \to 0+} \lim_{\delta \to 0+} \chi\big(X \cap B(x + (\varepsilon + \delta)n, \varepsilon)\big)\Big) \tag{5.3}$$

provided that the right hand side is determined, where χ denotes the Euler-Poincaré characteristic in the sense of singular homology.

In the proof of Theorem 5.6 below we will see that $i_X(x,n) = \mathbf{1}_{\mathrm{nor}\, X}(x,n)$ if reach $X > 0$ and, therefore Eq. (5.2) in this case corresponds to the additivity of the Euler characteristic. At the same time the latter will guarantee our additive extensions. In order to show the existence of the index function for $\mathcal{U}_{\mathcal{PR}}$-sets we will need the following propositions which are also of independent interest.

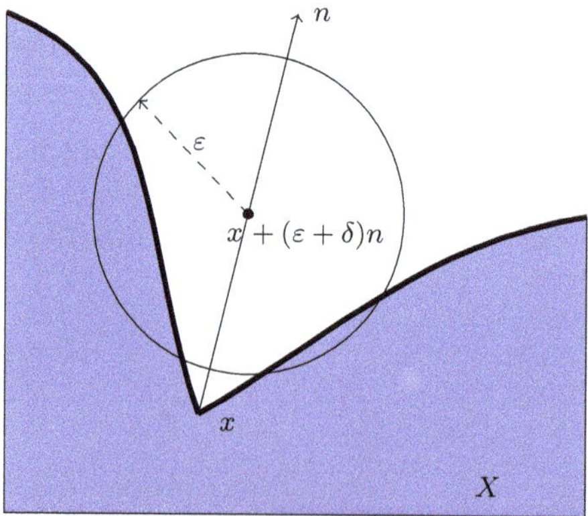

Fig. 5.1 The index function $i_X(x,n) = 1 - 2 = -1$

Proposition 5.3 *Any $X \in \mathcal{U}_{\mathcal{PR}}$ is locally contractible.*

Proof Let $X = \bigcup_{i=1}^{\infty} X^i$ be a $\mathcal{U}_{\mathcal{PR}}$-representation, $x \in X$, $I(x) := \{i : x \in X^i\}$ and $r > 0$ be such that reach $(\bigcap_{i \in J} X^i) > r$ for all $J \subset I(x)$ and $B(x, r) \cap X^i = \emptyset$ for any $i \notin I(x)$. We will show that $X \cap B(x, r)$ is contractible. The main point is the existence of a continuous mapping $\phi : B(x, r) \to \bigcup_{i \in I(x)} X^i$ which preserves points from $\bigcup_{i \in I(x)} X^i$. Then we can use the contraction given by

$$g(y, t) := \phi(tx + (1 - t)y), \quad y \in X \cap B(x, r), \quad t \in [0, 1]$$

(cf. the proof of Lemma 1.49 where the metric projection in case $X \in \mathcal{PR}$ was used). We will define ϕ so that $\phi(z) \in X^i$ if and only if $d_{X^i}(z) = d_X(z)$. For given $z \in B(x, r)$ denote $d_i := d_{X^i}(z)$ and suppose that $I(x) = \{1, \dots, N\}$ and $d_1 \le \dots \le d_N$. Let $\Pi_{1,\dots,i}$ be the metric projection onto $X^1 \cap \dots \cap X^i$. We introduce a sequence of points $(z_N, z_{N-1}, \dots, z_1)$ by $z_N := \Pi_{1,\dots,N}(z)$ and

$$z_i := \Pi_{1,\dots,i}\left(\left(1 - \frac{d_i}{d_{i+1}}\right)\Pi_{1,\dots,i}(z) + \frac{d_i}{d_{i+1}} z_{i+1}\right), \quad i = N - 1, \dots, 1$$

(setting $0/0 := 1$). Then put $\phi(z) := z_1$. The correctness of the definition of ϕ follows from the fact that if $d_{i+1} = d_i$ for some i then we obtain the same point z_i for both the orderings of $i, i + 1$. Continuity of the metric projection implies that of ϕ. □

Proposition 5.4 *If $X, Y, X \cap Y \in \mathcal{U}_{\mathcal{PR}}$ are compact, then the Euler-Poincaré characteristic of these sets and their union is defined and additive:*

$$\chi(X \cup Y) + \chi(X \cap Y) = \chi(X) + \chi(Y).$$

Proof Follows from Proposition 5.3 and Theorem 3.5. □

The following auxiliary relationship leads, in particular, to the values of the index function i_X for the special case of $\mathcal{PR}$-sets X.

Lemma 5.5 *If $0 < r <$ reach X, then for any ball $B(r)$ of radius r intersecting X we have $\chi(X \cap B(r)) = 1$.*

Proof By Proposition 4.20 for $Z := X \cap B(r)$ we obtain reach $Z > r$. Note that the convex hull conv Z of the intersection set Z lies in a distance not greater than r from Z. Therefore the metric projection from conv Z onto Z yields a homotopy between these sets. Consequently, $\chi(Z) = \chi(\operatorname{conv} Z) = 1$. □

Theorem 5.6 *The index $i_X(x, n)$ is determined for any $X \in \mathcal{U}_{\mathcal{PR}}$, $x \in \mathbb{R}^d$ and $n \in S^{d-1}$. For $X \in \mathcal{PR}$ we have the identity $i_X = \mathbf{1}_{\operatorname{nor} X}$. Moreover, the index function is locally bounded and satisfies the additivity property*

$$i_X + i_Y = i_{X \cup Y} + i_{X \cap Y}$$

whenever $X, Y, X \cap Y, X \cup Y \in \mathcal{U}_{\mathcal{PR}}$.

Proof First we show that $i_X = \mathbf{1}_{\operatorname{nor} X}$ for $X \in \mathcal{PR}$. If $(x, n) \in \operatorname{nor} X$, then $X \cap B(x + (\varepsilon + \delta)n, \varepsilon) = \emptyset$ for sufficiently small ε and δ by the unique nearest point property of sets with positive reach, hence $i_X(x, n) = 1$. Further, let $(x, n) \notin \operatorname{nor} X$ and $x \in X$ (the case $x \notin X$ is trivial). Then we have $X \cap B(x + (\varepsilon + \delta)n, \varepsilon) \neq \emptyset$ for all $\varepsilon > 0$ and sufficiently small δ (cf. Lemma 4.5). Therefore Lemma 5.5 implies that $\chi\big(X \cap B(x + (\varepsilon + \delta)n, \varepsilon)\big) = 1$ for $\varepsilon < \operatorname{reach} X$ and δ sufficiently small, hence $i_X(x, n) = 0$. The existence and additivity of i_X for general $X \in \mathcal{U}_{\mathcal{PR}}$ follows from this special case by means of Proposition 5.4, since the set $X \cap B(x + (\varepsilon + \delta)n, \varepsilon)$ has a finite $\mathcal{U}_{\mathcal{PR}}$-representation due to Proposition 4.20. The measurability of i_X for $X \in \mathcal{PR}$ follows by the measurability of $\operatorname{nor} X$, and for $X \in \mathcal{U}_{\mathcal{PR}}$ we use the additivity. The additivity and locality of the index function imply the following inclusion-exclusion formula valid for any (x, n):

$$i_X(x, n) = \sum_{\emptyset \neq I \subset I(x)} (-1)^{\operatorname{card}(I)-1} i_{\bigcap_{i \in I} X^i}(x, n), \tag{5.4}$$

where $I(x) = \{i : x \in X^i\}$ and X^i are the $\mathcal{PR}$-components from (5.1). Thus, $|i_X(x, n)| \leq 2^{\operatorname{card} I(x)}$ and i_X is locally bounded. □

The identity $i_X = \mathbf{1}_{\operatorname{nor} X}$ for $X \in \mathcal{PR}$ suggests the following generalization.

Definition 5.7 For $X \in \mathcal{U}_{\mathcal{PR}}$, the *unit normal bundle* is defined by

$$\operatorname{nor} X := \{(x, n) \in \mathbb{R}^d \times S^{d-1} : i_X(x, n) \neq 0\}.$$

Note that, in general, the unit normal bundle $\operatorname{nor} X$ need not be closed.

Proposition 5.8 *For any $X \in \mathcal{U}_{\mathcal{PR}}$,*

(i) $\operatorname{nor} X$ *is a locally $(d-1)$-rectifiable and $\mathcal{H}^{d-1}$-measurable subset of* $\mathbb{R}^d \times S^{d-1}$,
(ii) *The index function i_X is locally $\mathcal{H}^{d-1}$-integrable.*

Proof The measurability of $\operatorname{nor} X$ follows from the measurability of the index function. Let $X = \bigcup_{i=1}^{\infty} X^i$ be a $\mathcal{U}_{\mathcal{PR}}$-representation of X (5.1). Note that by the definition of the unit normal bundle and by the additivity of the index function we have

$$\operatorname{nor} X \subseteq \bigcup_{I \in \mathcal{P}_*(\mathbb{N})} \operatorname{nor}\Big(\bigcap_{i \in I} X^i\Big) \tag{5.5}$$

and the union on the right hand side is locally finite (i.e., the intersection of $\operatorname{nor} X$ with a bounded domain is covered by a finite union of unit normal bundles of sets of positive reach). The local rectifiability follows immediately from the same property of the unit normal bundles of sets with positive reach. In particular, $\operatorname{nor} X$ has locally finite $\mathcal{H}^{d-1}$-measure and since i_X is locally finite, it is locally $\mathcal{H}^{d-1}$-integrable. □

Definition 5.9 For $X \in \mathcal{U}_{\mathcal{PR}}$, the *normal cycle* $N_X \in \mathcal{D}_{d-1}(\mathbb{R}^d \times \mathbb{R}^d)$ is defined by

$$N_X = \left(H^{d-1} \llcorner \operatorname{nor} X\right) \wedge i_X a_X,$$

where a_X is the simple unit $(d-1)$-vector field associated with $\operatorname{Tan}^{d-1}(\operatorname{nor} X, \cdot)$ with orientation (sign) given by

$$\lim_{\varepsilon \to 0} \left\langle \frac{\bigwedge_{d-1}(\pi_0 + \varepsilon\pi_1) a_X(x,n)}{|\bigwedge_{d-1}(\pi_0 + \varepsilon\pi_1) a_X(x,n)|} \wedge n, \Omega_d \right\rangle > 0$$

for sufficiently small $\varepsilon > 0$.

Remark 5.10

1. In view of Lemma 4.44, the definition is consistent with that for $\mathcal{PR}$-sets.
2. By construction, N_X is locally $(d-1)$-rectifiable with support

$$\operatorname{spt} N_X \subset \overline{\operatorname{nor} X} \subset \partial X \times S^{d-1}.$$

3. Due to Proposition 1.19, for $\mathcal{H}^{d-1}$-almost all $(x,n) \in \operatorname{nor} X$, the approximate tangent space $\operatorname{Tan}^{d-1}(\operatorname{nor} X, (x,n))$ is a $(d-1)$-dimensional subspace and agrees with $\operatorname{Tan}^{d-1}(\operatorname{nor} \bigcap_I X^i, (x,n))$ for some $\emptyset \neq I \subset \mathbb{N}$ finite.

We will show that N_X is the desired additive extension of the normal cycle and that the main geometric properties of normal cycles of $\mathcal{PR}$-sets remain valid. (Recall Definition 4.39 for the contact 1-form α.)

Theorem 5.11 *If $X, Y, X \cap Y, X \cup Y \in \mathcal{U}_{\mathcal{PR}}$, then we have the following.*

(i) $N_X + N_Y = N_{X\cup Y} + N_{X\cap Y}$ *(additivity).*
(ii) $\partial N_X = 0$ *(N_X is a cycle).*
(iii) $N_X \llcorner \alpha = 0$ *(N_X is Legendrian).*

Proof Note that the additivity property (i) implies the inclusion-exclusion principle (5.2). Therefore, (ii) and (iii) are consequences of the corresponding properties of $\mathcal{PR}$-sets.

In order to show (i), we recall the additivity of the index function (Proposition 5.6). Hence, by the definition of the unit normal bundle,

$$\operatorname{nor}(X \cup Y) \subset \operatorname{nor} X \cup \operatorname{nor} Y \cup \operatorname{nor}(X \cap Y).$$

Consequently, in view of the local $(d-1)$-rectifiability of the unit normal bundles and due to Corollary 4.45, for $\mathcal{H}^{d-1}$-almost all $(x,n) \in \operatorname{nor}(X \cup Y)$ we have $a_{X\cup Y}(x,n) = a_Z(x,n)$ for some $Z \equiv Z(x,n) \in \{X, Y, X \cap Y\}$ and, moreover, $a_Z(x,n) = a_{Z'}(x,n)$ whenever $(x,n) \in \operatorname{nor} Z'$ for another $Z' \in \{X, Y, X \cap Y\}$.

From this we infer

$$
\begin{aligned}
N_{X\cup Y} &= \big(\mathcal{H}^{d-1} \llcorner \operatorname{nor}(X\cup Y)\big) \wedge i_{X\cup Y} a_{X\cup Y} \\
&= \big(\mathcal{H}^{d-1} \llcorner (\operatorname{nor} X \cup \operatorname{nor} Y \cup \operatorname{nor}(X\cap Y))\big) \wedge i_{X\cup Y} a_Z \\
&= \big(\mathcal{H}^{d-1} \llcorner (\operatorname{nor} X \cup \operatorname{nor} Y \cup \operatorname{nor}(X\cap Y))\big) \wedge (i_X + i_Y - i_{X\cap Y}) a_Z \\
&= \big(\mathcal{H}^{d-1} \llcorner \operatorname{nor} X\big) \wedge i_X a_Z + \big(\mathcal{H}^{d-1} \llcorner \operatorname{nor} Y\big) \wedge i_Y a_Z \\
&\quad - \big(\mathcal{H}^{d-1} \llcorner \operatorname{nor}(X\cap Y)\big) \wedge i_{X\cap Y} a_Z \\
&= \big(\mathcal{H}^{d-1} \llcorner \operatorname{nor} X\big) \wedge i_X a_X + \big(\mathcal{H}^{d-1} \llcorner \operatorname{nor} Y\big) \wedge i_Y a_Y \\
&\quad - \big(\mathcal{H}^{d-1} \llcorner \operatorname{nor}(X\cap Y)\big) \wedge i_{X\cap Y} a_{X\cap Y} \\
&= N_X + N_Y + N_{X\cap Y}.
\end{aligned}
$$

□

The index function is not defined uniquely in the literature. The following one was introduced by Fu [Fu89a]. We use the notation

$$H_{v,t} := \{y \in \mathbb{R}^d : y \cdot v \le t\}, \quad v \in S^{d-1},\ t \in \mathbb{R},$$

for halfspaces in $\mathbb{R}^d$.

Definition 5.12 For any $X \subset \mathbb{R}^d$, $x \in \mathbb{R}^d$ and $n \in S^{d-1}$, we define

$$\iota_X(x,n) := \lim_{\varepsilon\to 0}\lim_{\delta\to 0} \big(\chi(X\cap B(x,\varepsilon)\cap H_{-n,-x\cdot n+\delta}) - \chi\big(X\cap B(x,\varepsilon)\cap H_{-n,-x\cdot n-\delta}\big),$$

whenever the right hand side is determined (Fig. 5.2).

Remark 5.13

1. The index $\iota_X(x,n)$ need not be defined even if reach $X > 0$. As an example, consider X to be the graph of $f(t) = t^4 \sin\frac{1}{t}$, $t \in [-1,1]$, $x = (0,0)$ and $n = (0,1)$.
2. Fu [Fu89a, Corollary 6.7] showed that for a set $X \subset \mathbb{R}^d$ with positive reach, for $\mathcal{H}^{d-1}$-almost all $n \in S^{d-1}$ and all $x \in \mathbb{R}^d$,

$$\iota_X(x,n) = (-1)^{\lambda_X(x,n)},$$

   where $\lambda_X(x,n)$ is the number of negative principal curvatures of X at a regular point $(x,n) \in \operatorname{nor} X$ (cf. Theorem 4.54).
3. It is easy to see that $\iota_X(x,n)$ exists and agrees with $i_X(x,n)$ for all (x,n) if X is a compact polyconvex set. In such a case, we can also write it in the form

$$\iota_X(x,n) = 1 - \lim_{\varepsilon\to 0}\lim_{\delta\to 0} \chi(X\cap B(x,\varepsilon)\cap \partial H_{n,n\cdot x+\delta}), \tag{5.6}$$

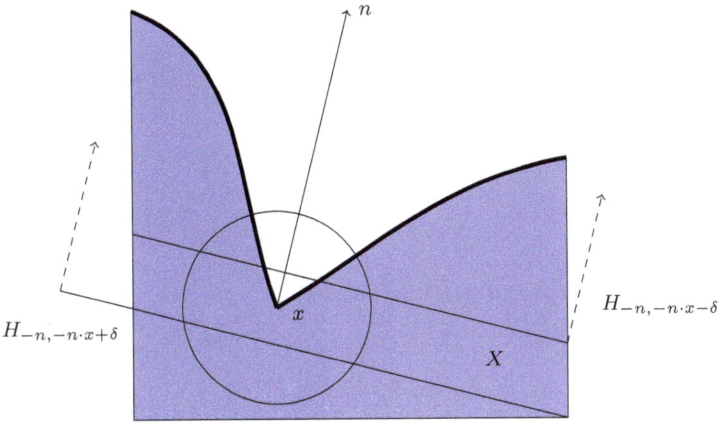

Fig. 5.2 The index function $\iota_X(x, n) = 1 - 2 = -1$

$x \in X, n \in S^{d-1}$. Indeed, decomposing

$$H_{-n,-n\cdot x+\delta} = H_{-n,-n\cdot x-\delta} \cup L_{n,\delta}(x)$$

with the layer $L_{n,\delta}(x) := \{y \in \mathbb{R}^d : -\delta \le (x - y)\cdot n \le \delta\}$, we can use additivity of the Euler characteristic on the convex ring and the fact that $\chi(X \cap B(x,\delta) \cap L_{n,\delta}(x)) = 1$ if $x \in X$ and ε, δ are small enough. In this version the index function has first been considered by Schneider [Sch80].

5.2 Curvature Measures, Generalized Principal Curvatures and Steiner Formula

Using the normal cycle (Definition 5.9) and the Lipschitz-Killing curvature forms from Definition 4.41 we now introduce similarly as for $\mathcal{PR}$-sets the associated curvature notions.

Definition 5.14 The *curvature-direction measure of order* $k \in \{0, \dots, d-1\}$ of $X \in \mathcal{U}_{\mathcal{PR}}$ is the signed Radon measure $\widetilde{C}_k(X, \cdot)$ on $\mathbb{R}^d \times \mathbb{R}^d$ given by

$$\begin{aligned}\widetilde{C}_k(X, E) &:= (N_X \llcorner \mathbf{1}_E)(\varphi_k)\\ &= \int_{E\cap \operatorname{nor} X} \langle i_X(x,n) a_X(x,n), \varphi_k(n)\rangle\, \mathcal{H}^{d-1}(\mathrm{d}(x,n)),\end{aligned}$$

for any bounded Borel set $E \subset \mathbb{R}^d \times \mathbb{R}^d$.

The corresponding *(Lipschitz-Killing) curvature measure* is the projection

$$C_k(X, B) := \widetilde{C}_k(B \times S^{d-1})$$

for any bounded Borel set $B \subset \mathbb{R}^d$.
For bounded ∂X the total curvature measure

$$\mathbf{C}_k(X) := C_k(X, \mathbb{R}^d)$$

is called *Lipschitz-Killing curvature of order k*.
We set again

$$C_d(X, \cdot) := \mathcal{L}^d(X \cap (\cdot))$$

and

$$\mathbf{C}_d(X) := C_d(X, \mathbb{R}^d) = \mathcal{L}^d(X)$$

when X is bounded.

The additivity of the normal cycle (cf. Theorem 5.11) implies the following.

Corollary 5.15 *The curvature direction-measures for $\mathcal{U}_{\mathcal{PR}}$-sets are the additive extensions of those for $\mathcal{PR}$-sets and we have*

$$\widetilde{C}_k(X \cup Y, \cdot) + \widetilde{C}_k(X \cap Y, \cdot) = \widetilde{C}_k(X, \cdot) + \widetilde{C}_k(Y, \cdot)$$

if $X, Y, X \cup Y, X \cap Y \in \mathcal{U}_{\mathcal{PR}}$.

In particular, we obtain the *additivity of the Lipschitz-Killing curvature measures*. The following important geometric properties of the curvature-direction measures are direct analogues to their variants for $\mathcal{PR}$-sets and follow again from the definition by means of the unit normal cycle: Recall the notation $\tilde{g}(x, n) = (g(x), O(n))$, $(x, n) \in \mathbb{R}^{2d}$, for a Euclidean motion $g \in \mathcal{G}_d$ with $g(x) = g(0) + O(x)$.

Proposition 5.16 *For any $X, Y \in \mathcal{U}_{\mathcal{PR}}$, $B \subset \mathbb{R}^d$ and $E \subset \mathbb{R}^{2d}$ bounded Borel sets, $\lambda > 0$ and $g \in \mathcal{G}_d$ we have*

(i) $\widetilde{C}_k(gX, \tilde{g}(E)) = \widetilde{C}_k(X, E)$ *(motion covariance),*
(ii) $C_k(\lambda X, \lambda B) = \lambda^k C_k(X, B)$ *(homogeneity),*
(iii) $\widetilde{C}_k(X, E) = \widetilde{C}_k(Y, E)$ *if $X, Y \in \mathcal{U}_{\mathcal{PR}}$ and $N_X \llcorner \mathbf{1}_E = N_Y \llcorner \mathbf{1}_E$ (locality).*

As a corollary of (iii) we get again $C_k(X, U) = C_k(Y, U)$ if U is a bounded open set in $\mathbb{R}^d$ and $X \cap U = Y \cap U$, i.e. the *locality of the Lipschitz-Killing curvature measures*.

As in the case of sets with positive reach, we can introduce (generalized) *principal curvatures* $\kappa_i(x, n)$ and *principal directions* $b_i(x, n)$ at almost all $(x, n) \in$

$\operatorname{nor} X$ of $\mathcal{U}_{\mathcal{PR}}$-sets X: Recall that if X has $\mathcal{U}_{\mathcal{PR}}$-representation (5.1), $\operatorname{nor} X$ is covered by the locally finite union

$$\operatorname{nor} X \subset \bigcup_{I \in \mathcal{P}_*(\mathbb{N})} \operatorname{nor}\left(\bigcap_{i \in I} X^i\right). \tag{5.7}$$

Let us call a point $(x, n) \in \operatorname{nor} X$ *regular* if (x, n) is regular in $\operatorname{nor} \bigcap_{i \in I} X^i$ whenever $(x, n) \in \operatorname{nor} \bigcap_{i \in I} X^i$. Clearly, $\mathcal{H}^{d-1}$-almost all $(x, n) \in \operatorname{nor} X$ are regular and due to Proposition 1.19 and Corollary 4.45, we have for regular points (x, n), $\operatorname{Tan}^{d-1}(\operatorname{nor} X, (x, n)) = \operatorname{Tan}^{d-1}(\operatorname{nor} Z, (x, n))$ and $a_X(x, n) = a_Z(x, n)$ whenever $Z = \bigcap_{i \in I} X^i$ and $(x, n) \in \operatorname{nor}(Z, (x, n))$.

Thus, we obtain from Proposition 4.23 and Lemma 4.24 the following.

Proposition 5.17 *If $X \in \mathcal{U}_{\mathcal{PR}}$ then for $\mathcal{H}$-almost all $(x, n) \in \operatorname{nor} X$, $\operatorname{Tan}^{d-1}(\operatorname{nor} X, (x, n))$ is a $(d-1)$-dimensional subspace and there exist vectors $b_1(x, n), \ldots, b_{d-1}(x, n)$ in $\mathbb{R}^d$ and numbers $\kappa_1(x, n), \ldots, \kappa_{d-1}(x, n) \in (-\infty, \infty]$ such that $b_1(x, n), \ldots, b_{d-1}(x, n), n$ form a positively oriented orthonormal basis of $\mathbb{R}^d$ and the vectors*

$$a_i(x, n) := \left(\frac{1}{\sqrt{1 + \kappa_i^2(x, n)}} b_i(x, n), \frac{\kappa_i(x, n)}{\sqrt{1 + \kappa_i^2(x, n)}} b_i(x, n)\right), \quad i = 1, \ldots, d-1,$$

form an orthonormal basis of $\operatorname{Tan}^{d-1}(\operatorname{nor} X, (x, n))$. (We set $\frac{1}{\sqrt{1+\infty^2}} = 0$ and $\frac{\infty}{\sqrt{1+\infty^2}} = 1$.) In particular, the $(d-1)$-vector field orienting $\operatorname{nor} X$ equals $a_X = a_1 \wedge \ldots \wedge a_{d-1}$.
The $\kappa_i(x, n)$ are uniquely determined up to their order and the subspace of $\mathbb{R}^d$ generated by the basis vectors $b_j(x, n)$ corresponding to one particular value $\kappa_i(x, n)$ is unique. The dimension of this subspace is given by the multiplicity of the number $\kappa_i(x, n)$ within $\{\kappa_1(x, n), \ldots, \kappa_{d-1}(x, n)\}$.

Then we get a representation of the curvature-direction measures in terms of the corresponding symmetric functions of principal curvatures $s_k(X; x, n)$, which are introduced as in Definition 4.25.

Theorem 5.18 *For $X \in \mathcal{U}_{\mathcal{PR}}$ we have*

$$\widetilde{C}_k(X, \cdot) = \frac{1}{(d-k)\omega_{d-k}} \int_{(\cdot) \cap \operatorname{nor} X} i_X(x, n) s_{d-1-k}(X; x, n)\, \mathcal{H}^{d-1}(\mathrm{d}(x, n)).$$

Proof This follows from the definition and property (4.14). □

If we take into regard the multiple points of the metric projection together with the index function we can also infer a *generalized Steiner formula* for $\mathcal{U}_{\mathcal{PR}}$-sets (cf.

Theorem 4.30 for $\mathcal{PR}$-sets). Recall that $\pi_0(x, n) = x$ is the projection mapping onto the first coordinate in $\mathbb{R}^d \times \mathbb{R}^d$.

Theorem 5.19 *For $X \in \mathcal{U}_{\mathcal{PR}}$, any bounded Borel set $E \subset \mathbb{R}^d \times \mathbb{R}^d$ and $r > 0$ we have*

$$\int_{\mathbb{R}^d} \sum_{x \in \partial X : |x-y| \le r} i_X\left(x, \frac{y-x}{|y-x|}\right) \mathbf{1}_E\left(x, \frac{y-x}{|y-x|}\right) \mathcal{L}^d(\mathrm{d}y) = \sum_{k=0}^{d-1} \omega_k r^k \widetilde{C}_{d-1-k}(X, E).$$

Proof As in Sect. 4.3 for sets of positive reach we apply the area formula to the mapping

$$f : \operatorname{nor} X \times [0, r] \to \mathbb{R}^d \text{ with } f(x, n, t) = x + tn$$

which is, in general, no more injective. Therefore we get instead of Eq. (4.10) for $g = i_X \mathbf{1}_E$,

$$\begin{aligned} &\int_0^r \int_{E \cap \operatorname{nor} X} i_X((x, n)\, J_d f(x, n, t)\, \mathcal{H}^{d-1}(\mathrm{d}(x, n))\, \mathrm{d}t \\ &= \int_{\mathbb{R}^d} \sum_{(x,n,t) \in \operatorname{nor} X \times [0,r] : x+tn=y} i_X(x, n) \mathbf{1}_E(x, n)\, \mathcal{L}^d(\mathrm{d}y) \\ &= \int_{\mathbb{R}^d} \sum_{x \in \partial X : |x-y| \le r} i_X\left(x, \frac{y-x}{|y-x|}\right) \mathbf{1}_E\left(x, \frac{y-x}{|y-x|}\right) \mathcal{L}^d(\mathrm{d}y), \end{aligned}$$

and the sum under the integral on the right hand side is finite for $\mathcal{L}^d$-a.a. y. The remaining calculations for the Jacobian function $J_d f$ leading to the polynomial representation of the left hand side with the curvature-direction measures as coefficients are the same as in Sects. 4.3 and 4.5 for sets of positive reach. □

Setting $E := B \times S^{d-1}$ for a bounded Borel set $B \subset \mathbb{R}^d$ we infer the *generalized Steiner formula for the Lipschitz-Killing curvature measures* of $\mathcal{U}_{\mathcal{PR}}$-sets X:

$$\int_{\mathbb{R}^d} \sum_{x \in E \cap \partial X : |x-y| \le r} i_X\left(x, \frac{y-x}{|y-x|}\right) \mathcal{L}^d(dy) = \sum_{k=0}^{d-1} \omega_k r^k C_{d-1-k}(X, E). \tag{5.8}$$

Note that in the special case $X \in \mathcal{PR}$ Eq. (5.8) extends the Steiner formula from Corollary 4.33, since we do not suppose now that $r < \operatorname{reach} X$.

5.3 Regular $\mathcal{U}_{\mathcal{PR}}$-Representations and the Reflection Principle for Normal Cycles

For $Y \subset \mathbb{R}^d$, denote the closure of the complement of Y by

$$\widetilde{Y} := \overline{Y^c}. \tag{5.9}$$

Recall that X_ε denotes the ε-parallel set of $X \subset \mathbb{R}^d$. The set $\widetilde{X_\varepsilon}$ for sufficiently small ε possesses in many cases much better measure geometric properties than the primary set X. Recall that by Corollary 4.22 for $X \in \mathcal{PR}$, $\widetilde{X_\varepsilon}$ is a closed $C^{1,1}$-domain. Fu [Fu85] showed that for *any compact* $X \subset \mathbb{R}^d$ with $d \leq 3$ and almost all $\varepsilon > 0$, the set $\widetilde{X_\varepsilon}$ has positive reach. In higher dimensions this is not true. However, we obtain such regularity for a large subclass of $\mathcal{U}_{\mathcal{PR}}$ sets.

Definition 5.20

(i) We say that the sets $X^1, \dots, X^k \in \mathcal{PR}$ *touch* (at a point x) if there are $n_1, \dots, n_k$ with $(x, n_1) \in \operatorname{nor} X^1, \dots, (x, n_k) \in \operatorname{nor} X^k$ and $t_1 n_1 + \dots + t_k n_k = 0$ for some $t_1, \dots, t_k \geq 0$ with $t_1 + \dots + t_k = 1$ (i.e., the convex hull of the unit normal vectors $n_1, \dots, n_k$ at the section point x contains the zero vector).

(ii) A $\mathcal{U}_{\mathcal{PR}}$-representation $X = \bigcup_{i=1}^{\infty} X^i$ (5.1) is called *regular* if $X^{i_1}, \dots, X^{i_k}$ do not touch for any $1 \leq i_1 < \dots < i_k < \infty$, $k \in \mathbb{N}$. In that case, we say that the $\mathcal{PR}$-sets $X^1, X^2, \dots$ are *completely non-touching*.

By Theorem 4.19 the finite intersection of non-touching compact $\mathcal{PR}$-sets has again positive reach. Therefore we conclude the following.

Proposition 5.21 *Any locally finite union of completely non-touching compact sets with positive reach provides a regular $\mathcal{U}_{\mathcal{PR}}$-representation.*

The main theorem of this section, the *reflection principle*, establishes a relationship between the unit normal cycle of X and that of $\widetilde{X_\varepsilon}$ for a subclass of $\mathcal{U}_{\mathcal{PR}}$-sets. We define the *normal reflection* on $\mathbb{R}^d \times \mathbb{R}^d$ by

$$\rho : (x, n) \mapsto (x, -n). \tag{5.10}$$

In the proof, we will use the following inclusion-exclusion principle for convex hulls. For the proof, see [KLZ17].

Lemma 5.22 *If $x_1, \dots, x_k \in \mathbb{R}^d$ and $x \in \operatorname{rel\,int} \operatorname{conv} \{x_1, \dots, x_k\}$ then*

$$\sum_{\emptyset \neq I \subset \{1, \dots, k\}} (-1)^{|I|-1} \mathbf{1}_{\operatorname{conv}\{x_i : i \in I\}}(x) = (-1)^{m-1},$$

where $m := \dim \operatorname{conv} \{x_1, \dots, x_k\}$.

Theorem 5.23 *Let* $X = \bigcup_{i=1}^N X^i$ *be a regular* $\mathcal{U}_{\mathcal{PR}}$*-representation with closed* $C^{1,1}$*-domains* $X^1, \dots, X^N$*. Then* $\widetilde{X}$ *has positive reach and*

$$N_X = \rho_\# N_{\widetilde{X}} . \tag{5.11}$$

Proof Since the X^i's are $C^{1,1}$-domains, there exists a unique unit normal vector $n_i(x) \in \operatorname{Nor}(X^i, x)$ at each $x \in \partial X^i$. For $x \in \partial X$, denote $I(x) := \{i \le N : x \in \partial X^i\}$, $k(x) := \operatorname{card} I(x)$ and let $\Delta(x)$ be the convex hull of $\{n_i(x) : i \in I(x)\}$. Denote

$$\eta(x) := \operatorname{dist}(\Delta(x), 0), \quad x \in \partial X.$$

Since $X^1, \dots, X^N$ do not touch we have $\eta(x) > 0$ for any $x \in \partial X$. Moreover, since nor X^i is closed for any $1 \le i \le N$ (see Lemma 4.11), the function η is lower semicontinuous on ∂X and we infer from the compactness that

$$\eta := \inf_{x \in \partial X} \eta(x) > 0. \tag{5.12}$$

Thus, reach $(\bigcap_{i=1}^N X^i) \ge r\eta^{N-1}$ by Theorem 4.19.

It is easy to see that each $\widetilde{X^i}$ is again a $C^{1,1}$ domain, that $-n_i(x)$ is the only unit outer normal to $\widetilde{X^i}$ at $x \in \partial X^i$ and, using the same reasoning as above, we obtain also reach $(\bigcap_{i=1}^N \widetilde{X^i}) \ge r\eta^{N-1}$. Since

$$\widetilde{X} = \overline{\left(\bigcup_{i=1}^N X^i\right)^c} = \overline{\bigcap_{i=1}^N (X^i)^c} = \bigcap_{i=1}^N \widetilde{X^i}, \tag{5.13}$$

we get reach $\widetilde{X} \ge r\eta^{N-1} > 0$.

We denote $C(x) := \operatorname{sco}\{n_i(x) : i \in I(x)\}$, where $\operatorname{sco} T$ denotes the spherical convex hull of a set $T \subset S^{d-1}$ (in fact, $C(x)$ agrees with the projection of $\Delta(x)$ onto S^{d-1} from the origin). Let $m(x) := \dim C(x)$. We show that for any $x \in \partial X$ and $n \in S^{d-1} \setminus \operatorname{rel\,bd} C(x)$,

$$i_X(x, n) = (-1)^{m(x)-1} \mathbf{1}_{C(x)}(n). \tag{5.14}$$

We use the inclusion-exclusion formula for the (additive) index function

$$i_X(x, n) = \sum_{\emptyset \ne I \subset \{1,\dots,N\}} (-1)^{|I|-1} \mathbf{1}_{\operatorname{Nor}(\bigcap_{i \in I} X^i, x)}(n).$$

Theorem 4.19 yields that

$$\operatorname{Nor}(\bigcap_{i \in I} X^i, x) \cap S^{d-1} = \begin{cases} \operatorname{sco}\{n_i(x) : i \in I\}, & I \subset I(x), \\ 0, & \text{otherwise}, \end{cases}$$

and we can apply Lemma 5.22 to obtain (5.14). (In fact, Lemma 5.22 is formulated in the Euclidean space and not on the sphere, nevertheless, the set $C(x)$ is contained in an open hemisphere by the non-touching assumption and we may project $C(x)$ onto a suitable hyperplane from the origin and apply Lemma 5.22 then.)

Theorem 4.19 together with (5.13) yield

$$\operatorname{nor} \widetilde{X} = \{(x, -n) : x \in \partial X,\ n \in C(x)\},$$

whereas by (5.14),

$$\{(x, n) : x \in \partial X,\ n \in \operatorname{rel\,int} C(x)\} \subset \operatorname{nor} X \subset \{(x, n) : x \in \partial X,\ n \in C(x)\}.$$

Thus, the symmetric difference of $\operatorname{nor} X$ and $\rho(\operatorname{nor} \widetilde{X})$ is contained in

$$D := \{(x, n) : x \in \partial X,\ n \in \operatorname{rel\,bd} C(x)\}.$$

If $(x, n) \in D$ then, clearly, $(x, n) \in \operatorname{nor} X_I$ with $I := I(x)$, and there exists a unit vector $u \perp n$ such that $-u \in \operatorname{Tan}(\operatorname{Nor}(X_I, x), n)$ and $u \in (\operatorname{Nor}(X_I, x))^o = \operatorname{Tan}(X_I, x)$ (recall Definition 2.6 for the polar cone). The first named property implies that $(0, -u) \in \operatorname{Tan}(\operatorname{nor} X_I, (x, n))$, whereas the second one yields $(0, u) \notin \operatorname{Tan}(\operatorname{nor} X_I, (x, n))$ by Lemma 5.24 below. Thus, $\operatorname{Tan}(\operatorname{nor} X_I, (x, n))$ is not a linear space and, hence, (x, n) is a non-regular point of $\operatorname{nor} X_I$, and the $\mathcal{H}^{d-1}$-measure of such points is zero. Hence, $\mathcal{H}^{d-1}(D) = 0$, which implies that

$$\operatorname{nor} X = \rho(\operatorname{nor} \widetilde{X}) \text{ up to a set of } \mathcal{H}^{d-1} - \text{measure zero.}$$

Theorem 1.49 yields the relation

$$\rho_{\#} N_{\widetilde{X}} = \left(\mathcal{H}^{d-1} \llcorner \operatorname{nor} X\right) \wedge \zeta,$$

where

$$\zeta(x, n) = (\textstyle\bigwedge_{d-1} \rho) a_{\widetilde{X}}(x, -n), \quad (x, n) \in \operatorname{nor} X.$$

In order to show (5.11), it will be enough to verify that $i_X(x, n) a_X(x, n) = \zeta(x, n)$ for $\mathcal{H}^{d-1}$-almost all $(x, n) \in \operatorname{nor} X$. Let $(x, n) \in \operatorname{nor} X$ be such that $(x, -n)$ is a regular point of $\operatorname{nor} \widetilde{X}$. Then, by Lemma 4.44,

$$\langle (\textstyle\bigwedge_{d-1}(\pi_0 + t\pi_1)) a_{\widetilde{X}}(x, -n) \wedge (-n), \Omega_d \rangle > 0$$

for sufficiently small $t > 0$. Hence,

$$\begin{aligned}
&\operatorname{sgn} \langle (\textstyle\bigwedge_{d-1}(\pi_0 + t\pi_1))(\textstyle\bigwedge_{d-1}\rho) a_{\widetilde{X}}(x, -n) \wedge n, \Omega_d \rangle \\
&= \operatorname{sgn} \langle (\textstyle\bigwedge_{d-1}(\pi_0 - t\pi_1)) a_{\widetilde{X}}(x, -n) \wedge n, \Omega_d \rangle \\
&= (-1)^{m(x)} \operatorname{sgn} \langle (\textstyle\bigwedge_{d-1}(\pi_0 + t\pi_1)) a_{\widetilde{X}}(x, -n) \wedge n, \Omega_d \rangle = (-1)^{m(x)-1},
\end{aligned}$$

since, for $\mathcal{H}^{d-1}$-almost all $(x, n) \in \operatorname{nor} X$, there are exactly $m(x) - 1$ infinite principal curvatures of $\widetilde{X}$ at $(x, -n)$. Indeed, if n lies in the relative interior of $C(x)$ then the principal curvatures are clearly infinite at all directions from the $(m(x)-1)$-dimensional space $\operatorname{Tan}(C(x), n)$, and are finite at any direction perpendicular to $\operatorname{Tan}(C(x), n)$, by Lemma 5.24 below. Thus, the proof is finished. □

Lemma 5.24 *Let* $X^1, \ldots, X^N$ *be as in Theorem* 5.23, $\emptyset \neq I \subset \{1, \ldots, N\}$, $(x, n) \in \operatorname{nor} X_I$ *and* $n \perp u \in \operatorname{Tan}(X_I, x) \cap S^{d-1}$, *where* $X_I := \bigcap_{i \in I} X^i$ *Then* $(0, u) \notin \operatorname{Tan}(\operatorname{nor} X_I, (x, n))$.

Proof Assume, for the contrary, that $(0, u) \in \operatorname{Tan}(\operatorname{nor} X_I, (x, n))$. Then there exist $(x_p, n_p) \in \operatorname{nor} X_I$ and $r_p > 0$ such that $x_p \to x$, $n_p \to n$, $r_p(x_p - x) \to 0$ and $r_p(n_p - n) \to u$, $p \to \infty$. In particular, we have

$$\frac{|x_p - x|}{(n_p - n) \cdot u} \to 0, \quad p \to \infty. \tag{5.15}$$

Due to Theorem 4.19, we can write n_p in the form $n_p = \sum_{i \in I} t_{p,i} n_i(x_p)$ with some $t_{p,i} \geq 0$. Passing to a subsequence if necessary, we can assume that $t_{p,i} \to t_i \geq 0$, $p \to \infty$, $i \in I$. Then, since $n_p \to n$, we get $n = \sum_{i \in I} t_i n_i(x)$. Denoting $J := \{i \in I : n_i(x) \cdot u = 0\}$, we have $t_i = 0$ for $i \in I \setminus J$, since $n \cdot u = 0$ by assumption. Choose $L > 0$ so that $x \mapsto n_i(x)$ is L-Lipschitz, $i \in I$. Then we have

$$n_i(x_p) \cdot u = ((n_i(x_p) - n_i(x)) + n_i(x)) \cdot u \leq L|x_p - x|, \quad i \in I,$$

since $n_i(x) \cdot u \leq 0$, $i \in I$, by assumption, and

$$|n_i(x_p) \cdot u| \leq |(n_i(x_p) - n_i(x)) \cdot u| + |n_i(x)) \cdot u| \leq L|x_p - x|, \quad i \in J.$$

Hence, denoting $\tilde{n}_p := \sum_{i \in I} t_i n_i(x_p)$, we get

$$\begin{aligned}
(n_p - \tilde{n}_p) \cdot u &= \sum_{i \in I} (t_{p,i} - t_i)(n_i(x_p) \cdot u) \\
&\leq \sum_{i \in J} |t_{p,i} - t_i||n_i(x_p) \cdot u| + \sum_{i \in I \setminus J} t_{p,i}(n_i(x_p) \cdot u) \\
&\leq \left(\sum_{i \in I} |t_{p,i} - t_i| \right) L|x_p - x| \\
&\leq L|x_p - x|
\end{aligned}$$

for sufficiently large p. We further have

$$|(\tilde{n}_p - n) \cdot u| \leq \left(\sum_{i \in I} t_i \right) |n_i(x_p) - n_i(x)| \leq \frac{L}{\eta} |x_p - x|,$$

since $\sum_{i\in I} t_i \le \eta^{-1}$ by (5.12). It follows that

$$(n_p - n)\cdot u \le (n_p - \tilde{n}_p)\cdot u + |(\tilde{n}_p - n)\cdot u| \le L(1+\eta^{-1})|x_p - x|$$

fur sufficiently large p, hence

$$\limsup_{p\to\infty} \frac{|n_p - n|}{|x_p - x|} \le L(1+\eta^{-1}),$$

which contradicts (5.15) and finishes the proof. □

The regular $\mathcal{U}_{\mathcal{PR}}$-representation property remains valid for the parallel sets of sufficiently small distance. We obtain a slightly sharper property:

Proposition 5.25 *For any regular $\mathcal{U}_{\mathcal{PR}}$-representation $X = \bigcup_{i=1}^N X^i$ of a compact set and all sufficiently small ε, $X_\varepsilon = \bigcup_{i=1}^N X^i_\varepsilon$ is also a regular $\mathcal{U}_{\mathcal{PR}}$-representation. Moreover, there exists an $r > 0$ such that* reach $(\bigcap_{j=1}^k X_\varepsilon^{i_j}) \ge r$ *whenever* $\{i_1,\dots,i_k\} \subseteq \{1,\dots,N\}$.

Proof Let $k \le N$, $1 \le i_1 < \cdots < i_k \le N$ be fixed. Use the notation for the simplex

$$\Delta_k = \{(t_1,\dots,t_k) \in [0,1]^k : t_1 + \cdots + t_k = 1\}$$

and consider the function Ψ defined on $\operatorname{nor} X^{i_1} \times \ldots \times \operatorname{nor} X^{i_k} \times \Delta_k$ by

$$\Psi : (x_1, n_1, \ldots x_k, n_k, t_1, \ldots, t_k) \mapsto \sum_{j=2}^k |x_j - x_1| + \left|\sum_{j=1}^k t_j n_j\right|.$$

In view of Definition 5.20, Ψ is positive. Since it is Lipschitz continuous, by the compactness assumption its infimum η is positive. Note that for all j, $\operatorname{nor} X_\varepsilon^{i_j}$ converge to $\operatorname{nor} X^{i_j}$ with $\varepsilon \to 0$ in the Hausdorff distance. Thus there exists an $\varepsilon_0 > 0$ such that

$$\Psi(y_1, n_1, \ldots, y_k, n_k, t_1, \ldots, t_k) \ge \eta/2$$

for all $(y_j, n_j) \in \operatorname{nor} X_\varepsilon^{i_j}$, $\varepsilon < \varepsilon_0$ and $(t_1,\dots,t_k) \in \Delta_k$. Consequently, again by Definition 5.20, $X_\varepsilon^{i_1},\dots,X_\varepsilon^{i_k}$ do not touch. Furthermore, since reach $X_\varepsilon^{i_j} \ge$ reach $X^{i_j} - \varepsilon$, Theorem 4.19 implies that

$$\operatorname{reach}\,(\bigcap_{j=1}^k X_\varepsilon^{i_j}) \ge (r_{\min} - \varepsilon_0)(\eta/2)^{k-1},$$

where $r_{\min} := \min_i \operatorname{reach} X^i$. □

As a corollary, we obtain the following result.

Theorem 5.26 *Let $X^1, \dots, X^N$ be compact sets of positive reach. Then there exists $\varepsilon_0 > 0$ such that whenever for some $0 < \varepsilon < \varepsilon_0$, $X_\varepsilon = \bigcup_{i=1}^N X_\varepsilon^i$ is a regular $\mathcal{U}_{\mathcal{PR}}$-representation, then $\widetilde{X_\varepsilon}$ has positive reach and*

$$N_{X_\varepsilon} = \rho_\# N_{\widetilde{X_\varepsilon}} .$$

Proof For all sufficiently small $\varepsilon > 0$, the parallel sets X_ε^i are $C^{1,1}$ domains by Corollary 4.22, and they do not touch by Proposition 5.25. Thus we can apply Theorem 5.23. □

Recall that the curvature measures of X_ε are given by

$$\begin{aligned} C_k(X, B) &= (N_X \llcorner \mathbf{1}_{B \times S^{d-1}})(\varphi_k) \\ &= \int_{\operatorname{nor} X} \mathbf{1}_B(x) \langle i_X(x,n) a_X(x,n), \varphi_k(n) \rangle \, \mathcal{H}^{d-1}(\mathrm{d}(x,n)) , \end{aligned}$$

for any bounded Borel set B in $\mathbb{R}^d$, where the φ_k are the Lipschitz-Killing curvature forms from Definition 4.41. Applying Theorem 5.26 to φ_k and regarding the change of the sign $(-1)^{d-1-k}$ under the mapping ρ we get the following.

Corollary 5.27 *Under the conditions of Theorem 5.26 we have*

$$C_k(X_\varepsilon, \cdot) = (-1)^{d-1-k} \, C_k(\widetilde{X_\varepsilon}, \cdot) , \ k = 0, \dots, d-1 .$$

Remark 5.28 In Chap. 7 (Theorem 7.3) we will show that the normal cycles of the parallel sets of small distance of any $\mathcal{U}_{\mathcal{PR}}$-set X converge in a suitable topology to that of X. In view of Theorem 5.26 we could also work with the sets $\widetilde{X_\varepsilon}$ instead of X_ε. The regularity condition on the $\mathcal{U}_{\mathcal{PR}}$-representation may be reformulated in terms of some regularity property of the distance function d_X in the sense of [Fu85]. Therefore such an approximation is the key for extending the notions of normal cycle and curvatures to geometric sets with a highly singular structure, in particular, to some fractals (see Chaps. 9, and 10).

5.4 Bibliographical Notes

1. Extensions of classical versions of the above curvatures to *piecewise flat subspaces* of Riemannian manifolds in differential geometry and Morse index theory have a long history:
   a. Allendoerfer and Weil [AW43] proved the Gauss-Bonnet theorem for Riemannian polyhedra.
   b. In mathematical physics, in particular in general relativity, the corresponding theory is known as Regge calculus, which goes back to [Reg61], where the

scalar curvature was involved. (The latter corresponds to C_{d-3} in the above case.)

c. Banchoff [Ban67] determined the Gauss curvature C_0 of Euclidean polyhedra via index theory.

d. Intrinsic constructions of the Lipschitz-Killing curvatures C_k for the Riemannian case were given in Wintgen [Win82], Cheeger [Che83], Cheeger, Müller and Schrader [CMS84]. The latter also obtained a version of the Steiner formula for such sets in [CMS86]. A purely combinatorial description of these curvatures in a more abstract setting can be found in [Bud89].

2. For parallel developments in convex geometry see Chap. 2. The closest reference is Schneider [Sch80] who extended curvature measures from convex bodies to the convex ring, using additivity and the index function i_X in the sense of Remark 5.13. He also considered a nonnegative extension of curvature measures to the convex ring.
3. $\mathcal{U}_{\mathcal{PR}}$-sets were introduced in [Zäh87], with an additional condition concerning the intersections of tangent cones of the $\mathcal{PR}$ components. In particular, the $\mathcal{U}_{\mathcal{PR}}$-representation (5.1) was assumed to fulfill for any $x \in \partial X$

$$\mathrm{Tan}\left(\bigcap_{i\in I} X^i, x\right) = \bigcap_{i\in I} \mathrm{Tan}(X^i, x), \quad \emptyset \neq I \subset \mathbb{N} \text{ finite.} \tag{5.16}$$

General $\mathcal{U}_{\mathcal{PR}}$-sets were investigated in [RZ01].

Chapter 6
Integral Geometric Formulas

Integral geometry, in general, is concerned with integrals of geometric characteristics with respect to invariant measures, usually under Euclidean motions. A classical and comprehensive reference to integral-geometric relations is the book of Santaló [San76]. In this chapter we derive integral-geometric formulas concerning curvature measures and their total values (total curvatures or intrinsic volumes). This is, in particular, the Principal kinematic formula expressing the curvature measures of the intersection of two bodies, one of them fixed and the other moving, in terms of those of the primary bodies. Similarly, the Crofton formula deals with the curvature measures of flat sections of a body, integrated with respect to the invariant measure over the sectioning flats. See Chap. 2, Theorems 2.4 and 2.5, for the versions of convex geometry.

Here we consider the formulas in the setting of sets with positive reach, though their validity is much broader. The reason is that the proof techniques can be well demonstrated on $\mathcal{PR}$-sets. Extensions to the $\mathcal{U}_{\mathcal{PR}}$-settings are mentioned in the bibliographic remarks at the end of the chapter. For the validity for other set classes see the remarks in Chap. 9.

We start now with the formulations of the main results. Recall that $\mathrm{d}g$ denotes integration with respect to the properly normalized motion invariant measure on the group $\mathcal{G}_d$ of all rigid motions in $\mathbb{R}^d$, see Sect. 2.3.

Theorem 6.1 (Principal kinematic formula for sets with positive reach) *Let $X, Y \subset \mathbb{R}^d$ be sets with positive reach and $A, B \subset \mathbb{R}^d$ be bounded Borel sets. Then* reach $(X \cap gY) > 0$ *for almost all motions $g \in \mathcal{G}_d$ and for any integer $0 \le k \le d$,*

$$\int_{\mathcal{G}_d} C_k\big(X \cap gY, A \cap gB\big)\mathrm{d}g = \sum_{\substack{1 \le r,s \le d \\ r+s=d+k}} \gamma(d,r,s) C_r(X,A) C_s(Y,B),$$

J. Rataj, M. Zähle, *Curvature Measures of Singular Sets*, Springer Monographs in Mathematics, https://doi.org/10.1007/978-3-030-18183-3_6

where

$$\gamma(d, r, s) = \frac{\Gamma((r+1)/2)\Gamma((s+1)/2)}{\Gamma((r+s-d+1)/2)\Gamma((d+1)/2)}.$$

The first proof was given by Federer in [Fed59] using approximation with smooth sets. A shorter and more direct proof was presented later by Rother and Zähle [RZ90b]. We will use here another approach based on the translative version as given in [RZ95]. (See Theorem 6.10 and Lemma 6.19.)

Before doing this, note that choosing for Y in Theorem 6.1 a j-flat, since $C_s(Y, B) = \mathcal{H}^j(Y \cap B)$ if $s = j$ and $C_s(Y, B) = 0$ otherwise, we get the Crofton formula stated as follows. Recall that μ_j^d is the properly normalized invariant measure on the set $\mathcal{A}_j^d$ of all j-flats in $\mathbb{R}^d$, see Chap. 2.

Theorem 6.2 (Crofton formula for sets with positive reach) *Let X be a set with positive reach in $\mathbb{R}^d$, $0 \le k \le j \le d$, $A \subset \mathbb{R}^d$ be a bounded Borel set. Then*

$$\int_{\mathcal{A}(d,j)} C_k(X \cap E, A \cap E)\, \mu_j^d(\mathrm{d}E) = \gamma(d, d+k-j, j) C_{d+k-j}(X, A).$$

Note that if X, Y are moreover compact we can relax the boundedness assumption for A, B in Theorems 6.1 and 6.2. In particular, we obtain:

Corollary 6.3 *Let $X, Y \subset \mathbb{R}^d$ be compact sets with positive reach. Then*

$$\int_{\mathcal{G}_d} \mathbf{C}_k(X \cap gY)\mathrm{d}g = \sum_{\substack{1 \le r,s \le d \\ r+s=d+k}} \gamma(d, r, s)\mathbf{C}_r(X)\mathbf{C}_s(Y),$$

$$\int_{\mathcal{A}(d,j)} \mathbf{C}_k(X \cap E)\, \mu_j^d(\mathrm{d}E) = \gamma(d, d+k-j, j)\mathbf{C}_{d+k-j}(X).$$

6.1 Translative and Kinematic Integral Formulas for Curvature Measures

If X, Y are two sets with positive reach that do not *touch*, i.e., there is no intersection point of the boundaries $x \in \partial X \cap \partial Y$ with opposite normal vectors, $0 \neq n \in \mathrm{Nor}(X, x)$, $-n \in \mathrm{Nor}(Y, x)$, then the unit normal bundle (and, hence, also the curvature measures) is completely described by means of the relation

$$\mathrm{Nor}(X \cap Y, x) = \mathrm{Nor}(X, x) + \mathrm{Nor}(Y, x)$$

from Proposition 4.17. If X and Y *do* touch at some point x then we cannot say in general anything about the local behaviour of the intersection at x. It can even happen that $X \cap Y$ does not have positive reach (see Example 4.16).

We will derive in this section a formula for the intersection

$$\int_{\mathbb{R}^d} C_k(X \cap (Y+z); \cdot)\, \mathcal{L}^d(\mathrm{d}z).$$

In order to guarantee the regular behaviour of the intersection for almost all shifts z, we assume that

$$\mathcal{L}^d\{z \in \mathbb{R}^d : X \text{ and } Y+z \text{ touch}\} = 0. \tag{6.1}$$

The validity of this assumption will be discussed later.

In order to describe the unit normal bundles of $X \cap (Y+z)$, we introduce the following mappings. Let $m, n \in \mathbb{R}^d$ be two nonzero vectors with angle $\angle(m,n) < \pi$, let $t \in [0,1]$, and set

$$u(m,n,t) := \frac{\sin((1-t)\angle(m,n))}{\sin \angle(m,n)} m + \frac{\sin(t\angle(m,n))}{\sin \angle(m,n)} n. \tag{6.2}$$

(The values of the fractions at $\angle(m,n) = 0$ are defined by the limits as $\angle(m,n) \to 0$.) Note that if m, n are unit vectors then $u(m,n,\cdot)$ is a unit-speed parametrization of the geodesic arc on S^{d-1} with endpoints m, n. Denote further

$$R^0 := \{(x,m,y,n) \in (\mathbb{R}^d)^4 : m, n \neq 0,\ \angle(m,n) < \pi\}$$

and consider the mapping

$$\begin{aligned} F :\ & R^0 \times [0,1] \to \mathbb{R}^{2d} \times S^{d-1}, \\ & (x,m,y,n,t) \mapsto (x,y,u(m,n,t)). \end{aligned} \tag{6.3}$$

Note that F is differentiable and, hence, locally Lipschitz, on $R^0 \times [0,1]$. We are going to push forward by F the product current (cf. Sect. 1.3.3)

$$\Big((N_X \times N_Y) \llcorner R^0\Big) \times [0,1]$$

(here $[0,1]$ denotes simultaneously the unit interval and also the 1-current $(\mathcal{L}^1 \llcorner [0,1]) \wedge 1$). We first define the measure

$$\|N_{X,Y}\| := \|F_\# \Big(((N_X \times N_Y) \llcorner R^0) \times [0,1]\Big)\|$$

as in Definition 1.48 and, provided that

$$\|N_{X,Y}\| \text{ is locally finite,} \tag{6.4}$$

also the current

$$N_{X,Y} := F_\#((N_X \times N_Y) \llcorner R^0 \times [0,1])$$

on $\mathbb{R}^{3d}$ called later *joint unit normal current of* X and Y. Its career will be denoted by

$$\operatorname{nor}(X,Y) := F\left(((\operatorname{nor} X \times \operatorname{nor} Y) \cap R^0) \times [0,1]\right).$$

Using (1.26), we can write

$$\begin{aligned}&\left((N_X \times N_Y) \llcorner R^0\right) \times [0,1]\\&= \left(\mathcal{H}^{2d-1} \llcorner ((\operatorname{nor} X \times \operatorname{nor} Y) \cap R^0) \times [0,1]\right) \wedge (a_X \oslash a_Y \oslash 1)\end{aligned}$$

(see Sect. 1.3.3 for the definition of the product of currents and of the symbol $\oslash$). Thus, condition (6.4) can be rewritten equivalently as

$$\int_{(\operatorname{nor} X \times \operatorname{nor} Y) \cap (R^0 \cap K) \times [0,1]} |\tilde{a}_{X,Y}| \, d\mathcal{H}^{2d-1} < \infty, \quad K \subset \mathbb{R}^{4d} \text{ compact}, \tag{6.5}$$

where

$$\tilde{a}_{X,Y} := (\textstyle\bigwedge_{2d-1} DF)(a_X \oslash a_Y \oslash 1). \tag{6.6}$$

Remarks on the validity of (6.4) will be given later.

In order to get an integral representation of $N_{X,Y}$, we need the following auxiliary result. Let f denote the restriction of F to $((\operatorname{nor} X \times \operatorname{nor} Y) \cap R^0) \times [0,1]$.

Lemma 6.4 *For $\mathcal{H}^{2d-1}$-almost all $(x,y,u) \in \operatorname{nor}(X,Y)$,*

$$\operatorname{card} f^{-1}\{(x,y,u)\} = 1.$$

Consequently,

$$N_{X,Y} = \left(\mathcal{H}^{2d-1} \llcorner \operatorname{nor}(X,Y)\right) \wedge a_{X,Y},$$

where the unit simple $(2d-1)$-vector field $a_{X,Y}$ is given by

$$a_{X,Y} := \frac{\tilde{a}_{X,Y} \circ f^{-1}}{|\tilde{a}_{X,Y} \circ f^{-1}|}.$$

Proof Let

$$N := \{x, y, u) \in \operatorname{nor}(X, Y) : \operatorname{card} f^{-1}\{(x, y, u)\} > 1\}.$$

By the Area formula (Theorem 1.14),

$$\int_{F^{-1}(N)} \operatorname{ap} J_{2d-1} f \, d\mathcal{H}^{2d-1} = \int_N \operatorname{card} F^{-1}\{(x, y, u)\} \, \mathcal{H}^{2d-1}(d(x, y, u)).$$

We will show that the approximate Jacobian $J_{2d-1} f$ vanishes on $f^{-1}(N)$, which implies the first statement.

Take a point $(x, m, y, n, t) \in f^{-1}(N)$. By the definition of N, there exists $(m', n', t') \neq (m, n, t)$ with $(x, m') \in \operatorname{nor} X$, $(y, n') \in \operatorname{nor} Y$ and $u = u(m, n, t) = u(m', n', t')$. Since the normal cones $\operatorname{Nor}(X, x)$ and $\operatorname{Nor}(Y, y)$ are convex, they contain the cones spanned by m, m' and n, n', respectively, and we have $(0, a) \in \operatorname{Tan}^{d-1}(\operatorname{nor} X, (x, m))$ and $(0, b) \in \operatorname{Tan}^{d-1}(\operatorname{nor} Y, (y, n))$ with $a = m' - m$ and $b = n' - n$. Thus, the linear hull L of $(0, a, 0, 0, 0)$, $(0, 0, 0, b, 0)$ and $(0, 0, 0, 0, 1)$ is a subspace of $\operatorname{Tan}^{2d-1}(\operatorname{nor} X \times \operatorname{nor} Y \times [0, 1])$. It is further clear that $\operatorname{ap} Df(x, m, y, n, t)$ maps all the vectors from L to $\{0\} \times M$ for the linear hull $M = \operatorname{Lin}\{m, n, m', n'\}$ and, at the same time, since u maps to the unit sphere, the image of its differential lies in $u^\perp$. The four vectors m, m', n, n' are linearly dependent (since $u = u(m, n, t) = u(m', n', t')$), thus, $\dim M \leq 3$ and $\dim M \cap u^\perp \leq 2$. It follows that $\operatorname{ap} Df(x, m, y, n, t)$ cannot be injective at points from $f^{-1}(N)$. Consequently, $J_{2d-1} f$ vanishes at $f^{-1}(N)$.

The second assertion is the Area formula for currents (Theorem 1.49). □

For any $z \in \mathbb{R}^d$, we can decompose the normal current of $X \cap (Y + z)$ into three parts,

$$N_{X \cap (Y+z)} = N_z^1 + N_z^2 + N_z^3, \tag{6.7}$$

where

$$N_z^1 = N_{X \cap (Y+z)} \llcorner \pi_0^{-1} \operatorname{int}(Y + z) = N_X \llcorner \pi_0^{-1} \operatorname{int}(Y + z),$$

$$N_z^2 = N_{X \cap (Y+z)} \llcorner \pi_0^{-1} \operatorname{int} X = N_{Y+z} \llcorner \pi_0^{-1} \operatorname{int} X$$

and

$$N_z^3 = N_{X \cap (Y+z)} \llcorner \pi_0^{-1} (\partial X \cap \partial(Y + z)).$$

We obtain the most intricate part, N_z^3, by sectioning $N_{X,Y}$ with the mapping

$$G : (x, y, u) \mapsto x - y$$

defined on $\mathbb{R}^{3d}$, with restriction

$$g := G \mid \operatorname{nor}(X, Y).$$

(See Sect. 1.3.7 for the definition of slicing.) Denote the projection

$$\hat{\pi} : (x, y, u) \mapsto (x, u).$$

Proposition 6.5 *Let two sets* X, Y *have positive reach and satisfy assumptions* (6.1) *and* (6.4). *Then*

$$N_z^3 = (-1)^d \hat{\pi}_\# \langle N_{X,Y}, G, z\rangle$$

for $\mathcal{L}^d$*-almost all* $z \in \mathbb{R}^d$.

Proof First, we apply the Coarea formula for currents (Theorem 1.65) to get an expression for the section current. We obtain that for $\mathcal{L}^d$-almost all $z \in \mathbb{R}^d$,

$$\langle N_{X,Y}, g, z\rangle = \left(\mathcal{H}^{d-1} \llcorner g^{-1}\{z\}\right) \wedge \zeta$$

with

$$\zeta(x, y, u) = \frac{a_{X,Y}(x, y, u) \llcorner G^\# \Omega_d}{J_d g(x, y, u)},$$

a unit simple $(2d-1)$-vector field associated with $g^{-1}\{z\}$. Further, note that by Lemma 6.4, $\hat{\pi}$ is one-to-one $\mathcal{H}^{2d-1}$-almost everywhere on $\operatorname{nor}(X, Y)$. Thus, the Area formula for currents (Theorem 1.49) yields

$$\hat{\pi}_\# \langle N_{X,Y}, g, z\rangle = \left(\mathcal{H}^{d-1} \llcorner \hat{\pi}(g^{-1}\{z\})\right) \wedge \eta$$

with unit simple $(d-1)$-vector field

$$\eta(x, u) = \frac{\left(\bigwedge_{d-1} \hat{\pi}(\hat{\pi}^{-1}(x, u))\right) \zeta(\hat{\pi}^{-1}(x, u))}{J_{d-1}(\hat{\pi} \mid \operatorname{nor}(X, Y))(\hat{\pi}^{-1}(x, u))}.$$

Due to assumption (6.1) and Proposition 4.17, we know that $\hat{\pi}(g^{-1}\{z\})$ agrees with N_z^3 for almost all z. It is thus sufficient to verify that the vector fields $(-1)^d \eta$ and $a_{X \cap (Y+z)}$ coincide $\mathcal{H}^{d-1}$-almost everywhere on N_z^3, for $\mathcal{L}^d$-almost all $z \in \mathbb{R}^d$. Since both are unit simple vector fields associated with the same vector space, it suffices to verify that they have the same orientation. We have

$$\begin{aligned}(-1)^d \langle \eta, \varphi_k\rangle &= \alpha \langle a_{X,Y} \llcorner G^\# \Omega_d, \hat{\pi}^\# \varphi_k\rangle \\ &= \alpha \langle a_{X,Y}, G^\# \Omega_d \wedge \hat{\pi}^\# \varphi_k\rangle,\end{aligned}$$

with a positive factor α. It will follow from Lemma 6.15 that

$$\langle a_{X,Y}, G^{\#}\Omega_d \wedge \hat{\pi}^{\#}\varphi_{k_0}\rangle > 0, \tag{6.8}$$

where k_0 is the largest index $\le d-1$ (depending on (x,m,y,n,t)) for which $\langle a_{X,Y}, G^{\#}\Omega_d \wedge \hat{\pi}^{\#}\varphi_{k_0}\rangle$ is nonzero. According to Remark 4.43, this proves already that η has the same orientation as $a_{X\cap(Y+z)}$. □

Before formulating the translative intersection formula, we need to define the mixed curvature measures.

Definition 6.6 (3-product of multivectors) Let α, β, γ be multivectors in $\mathbb{R}^d$ of multiplicities r, s, q, respectively, and assume that $r+s+q=2d$. The *3-product* of α, β and γ is defined by

$$\langle \alpha, \beta, \gamma\rangle := \langle \star\alpha \wedge \star\beta \wedge \star\gamma, \Omega_d\rangle$$

(for the definition of the Hodge star operator, see Definition 1.29).

In the sequel, we will need the following auxiliary result. Recall Definition 1.31 of the bracket $[\cdot,\cdot]$ for subspaces. If ξ, η are simple multivectors, we will often write briefly $[\xi,\eta]$ instead of $[L(\xi), L(\eta)]$ ($L(\xi)$ is the linear subspace associated with ξ).

Lemma 6.7 *If $\{a_1,\dots,a_d\}$, $\{b_1,\dots,b_d\}$ are two positively oriented orthonormal bases of $\mathbb{R}^d$, then for all $1\le r,s\le d-1$ with $r+s\ge d$,*

$$\left\langle \bigwedge_{i=1}^{r} a_i, \bigwedge_{j=1}^{s} b_j, \bigwedge_{i=r+1}^{d} a_i \wedge \bigwedge_{j=s+1}^{d} b_j \right\rangle = \left[\bigwedge_{i=1}^{r} a_i, \bigwedge_{j=1}^{s} b_j\right]^2.$$

Proof Denote $A := \operatorname{Lin}\{a_1,\dots,a_r\}$, $B := \operatorname{Lin}\{b_1,\dots,b_s\}$, $k := \dim(A\cap B)$. If $k > r+s-d$ then $A+B \subsetneq \mathbb{R}^d$ and both sides of the equation vanish. Assume thus that $k = r+s-d$.

We can assume without loss of generality that $a_i = b_i$ for $i = 1,\dots,k$. We will use the identity

$$\bigwedge_{i=r+1}^{d} a_i \wedge \bigwedge_{j=s+1}^{d} b_j = \left\langle \bigwedge_{i=1}^{k} a_i \wedge \bigwedge_{i=r+1}^{d} a_i \wedge \bigwedge_{j=s+1}^{d} b_j, \Omega \right\rangle \bigwedge_{i=k+1}^{d} a_i. \tag{6.9}$$

To see this, note that on both sides there are simple multivectors associated with the same linear subspace, $(A\cap B)^{\perp}$, and that the wedge product with $\bigwedge_1^k a_i$ from left yields the same d-vector on both sides. Using this identity and the definition of the

3-product, we obtain

$$
\begin{aligned}
&\left\langle \bigwedge_{i=1}^{r} a_i, \bigwedge_{j=1}^{s} b_j, \bigwedge_{i=r+1}^{d} a_i \wedge \bigwedge_{j=s+1}^{d} b_j \right\rangle \\
&= \left\langle \bigwedge_{i=1}^{k} a_i \wedge \bigwedge_{i=r+1}^{d} a_i \wedge \bigwedge_{j=s+1}^{d} b_j, \Omega \right\rangle \left\langle \bigwedge_{i=r+1}^{d} a_i \wedge \bigwedge_{j=s+1}^{d} b_j \wedge (-1)^{k(d-k)} \bigwedge_{i=1}^{k} a_i, \Omega \right\rangle \\
&= \left\langle \bigwedge_{i=1}^{k} a_i \wedge \bigwedge_{i=r+1}^{d} a_i \wedge \bigwedge_{j=s+1}^{d} b_j, \Omega \right\rangle^2 \\
&= \left[\bigwedge_{i=1}^{r} a_i, \bigwedge_{j=1}^{s} b_j \right]^2 .
\end{aligned}
$$

□

Definition 6.8 (**$\psi_{r,s}$-forms**) Given $r, s \in \{1, \ldots, d\}, d \leq r+s \leq 2d-1$, we define the $(2d-1)$-form $\psi_{r,s}$ on $\mathbb{R}^{3d}$ by

$$
\begin{aligned}
&\left\langle \bigwedge_{i=1}^{2d-1} (a_i, b_i, c_i), \psi_{r,s}(x, y, u) \right\rangle \\
&= (-1)^{d+s(d-r)} \mathcal{O}_q^{-1} \sum_{\sigma \in \mathrm{Sh}(r,s,q)} \operatorname{sgn} \sigma \left\langle \bigwedge_{i=1}^{r} a_{\sigma_i}, \bigwedge_{j=r+1}^{r+s} b_{\sigma_j}, \bigwedge_{k=r+s+1}^{2d-1} c_{\sigma_k} \wedge u \right\rangle,
\end{aligned}
$$

where $q = 2d - r - s - 1$,

$$\mathcal{O}_q := \mathcal{H}^q(S^q) = (q+1)\omega_{q+1}$$

is the Hausdorff measure of the unit q-sphere in R^{q+1}, and $\mathrm{Sh}(r, s, q)$ is the set of all permutations σ of $\{1, \ldots, 2d-1\}$ which are increasing on $\{1, \ldots, r\}$, $\{r+1, \ldots, r+s\}$ and $\{r+s+1, r+s+q\}$.

Since $\psi_{r,s}(x, y, u)$ depends only on the third component u, we will write $\psi_{r,s}(u)$ in the sequel.

Definition 6.9 (Mixed curvature measures) Let X, Y be two sets in $\mathbb{R}^d$ with positive reach and assume that (6.4) is fulfilled. We define the mixed curvature measure $\widetilde{C}_{r,s}(X, Y, \cdot)$ of sets X, Y and orders $r, s \in \{1, \ldots, d\}, d \leq r+s \leq 2d-1$, as follows. If $r = d$ we set

$$\widetilde{C}_{d,s}(X, Y, \cdot) := C_d(X, \cdot) \otimes \widetilde{C}_s(Y, \cdot),$$

and if $s = d$,

$$\widetilde{C}_{r,d}(X, Y, \cdot) := \widetilde{C}_r(X, \cdot) \otimes C_s(Y, \cdot).$$

If $r, s < d$ we set

$$\widetilde{C}_{r,s}(X, Y, E) := (N_{X,Y} \llcorner \mathbf{1}_E)(\psi_{r,s}),$$

for any bounded Borel subset $E \subset \mathbb{R}^{3d}$. By $C_{r,s}(X, Y, \cdot) = \widetilde{C}_{r,s}(X, Y, \cdot \times \mathbb{R}^d)$ we denote the projection of $\widetilde{C}_{r,s}(X, Y, \cdot)$ into the first two vector coordinates.

Theorem 6.10 (Translative intersection formula) *Let X, Y be two subsets of $\mathbb{R}^d$ with positive reach and assume that* (6.1) *and* (6.4) *are fulfilled. Then, for any bounded measurable function h on $(\mathbb{R}^d)^3$ such that* $\operatorname{spt} h \cap ((X - Y) \times X \times S^{d-1})$ *is bounded, and for any $k \in \{0, \dots, d-1\}$,*

$$\int \int h(z, x, u)\, \widetilde{C}_k(X \cap (Y + z), \mathrm{d}(x, u))\, \mathcal{L}^d(\mathrm{d}z)$$
$$= \sum_{r+s=k+d} \int h(x - y, x, u)\, \widetilde{C}_{r,s}(X, Y, \mathrm{d}(x, y, u)).$$

Proof Using the definition of $\widetilde{C}_k$ and the decomposition (6.7), we can write the left side as

$$\int \left(N_{X\cap(Y+z)} \llcorner h(z, \cdot, \cdot)\right)(\varphi_k)\, \mathcal{L}^d(\mathrm{d}z) = \sum_{i=1}^{3} \int \left(N_z^i \llcorner h(z, \cdot, \cdot)\right)(\varphi_k)\, \mathcal{L}^d(\mathrm{d}z).$$

The first summand can be calculated as

$$\begin{aligned}
&\int \left(N_z^1 \llcorner h(z, \cdot, \cdot)\right)(\varphi_k)\, \mathcal{L}^d(\mathrm{d}z)\\
&= \int \left(N_X \llcorner (\mathbf{1}_{\pi_0^{-1}\operatorname{int}(Y+z)} h(z, \cdot, \cdot))\right)(\varphi_k)\, \mathcal{L}^d(\mathrm{d}z)\\
&= \int \int \mathbf{1}_{\operatorname{int}(Y+z)}(x) h(z, x, u)\, \widetilde{C}_k(X, \mathrm{d}(x, u))\, \mathcal{L}^d(\mathrm{d}z)\\
&= \int \int \mathbf{1}_{\operatorname{int} Y}(y) h(x - y, x, u)\, \widetilde{C}_k(X, \mathrm{d}(x, u))\, \mathcal{L}^d(\mathrm{d}y)\\
&= \int \int h(x - y, x, u)\, \widetilde{C}_k(X, \mathrm{d}(x, u))\, C_d(Y, \mathrm{d}y),
\end{aligned}$$

where we have used the substitution $z = x - y$, $\mathcal{L}^d(\mathrm{d}z) = \mathcal{L}^d(\mathrm{d}y)$ for the outer integral. Similarly we obtain

$$\int \left(N_z^2 \llcorner h(z,\cdot,\cdot)\right)(\varphi_k)\,\mathcal{L}^d(\mathrm{d}z)$$
$$= \int\int h(x-y,x,u)\,C_d(X,\mathrm{d}x)\,\widetilde{C}_k(Y,\mathrm{d}(y,u)).$$

For the third summand, we apply Proposition 6.5 and get

$$\int \left(N_z^3 \llcorner h(z,\cdot,\cdot)\right)(\varphi_k)\,\mathcal{L}^d(\mathrm{d}z)$$
$$= (-1)^d \int \left(\hat{\pi}_\#\langle N_{X,Y},G,z\rangle \llcorner h(z,\cdot,\cdot)\right)(\varphi_k)\,\mathcal{L}^d(\mathrm{d}z)$$
$$\overset{(1.30)}{=} (-1)^d \int \hat{\pi}_\#\left(\langle N_{X,Y},G,z\rangle \llcorner \hat{\pi}^\# h(z,\cdot,\cdot)\right)(\varphi_k)\,\mathcal{L}^d(\mathrm{d}z)$$
$$\overset{(1.27)}{=} (-1)^d \int \langle N_{X,Y} \llcorner h\circ\Gamma, G, z\rangle(\hat{\pi}^\#\varphi_k)\,\mathcal{L}^d(\mathrm{d}z)$$
$$\overset{(1.33)}{=} (-1)^d((N_{X,Y} \llcorner h\circ\Gamma) \llcorner G^\#\Omega_d)(\hat{\pi}^\#\varphi_k)$$
$$\overset{(1.22)}{=} (-1)^d(N_{X,Y} \llcorner h\circ\Gamma)(G^\#\Omega_d \wedge \hat{\pi}^\#\varphi_k),$$

with $\Gamma : (x,y,u) \mapsto (x-y,x,u)$. The right side of the Translative integral formula to be proved can be rewritten, using the definition of mixed curvature measures, as

$$\sum_{\substack{r+s=k+d\\1\le r,s\le d-1}} (N_{X,Y} \llcorner h\circ\Gamma)(\psi_{r,s}).$$

The proof is thus finished by Lemma 6.11. □

Lemma 6.11 *For all* $u \in S^{d-1}$,

$$G^\#\Omega_d \wedge \hat{\pi}^\#\varphi_k(u) = (-1)^d \sum_{\substack{r+s=k+d\\1\le r,s\le d-1}} \psi_{r,s}(u).$$

Proof Consider a multivector

$$\tau = \bigwedge_{i=1}^{r}(a_i,0,0) \wedge \bigwedge_{j=1}^{s}(0,b_j,0) \wedge \bigwedge_{l=1}^{q}(0,0,c_l)$$

with $0 \leq r, s \leq d$, $q := 2d - r - s - 1$ and vectors a_i, b_j, c_k from a positively oriented orthonormal basis $\{e_1, \dots, e_d\}$ of $\mathbb{R}^d$ such that $u = e_d$. Since such multivectors form a basis of $\bigwedge_{2d-1}(\mathbb{R}^d)^3$, it is sufficient to show that

$$\langle \tau, G^{\#}\Omega_d \wedge \hat{\pi}^{\#}\varphi_k(u)\rangle = (-1)^d \sum_{\substack{r'+s'=k+d \\ 1\leq r',s'\leq d-1}} \langle \tau, \psi_{r',s'}(u)\rangle. \tag{6.10}$$

Due to the special form of τ, the left side of (6.10) equals $\langle \tau, \psi_{r,s}(u)\rangle$ if $1 \leq r, s \leq d$ and $r + s = d + k$, and zero otherwise. If $r + s \neq d + k$ then both sides of (6.10) vanish. If $r + s = d + k$ we can assume without loss of generality that $a_i = b_i$ for $1 \leq i \leq k$ (this can be achieved by reordering the components). Since

$$(u, 0, 0) \wedge (0, u, 0) = \left(\frac{u}{\sqrt{2}}, -\frac{u}{\sqrt{2}}, 0\right) \wedge \left(\frac{u}{\sqrt{2}}, \frac{u}{\sqrt{2}}, 0\right)$$

for any vector u, we can rewrite τ in the form

$$\tau = \bigwedge_{i=1}^{k}\left(\frac{a_i}{\sqrt{2}}, -\frac{a_i}{\sqrt{2}}, 0\right) \wedge \bigwedge_{i=k+1}^{r} (a_i, 0, 0) \wedge \bigwedge_{j=1}^{k}\left(\frac{a_j}{\sqrt{2}}, \frac{a_j}{\sqrt{2}}, 0\right) \wedge \bigwedge_{j=k+1}^{s} (0, b_j, 0) \wedge \bigwedge_{l=1}^{q}(0, 0, c_l).$$

In this expression, exactly d wedge factors do not lie in the kernel of G. Thus, by the shuffle formula (1.3)

$$\begin{aligned}
&\langle \tau, G^{\#}\Omega_d \wedge \hat{\pi}^{\#}\varphi_k(u)\rangle \\
&= (-1)^{k(s-k)} \left\langle \bigwedge_{i=1}^{k}\left(\frac{a_i}{\sqrt{2}}, -\frac{a_i}{\sqrt{2}}, 0\right) \wedge \bigwedge_{i=k+1}^{r} (a_i, 0, 0) \wedge \bigwedge_{j=k+1}^{s} (0, b_j, 0), G^{\#}\Omega_d \right\rangle \\
&\qquad \times \left\langle \bigwedge_{j=1}^{k}\left(\frac{a_j}{\sqrt{2}}, \frac{a_j}{\sqrt{2}}, 0\right) \wedge \bigwedge_{l=1}^{q}(0, 0, c_l), \hat{\pi}^{\#}\varphi_k(u) \right\rangle \\
&= (-1)^{k(s-k)} 2^{k/2} (-1)^{s-k} \left\langle \bigwedge_{i=1}^{r} a_i \wedge \bigwedge_{j=k+1}^{s} b_j, \Omega_d \right\rangle \\
&\qquad \times \left\langle \bigwedge_{j=1}^{k} 2^{-k/2}(a_j, 0) \wedge \bigwedge_{l=1}^{q}(0, c_l), \varphi_k(u) \right\rangle \\
&= (-1)^{s(d-r)} \mathcal{O}_q^{-1} \left\langle \bigwedge_{i=1}^{r} a_i \wedge \bigwedge_{j=k+1}^{s} b_j, \Omega_d \right\rangle \left\langle \bigwedge_{j=1}^{k} a_j \wedge \bigwedge_{l=1}^{q} c_l \wedge u, \Omega_d \right\rangle.
\end{aligned}$$

If $c_l = u$ for some $l \le q$ then the last expression vanishes, but then, $\langle \tau, \psi_{r,s}(u)\rangle = 0$ as well. Assume thus that $c_l \neq u$ for all $l \le q$, and complete the vector sequences (a_i), (b_j) and (c_l) to positively oriented orthonormal bases of $\mathbb{R}^d$ such that $c_d = u$. We have

$$\left\langle \bigwedge_{j=1}^{k} a_j \wedge \bigwedge_{l=1}^{q} c_l \wedge u, \Omega_d \right\rangle = \left\langle \bigwedge_{j=k+1}^{d} a_j \wedge \bigwedge_{l=q+1}^{d-1} c_l, \Omega_d \right\rangle.$$

Moreover, applying (6.9) and the fact that

$$\left\langle \bigwedge_{i=1}^{r} a_i \wedge \bigwedge_{j=k+1}^{s} b_j, \Omega_d \right\rangle = \left\langle \bigwedge_{i=1}^{k} a_i \wedge \bigwedge_{i=r+1}^{d} a_i \wedge \bigwedge_{j=s+1}^{d} b_j, \Omega_d \right\rangle,$$

we get

$$\begin{aligned}
&\langle \tau, G^{\#}\Omega_d \wedge \hat{\pi}^{\#}\varphi_k(u)\rangle \\
&= (-1)^{s(d-r)}\mathcal{O}_q^{-1} \left\langle \bigwedge_{i=r+1}^{d} a_i \wedge \bigwedge_{j=s+1}^{d} b_j \wedge \bigwedge_{l=q+1}^{d-1} c_l, \Omega_d \right\rangle \\
&= (-1)^{s(d-r)}\mathcal{O}_q^{-1} \left\langle \bigwedge_{i=1}^{r} a_i, \bigwedge_{j=1}^{s} b_j, \bigwedge_{l=1}^{q} c_l \wedge u \right\rangle \\
&= (-1)^d \langle \tau, \psi_{r,s}(u)\rangle,
\end{aligned}$$

as required. □

For the particular choice of the function h

$$h(z, x, u) = \mathbf{1}_A(x)\mathbf{1}_B(x - z)$$

we infer the following.

Corollary 6.12 *Let X, Y be two subsets of $\mathbb{R}^d$ with positive reach and assume that* (6.1) *and* (6.4) *hold. Then, for any bounded Borel sets $A, B \subset \mathbb{R}^d$ and any $k \in \{0, \dots, d-1\}$,*

$$\int C_k(X \cap (Y + z), A \cap (B + z))\, \mathcal{L}^d(\mathrm{d}z) = \sum_{r+s=d+k} C_{r,s}(X, Y, A \times B).$$

We will give now some sufficient conditions for (6.1). (Sufficient conditions for (6.4) will be given later.)

Proposition 6.13

(i) *Any two closed convex sets* $X, Y \subset \mathbb{R}^d$ *satisfy* (6.1).
(ii) *Any two sets* $X, Y \subset \mathbb{R}^2$ *with positive reach satisfy* (6.1).
(iii) *Any two sets* $X, Y \subset \mathbb{R}^d$ *with positive reach such that* $\partial X, \partial Y$ *are smooth submanifolds of differentiability class* d *satisfy* (6.1).
(iv) *Let* $X, Y \subset \mathbb{R}^d$ *have positive reach. Then* $X, \rho Y$ *satisfy* (6.1) *for almost all* $\rho \in SO(d)$.

Proof Denote $N = \{z \in \mathbb{R}^d : X \text{ and } Y + z \text{ touch}\}$.

(i) If X, Y are convex then the difference set $X - Y = \{x - y : x \in X, y \in Y\}$ is convex as well. Since $N \subset \partial(X - Y)$, we have $\mathcal{L}^d(N) \le \mathcal{L}^d(\partial(X - Y)) = 0$.
(ii) Consider first a general dimension d and define the mapping

$$\Psi : (x, m, y, n) \mapsto x - y, \quad (x, m, y, n) \in \operatorname{nor} X \times \operatorname{nor} Y.$$

Take any $(x, m, y, n) \in W^0 := (\operatorname{nor} X \times \operatorname{nor} Y) \setminus R^0$ (for the definition of R^0, see the beginning of Sect. 6.1) and $(a, a', b, b') \in \operatorname{Tan}^{2d-2}(\operatorname{nor} X \times \operatorname{nor} Y, (x, m, y, n))$. Then $a \perp m = -n \perp b$, hence, $a - b \perp m$. Consequently, $\operatorname{ap} J_d \Psi(x, m, y, n) = 0$. If $d = 2$ then the area formula gives

$$0 = \int_{W^0} \operatorname{ap} J_2 \Psi \, d\mathcal{H}^2 = \int_{\Psi(W^0)} \mathcal{H}^0(\Psi^{-1}\{z\}) \, \mathcal{L}^2(dz),$$

hence, $\mathcal{L}^2(N) = 0$ since $N = \Psi(W^0)$.
(iii) The assumption of (iii) implies that $\operatorname{nor} X \times \operatorname{nor} Y$ is of differentiability class $l = d-1$. Thus, applying the Sard-type theorem [Fed69, §3.4.3] to the mapping Ψ, we get $\mathcal{H}^{d-1+(d-1)/l}(\Psi(W^0)) = 0$, which yields the assertion.
(iv) We apply Lemma 1.73 with $Z = \operatorname{nor} X \times \operatorname{nor} Y$ and $p = 2d - 1$ and obtain that $\mathcal{H}^q(A) = 0$, where

$$A = \{(x, m, y, n, \rho) \in \operatorname{nor} X \times \operatorname{nor} Y \times SO(d) : m + \rho n = 0\}$$

and

$$q = 2d - 1 + \frac{(d-1)(d-2)}{2} = d + \frac{d(d-1)}{2}.$$

The mapping $h : (x, m, y, n, \rho) \mapsto (x - \rho y, \rho)$ is Lipschitz and maps A onto

$$h(A) = \{(z, \rho) \in \mathbb{R}^d \times SO(d) : X \text{ and } z + \rho Y \text{ touch}\}.$$

We have $\mathcal{H}^q(h(A)) = 0$ by Lemma 1.10 and, using the Fubini theorem, we get

$$0 = \mathcal{H}^q(h(A)) = \int_{\mathrm{SO}(d)} \mathcal{L}^d\{z \in \mathbb{R}^d : X \text{ and } z + \rho Y \text{ touch}\}\, \mathcal{H}^{d(d-1)/2}(\mathrm{d}\rho),$$

which yields the result.

□

6.2 Local Representation of Mixed Curvature Measures

In this section we derive a representation of mixed curvature measures as integrals over the product of the two unit normal bundles. This will give a better insight in the Translative integral formula and, in particular, we will derive the Principal kinematic formula as a corollary.

The basic idea is to apply the Area formula to the mapping F from (6.3). In order to do this, we need to know the differential. Recall the mapping u defined in (6.2).

Lemma 6.14 *Let* $(x, m, y, n) \in (\mathbb{R}^d \times S^{d-1})^2 \cap R^0$ *be fixed. Then, for any vectors* $v, w \in \mathbb{R}^d$, *we have*

$$\begin{aligned} Du(m, n, t)(v, 0, 0) &= \frac{\sin((1-t)\theta)}{\sin\theta} v + p, \\ Du(m, n, t)(0, w, 0) &= \frac{\sin t\theta}{\sin\theta} w + q, \\ Du(m, n, t)(0, 0, 1) &= \theta u^*, \end{aligned}$$

for certain vectors p, q *from* $\mathrm{Lin}\{m, n\}$, *where* $\theta := \angle(m, n)$, $u := u(m, n, t)$, $u^* := p_{u^\perp}(n - m)/|p_{u^\perp}(n - m)|$ *if* $m \neq n$ *and* $u^* = 0$ *otherwise* ($p_{u\perp}$ *denotes the orthogonal projection to* $u^\perp$).

Proof The first two equations are obtained by using the product rule for differentiation. The differential at $(0, 0, 1)$ is obtained from the fact that $t \mapsto u(m, n, t)$ is a unit-speed parametrization of the shortest path on S^{d-1} from m to n. □

We find now a representation of the simple multivector $\tilde{a}_{X,Y}$ defined in (6.6). Let $(x, m) \in \operatorname{nor} X$ and $(x, n) \in \operatorname{nor} Y$ be regular points and $t \in [0, 1]$. We can write $a_X \circledwedge a_Y \circledwedge 1(x, m, y, n, t)$ in the form

$$a_X \circledwedge a_Y \circledwedge 1 = (KL)^{-1} \bigwedge_{i=1}^{d-1} (a_i, \kappa_i a_i, 0, 0, 0) \wedge \bigwedge_{j=1}^{d-1} (0, 0, b_j, \lambda_j b_j, 0) \wedge (0, 0, 0, 0, 1),$$

where $\kappa_i = \kappa_i(x, m)$, $\lambda_j = \lambda_j(y, n)$ are the principal curvatures and $a_i = a_i(x, m)$, $b_j = b_j(y, n)$ the corresponding principal directions of X, Y at (x, m), (y, n), respectively, and

$$K = \prod_{i=1}^{d-1} \sqrt{1+\kappa_i^2}, \quad L = \prod_{j=1}^{d-1} \sqrt{1+\lambda_j^2} \tag{6.11}$$

(the case $\kappa_i, \lambda_j = \infty$ is treated as usually). Now we have, using Lemma 6.14,

$$\begin{aligned}\tilde{a}_{X,Y} = \frac{1}{KL} \bigwedge_{i=1}^{d-1} \left(a_i, 0, \kappa_i \frac{\sin((1-t)\theta)}{\sin\theta} a_i + p_i \right) \\ \wedge \bigwedge_{j=1}^{d-1} \left(0, b_j, \lambda_j \frac{\sin(t\theta)}{\sin\theta} b_j + q_j \right) \wedge (0, 0, \theta u^*)\end{aligned} \tag{6.12}$$

with $p_i, q_j \in \mathrm{Lin}\,\{m, n\}$ and, using the fact that $p_i \wedge u \wedge u^* = q_j \wedge u \wedge u^* = 0$ for all i, j, we get

$$\begin{aligned}\langle \tilde{a}_{X,Y}, \psi_{r,s} \rangle = \mathcal{O}_{2d-r-s-1}^{-1} \theta \left(\frac{\sin((1-t)\theta)}{\sin\theta} \right)^{d-1-r} \left(\frac{\sin t\theta}{\sin\theta} \right)^{d-1-s} \\ \times \frac{1}{KL} \sum_{|I|=r} \sum_{|J|=s} \left(\prod_{i\in I^c} \kappa_i \right) \left(\prod_{j \in J^c} \lambda_j \right) \\ \times (-1)^{d+s(d-r)} \operatorname{sgn}\sigma \left\langle \bigwedge_{i\in I} a_i, \bigwedge_{j\in J} b_j, \bigwedge_{i\in I^c} a_i \wedge \bigwedge_{j \in J^c} b_j \wedge u^* \wedge u \right\rangle,\end{aligned}$$

where the summation is carried over subsets I, J of $\{1, \dots, d-1\}$ of given cardinalities, I^C, J^C denote the complements of I, J, respectively, in $\{1, \dots, d-1\}$, and σ is the permutation of $\{1, \dots, 2d-1\}$ that maps the first r numbers increasingly onto I, the following s numbers increasingly onto $d-1+J$, and the remaining numbers again increasingly onto the remaining numbers. An easy calculation yields

$$\operatorname{sgn}\sigma = (\operatorname{sgn} I)(\operatorname{sgn} J)(-1)^{(d-1-r)s},$$

hence,

$$(-1)^{d+s(d-r)} \operatorname{sgn}\sigma = (-1)^{d-s} (\operatorname{sgn} I)(\operatorname{sgn} J),$$

where sgn I, sgn J denotes the sign of the permutation of $\{1, \ldots, d-1\}$ mapping the first r, s elements increasingly into I, J, and the remaining elements again increasingly onto I^C, J^C, respectively. If $\theta \neq 0$, we can substitute $u^* \wedge u = -(\sin\theta)^{-1} m \wedge n$ and, using Lemma 6.7, we obtain

$$\begin{aligned}
&\left\langle \bigwedge_{i\in I} a_i, \bigwedge_{j\in J} b_j, \bigwedge_{i\in I^c} a_i \wedge \bigwedge_{j\in J^c} b_j \wedge u^* \wedge u \right\rangle \\
&= -\frac{1}{\sin\theta}(-1)^{d-1-s} \left\langle \bigwedge_{i\in I} a_i, \bigwedge_{j\in J} b_j, \bigwedge_{i\in I^C} a_i \wedge m \wedge \bigwedge_{j\in J^C} b_j \wedge n \right\rangle \\
&= (-1)^{d-s} (\operatorname{sgn} I)(\operatorname{sgn} J) \frac{1}{\sin\theta} \left[\bigwedge_{i\in I} a_i, \bigwedge_{j\in J} b_j \right]^2 .
\end{aligned}$$

Lemma 6.15 *Let X, Y be sets with positive reach. Then we have for $\mathcal{H}^{2d-1}$-almost all $(x, m, y, n, t) \in ((\operatorname{nor} X \times \operatorname{nor} Y) \cap R^0) \times [0, 1]$, $\tilde{a}_{X,Y} = 0$ if $m = n$ and*

$$\begin{aligned}
\langle \tilde{a}_{X,Y}, \psi_{r,s} \rangle &= \mathcal{O}_{2d-r-s-1}^{-1} \frac{\theta}{\sin\theta} \left(\frac{\sin((1-t)\theta)}{\sin\theta} \right)^{d-1-r} \left(\frac{\sin t\theta}{\sin\theta} \right)^{d-1-s} \\
&\quad \times \frac{1}{KL} \sum_{|I|=r} \sum_{|J|=s} \left(\prod_{i\in I^C} \kappa_i \right) \left(\prod_{j\in J^C} \lambda_j \right) \left[\bigwedge_{i\in I} a_i, \bigwedge_{j\in J} b_j \right]^2
\end{aligned}$$

otherwise, where $\theta = \angle(m, n)$. In both cases, we have

$$|\tilde{a}_{X,Y}| \le h(\theta) := \begin{cases} \binom{2d-2}{d} \theta \sin\theta, & 0 \le \theta \le \frac{\pi}{2}, \\ \binom{2d-2}{d} \theta \sin^{3-d}\theta, & \frac{\pi}{2} < \theta < \pi. \end{cases}$$

Proof The equality follows from the above computation. In order to show the inequality, we use (6.12). Note that $\tilde{a}_{X,Y}$ is orthogonal to $(0, 0, u)$, thus, its norm does not change when being multiplied by $(0, 0, u)$. Using again the identity $u^* \wedge u = -(\sin\theta)^{-1} m \wedge n$ for $m \neq n$, we get

$$|\tilde{a}_{X,Y}| = \frac{1}{KL} \frac{\theta}{\sin\theta} \left| \bigwedge_{i=1}^{d-1} (a_i, 0, \alpha\kappa_i a_i) \wedge (0, 0, m) \wedge \bigwedge_{j=1}^{d-1} \left(0, b_j, \beta\lambda_j b_j\right) \wedge (0, 0, n) \right|$$

with

$$\alpha := \frac{\sin((1-t)\theta)}{\sin\theta} \text{ and } \beta := \frac{\sin(t\theta)}{\sin\theta}.$$

Since the mappings $P : (x, y) \mapsto (x, 0, y)$ and $Q : (x, y) \mapsto (0, x, y)$ are isometric immersions, the mapping

$$\bigwedge_{d-1} P \wedge \bigwedge_{d-1} Q : \bigwedge_{2d}(\mathbb{R}^{2d}) \to \bigwedge_{2d}(\mathbb{R}^{3d})$$

is 1-Lipschitz and we have

$$|\tilde{a}_{X,Y}| \leq \frac{1}{KL} \frac{\theta}{\sin\theta} \left| \bigwedge_{i=1}^{d-1} (a_i, \kappa_i \alpha a_i) \wedge (0, m) \wedge \bigwedge_{j=1}^{d-1} (b_j, \lambda_j \beta b_j) \wedge (0, n) \right| .$$

We can apply the shuffle formula for the last norm and obtain

$$|\tilde{a}_{X,Y}| \leq \frac{1}{KL} \frac{\theta}{\sin\theta} \sum_{|I|+|J|=d} \left| \bigwedge_{i\in I} a_i \wedge \bigwedge_{j\in J} b_j \right|^2 \left| \prod_{i\in I^c} \kappa_i \alpha \right| \left| \prod_{j\in J^c} \lambda_j \beta \right|$$

(here, as above, $I, J \subset \{0, \dots, d-1\}$). Since $| \bigwedge_{i\in I} a_i \wedge \bigwedge_{j\in J} b_j| \leq |m \wedge n| = \sin\theta$, $\alpha, \beta \leq 1$ if $\theta \leq \frac{\pi}{2}$, $\alpha, \beta \leq (\sin\theta)^{-1}$ if $\theta > \frac{\pi}{2}$ and $\left|\prod_{i\in I^c} \kappa_i\right| \leq K$, $\left|\prod_{j\in J^c} \lambda_j\right| \leq L$ by (6.11), the result follows. □

Remark 6.16 For given $(x, m, y, n, t) \in \operatorname{nor} X \times \operatorname{nor} Y \times (0, 1)$, let r_0, s_0 be the number of finite principal curvatures of X, Y at (x, m), (y, n), respectively (assuming, of course, that the principal curvatures exist). It follows from Lemma 6.15 that $(-1)^d \langle a_{X,Y}, \psi_{r_0,s_0} \rangle > 0$, and that $(-1)^d \langle a_{X,Y}, \psi_{r,s} \rangle \geq 0$ whenever $r + s \geq r_0 + s_0$. Thus, using Lemma 6.11, we get (6.8), which completes the proof of Proposition 6.5.

Proposition 6.17 *If X, Y are sets with positive reach satisfying* (6.4) *and* $1 \leq r, s \leq d-1$, $r+s \geq d$, *then, for any bounded Borel subsets $A \subset \mathbb{R}^d \times \mathbb{R}^d$ and $B \subset S^{d-1}$,*

$$\begin{aligned} \widetilde{C}_{r,s}(X, Y, A \times B) = &\int_{(\operatorname{nor} X \times \operatorname{nor} Y)\cap R^0} \mathbf{1}_A(x, y)\mu(m, n; B) \\ &\times \prod_{i=1}^{d-1} \frac{1}{\sqrt{1+\kappa_i^2}\sqrt{1+\lambda_i^2}} \sum_{|I|=r} \sum_{|J|=s} \prod_{i\in I^c} \kappa_i \prod_{j\in J^c} \lambda_j \\ &\times \left[\bigwedge_{i\in I} a_i, \bigwedge_{j\in J} b_j \right]^2 \mathcal{H}^{2d-2}(\mathrm{d}(x, m, y, n)), \end{aligned}$$

where

$$\mu(m,n;B) = \mathcal{O}_{2d-r-s-1}^{-1}\frac{\theta}{\sin\theta} \times \int_0^1 \mathbf{1}_B(u(m,n,t))\left(\frac{\sin((1-t)\theta)}{\sin\theta}\right)^{d-1-r}\left(\frac{\sin t\theta}{\sin\theta}\right)^{d-1-s} \mathrm{d}t$$

with $\theta = \angle(m,n)$.

Proof We have by definition,

$$\begin{aligned}\widetilde{C}_{r,s}(X,Y,A\times B) &= (N_{X,Y} \llcorner \mathbf{1}_{A\times B})(\psi_{r,s}) \\ &= F_\#\left(((N_X \times N_Y) \llcorner \mathbf{1}_{R^0}) \times [0,1]\right)(\psi_{r,s} \wedge \mathbf{1}_{A\times B}) \\ &= \left(((N_X \times N_Y) \llcorner \mathbf{1}_{R^0}) \times [0,1]\right)(F^\#(\psi_{r,s} \wedge \mathbf{1}_{A\times B})) \\ &= \int_{(\mathrm{nor}\, X \times \mathrm{nor}\, Y)\cap R^0}\int_0^1 \langle \tilde{a}_{X,Y}, \psi_{r,s} \wedge \mathbf{1}_{A\times B}\rangle \,\mathrm{d}\mathcal{L}^1\mathrm{d}\mathcal{H}^{2d-2},\end{aligned}$$

and the result follows from Lemma 6.15. □

Now we will prove that condition (6.4) is fulfilled for almost all rotations or reflections of one of the sets with positive reach.

Lemma 6.18 *Let X, Y be sets with positive reach, let $A, B \subset \mathbb{R}^d$ are Borel sets and let h be the function from Lemma* 6.15*. Then,*

$$\|N_{X,Y}\|(A \times B \times S^{d-1}) \le \int_{(\mathrm{nor}\, X \times \mathrm{nor}\, Y)\cap R^0} \mathbf{1}_A(x)\mathbf{1}_B(y)\, h(\angle(m,n))\, \mathcal{H}^{2d-2}(\mathrm{d}(x,m,y,n)). \tag{6.13}$$

Further, there exists a finite constant C such that

$$\int_{\mathrm{SO}(d)} \|N_{X,\rho Y}\|(A \times \rho B \times S^{d-1})\, \vartheta_d(\mathrm{d}\rho) \le C\mathcal{H}^{d-1}(\mathrm{nor}\, X \cap \pi_0^{-1}(A))\mathcal{H}^{d-1}(\mathrm{nor}\, Y \cap \pi_0^{-1}(B)). \tag{6.14}$$

Consequently, the pair $X, \rho Y$ satisfies (6.4) *for ϑ_d-almost all $\rho \in \mathrm{SO}(d)$.*

Proof Since

$$\|N_{X,Y}\|(\cdot) = \int_{F^{-1}(\cdot)} |a_{X,Y}|\, \mathrm{d}\|((N_X \times N_Y) \llcorner R^0) \times [0,1]\|,$$

Equation (6.13) follows from the inequality in Lemma 6.15. Using the fact that $(y, n) \mapsto (\rho y, \rho n)$ is an isometry between nor Y and nor ρY, we get

$$\|N_{X,\rho Y}\|(A \times \rho B \times S^{d-1})$$
$$\leq \int_{(\operatorname{nor} X \times \operatorname{nor} Y) \cap R^0} \mathbf{1}_A(x)\mathbf{1}_B(y)\, h(\angle(m, \rho^{-1}n))\, \mathcal{H}^{2d-2}(\mathrm{d}(x, m, y, n)).$$

Integrating over SO(d) and applying the Fubini theorem, we get (6.14), since $h(\theta) \leq \sin^{3-d}\theta$ and

$$\int_{\mathrm{SO}(d)} \sin^{3-d} \angle(m, \rho^{-1}n)\, \vartheta_d(\mathrm{d}\rho) = \mathcal{O}_{d-1}^{-1} \int_{S^{d-1}} \sin^{3-d} \angle(u, v)\, \mathcal{H}^{d-1}(\mathrm{d}v),$$

and the last integral can be shown to be finite by direct computation. □

We can compute now the rotational integral of the mixed curvature measures which will imply, together with Theorem 6.10, the Principal kinematic formula (Theorem 6.1).

Lemma 6.19 *For any sets $X, Y \subset \mathbb{R}^d$ with positive reach, for any $1 \leq r, s \leq d-1$, $r + s \geq d$, and for any bounded measurable sets $A, B \subset \mathbb{R}^d$, we have*

$$\int_{\mathrm{SO}(d)} C_{r,s}(X, \rho Y, A \cap \rho B)\, \vartheta_d(\mathrm{d}\rho) = \gamma(d, r, s) C_r(X, A) C_s(Y, B).$$

Proof Using Lemma 6.18 and Proposition 6.13(iv), we can write

$$\int_{SO_d} C_{r,s}(X, \rho Y, A \times \rho B) \vartheta_d(\mathrm{d}\rho)$$
$$= \int_{\operatorname{nor} X \times \operatorname{nor} \rho Y} \mathbf{1}_A(x) \mathbf{1}_{\rho B}(y) \mu(m, n; S^{d-1})$$
$$\times \frac{\sum_{|I|=r} \sum_{|J|=s} \prod_{i \in I^c} \kappa_i(X; x, m) \prod_{j \in J^c} \lambda_j(\rho Y; y, n)}{\prod_{i=1}^{d-1} \sqrt{1 + \kappa_i(X; x, m)^2} \sqrt{1 + \lambda_i(\rho Y; y, n)^2}}$$
$$\times \left[\bigwedge_{i \in I} a_i(X; x, m), \bigwedge_{j \in J} b_j(\rho Y; y, n) \right]^2 \mathcal{H}^{2d-2}(\mathrm{d}(x, m, y, n)) \vartheta_d(\mathrm{d}\rho).$$

Since the principal curvatures are rotation invariant and the principal directions are rotation covariant, we have

$$\lambda_j(\rho Y; y, n) = \lambda_j(Y; \rho^{-1}y, \rho^{-1}n)$$

and

$$b_j(\rho Y; y, n) = \rho b_j(Y; \rho^{-1}y, \rho^{-1}n).$$

Thus, using the substitution $\tilde{y} = \rho^{-1}y$, $\tilde{n} = \rho^{-1}n$, and the rotational invariance of the Hausdorff measure, we get

$$\begin{aligned}
&\int_{SO_d} C_{r,s}(X, \rho Y, A \times \rho B)\vartheta_d(\mathrm{d}\rho) \\
&= \int_{\operatorname{nor} X \times \operatorname{nor} Y} \mathbf{1}_A(x)\mathbf{1}_B(y) \\
&\quad \times \frac{\sum_{|I|=r}\sum_{|J|=s}\prod_{i\in I^C}\kappa_i(X; x, m)\prod_{j\in J^C}\lambda_j(Y; \tilde{y}, \tilde{n})}{\prod_{i=1}^{d-1}\sqrt{1+\kappa_i(X; x, m)^2}\sqrt{1+\lambda_i(Y; \tilde{y}, \tilde{n})^2}} \\
&\quad \times \int_{SO_d} \mu(m, \rho\tilde{n}; S^{d-1}) \left[\bigwedge_{i\in I} a_i(X; x, m), \bigwedge_{j\in J} \rho b_j(Y; \tilde{y}, \tilde{n})\right]^2 \vartheta_d(\mathrm{d}\rho) \\
&\quad \times \mathcal{H}^{2d-2}(\mathrm{d}(x, m, \tilde{y}, \tilde{n})).
\end{aligned}$$

We claim that the inner integral over SO(d) is a constant, say $\bar{\gamma}(d, r, s)$, depending on d, r, s only. Indeed we can write it in the form $\int \Phi(\mathcal{B}_1, \rho\mathcal{B}_2)\,\vartheta_d(d\rho)$ with two positively oriented orthonormal bases $\mathcal{B}_1 = (a_1, \ldots, a_{d-1}, m)$, $\mathcal{B}_2 = (b_1, \ldots, b_{d-1}, \tilde{n})$ and invariant function Φ ($\Phi(\rho\mathcal{B}_1, \rho\mathcal{B}_2) = \Phi(\mathcal{B}_1, \mathcal{B}_2)$), and the constancy of the integral follows from the invariance property of ϑ_d.

Therefore, Definitions 4.28 and 4.31 imply that

$$\int_{\mathrm{SO}(d)} C_{r,s}(X, Y, A \cap \rho B)\,\vartheta_d(\mathrm{d}\rho) = \tilde{\gamma}(d, r, s)C_r(X, A)C_s(Y, B)$$

for another constant $\tilde{\gamma}(d, r, s)$. Applying now Theorem 6.10, we get the Principal kinematic formula with coefficients $\tilde{\gamma}(d, r, s)$. Applying this to two unit balls gives the final result with $\tilde{\gamma}(d, r, s) = \gamma(d, r, s)$. □

Finally, we will derive a translative version of the Crofton formula (Theorem 6.2) as a corollary of Theorem 6.10. As above, κ_i and a_i, $i = 1, \ldots, d-1$, denote the principal curvatures and principal directions of a set X with positive reach. If $X \subset \mathbb{R}^d$ has positive reach, $F \in \mathcal{A}(d, j)$ and $X \cap F$ has also positive reach, the curvature-direction measure of $X \cap F$ can be considered either in the whole $\mathbb{R}^d$, i.e., as a measure on $\mathbb{R}^d \times S^{d-1}$, or relatively in F, i.e., as a measure on $F \times S^{j-1}$. where S^{j-1} stands for the unit sphere in the j-space parallel to F. In the second case, we will use the notation $\widetilde{C}_k^{(F)}(X \cap F, \cdot)$.

Recal that ν_j^d denotes the normalized invariant measure on the Grassmannian $G(d, j)$, and for $L \in G(d, j)$, $\pi_L : v \mapsto \frac{v}{|v|}$, $v \not\perp L$, is the spherical projection onto L.

Theorem 6.20 *Let $X \subset \mathbb{R}^d$ be a set with positive reach and $0 \le k < j < d$ be integers. Then, for ν_j^d-almost all $L \in G(d, j)$ and for any bounded measurable function g defined on $\mathbb{R}^d \times (L \cap S^{d-1})$ and with compact support, we have*

$$\int_{L^\perp} g(x,v)\,\widetilde{C}_k^{(L+z)}(X \cap (L+z), \mathrm{d}(x,v))\,\mathcal{L}^{d-j}(\mathrm{d}z)$$

$$= \mathcal{O}_{j-1-k}^{-1} \int_{\operatorname{nor} X} \frac{g(x, \pi_L(n))}{|p_L n|^{j-k}} \frac{\sum_{|I|=d+k-j} \left(\prod_{i \in I^c} \kappa_i\right) \left[\bigwedge_{i \in I} a_i, L\right]^2}{\prod_{i=1}^{d-1} \sqrt{1+\kappa_i^2}} \mathcal{H}^{d-1}(\mathrm{d}(x,n)).$$

In particular, if $A \subset \mathbb{R}^d$ is a bounded Borel set then

$$\int_{L^\perp} C_k(X \cap (L+z), A \cap (L+z))\,\mathcal{L}^{d-j}(\mathrm{d}z)$$

$$= \mathcal{O}_{j-1-k}^{-1} \int_{\operatorname{nor} X} \frac{\mathbf{1}_A(x)}{|p_L n|^{j-k}} \frac{\sum_{|I|=d+k-j} \left(\prod_{i \in I^c} \kappa_i\right) \left[\bigwedge_{i \in I} a_i, L\right]^2}{\prod_{i=1}^{d-1} \sqrt{1+\kappa_i^2}} \mathcal{H}^{d-1}(\mathrm{d}(x,n)).$$

Proof Let B_1 be a ball of unit j-volume in L. We apply Theorem 6.10 with X, $Y = L$ and

$$h(\tilde{z}, x, u) = \mathbf{1}_{B_1}(p_L \tilde{z}) g(x, \pi_L(n)).$$

Note that, due to Proposition 6.13(iv) and Lemma 6.18, the assumptions of Theorem 6.10 are satisfied for almost all $L \in G(d, j)$ and, for such subspaces L and for any $(x, n) \in \operatorname{nor} X$, $n \not\perp L$ by (6.1) and, hence $\pi_L(n) = \frac{p_L n}{|p_L n|}$ is defined.

It follows from Theorem 6.17 that $\widetilde{C}_{r,s}(X, L, \cdot)$ vanishes unless $s = j$ (note that, at each point of L, exactly j principal curvatures are zero, and the remaining $d-1-j$ are infinite). Hence we obtain, using Theorem 6.10 and Proposition 6.17,

$$\int_{L^\perp} g(x,v)\,\widetilde{C}_k^{(L+z)}(X \cap (L+z), \mathrm{d}(x,v))\,\mathcal{L}^{d-j}(\mathrm{d}z)$$

$$= \int_{L^\perp} g(x, \pi_L(n))\,\widetilde{C}_k(X \cap (L+z), \mathrm{d}(x,n))\,\mathcal{L}^{d-j}(\mathrm{d}z)$$

$$= \int_{\mathbb{R}^d} h(\tilde{z}, x, u)\,\widetilde{C}_k(X \cap (L+\tilde{z}), \mathrm{d}(x,u))\,\mathcal{L}^d(\mathrm{d}\tilde{z})$$

$$= \int h(x-y,x,u)\,\widetilde{C}_{d+k-j,j}(X,L,\mathrm{d}(x,y,u))$$

$$= \int_{\operatorname{nor} X} g(x,\pi_L(n)) \int_L \int_{L^\perp\cap S^{d-1}} \int_0^1 \mathbf{1}_{\{m+n\neq 0\}} \mathcal{O}_{d-1-k}^{-1}$$

$$\times \frac{\theta}{\sin\theta}\left(\frac{\sin((1-t)\theta)}{\sin\theta}\right)^{j-1-k}\left(\frac{\sin t\theta}{\sin\theta}\right)^{d-1-j} \mathrm{d}t\,\mathcal{H}^{d-1-j}(\mathrm{d}n)\,\mathcal{L}^j(\mathrm{d}y)$$

$$\times \frac{\sum_{|I|=d+k-j}\left(\prod_{I^C}\kappa_i\right)\left[\bigwedge_I a_i, L\right]^2}{\prod_{i=1}^{d-1}\sqrt{1+\kappa_i^2}}\mathcal{H}^{d-1}(\mathrm{d}(x,m))$$

(recall that $\theta = \angle(m,n)$). Note that if $m+n=0$ then $\left[\bigwedge_I a_i, L\right] = 0$. Thus, we can omit the indicator function of $\{m+n\neq 0\}$ in the last integral. We will show that

$$\int_{L^\perp\cap S^{d-1}} \frac{\theta}{\sin\theta}\int_0^1 \left(\frac{\sin((1-t)\theta)}{\sin\theta}\right)^{j-1-k}\left(\frac{\sin t\theta}{\sin\theta}\right)^{d-1-j} \mathrm{d}t\,\mathcal{H}^{d-1-j}(\mathrm{d}n)$$
$$= \frac{\mathcal{O}_{d-1-k}}{\mathcal{O}_{j-1-k}}|p_L m|^{k-j}. \tag{6.15}$$

This will complete the proof.

Fix $m\in S^{d-1}\setminus L^\perp$ and consider the mapping

$$\Phi : (n,t)\mapsto u(m,n,t).$$

The differential of Φ is (see Lemma 6.14)

$$D\Phi(n,t)(w,0) = \frac{\sin t\theta}{\sin\theta}w + q, \quad D\Phi(n,t)(0,1) = \theta u^*,$$

with some $q\in \operatorname{Lin}\{m,n\}$. Let ϕ be the restriction of Φ to $(L^\perp\cap S^{d-1})\times(0,1)$ and let $\{b_1,\dots,b_{d-1-j},n\}$ be an orthonormal basis of $L^\perp$. Then we have with some $q_1,\dots,q_{d-1}\in\operatorname{Lin}\{m,n\}$,

$$J_{d-j}\phi(n,t) = \left|\left(\bigwedge\nolimits_{d-j} D\Phi(n,t)\right)\left(\bigwedge_{i=1}^{d-1-j}(b_i,0)\wedge(0,1)\right)\right|$$

$$= \left|\bigwedge_{i=1}^{d-j-1}\left(\frac{\sin t\theta}{\sin\theta}b_i + q_i\right)\wedge\theta u^*\right|$$

$$= \left| \bigwedge_{i=1}^{d-j-1} \left(\frac{\sin t\theta}{\sin\theta} b_i + q_i \right) \wedge \theta u^* \wedge u \right|$$

$$= \theta \left(\frac{\sin t\theta}{\sin\theta} \right)^{d-1-j} \left| \bigwedge_{i=1}^{d-1-j} b_i \wedge u^* \wedge u \right|$$

(we have used the fact that the image of the differential of Φ is always perpendicular to u). Using again the fact that $u \wedge u^* = (m \wedge n)/|m \wedge n|$, we get

$$|b_1 \wedge \cdots \wedge b_{d-1-j} \wedge u^* \wedge u| = \frac{|b_1 \wedge \cdots \wedge b_{d-1-j} \wedge m \wedge n|}{|m \wedge n|} = \frac{|p_L m|}{\sin\theta}.$$

Hence,

$$J_{d-j}\phi(n,t) = \theta \left(\frac{\sin t\theta}{\sin\theta} \right)^{d-1-j} \frac{|p_L m|}{\sin\theta}.$$

Observing that

$$\frac{\sin(1-t)\theta}{\sin\theta} = \frac{u \cdot p_L m}{m \cdot p_L m} = \frac{u \cdot p_L m}{|p_L m|^2},$$

and applying the Area formula to ϕ, we get

$$\int_{L^\perp \cap S^{d-1}} \frac{\theta}{\sin\theta} \int_0^1 \left(\frac{\sin((1-t)\theta)}{\sin\theta} \right)^{j-1-k} \left(\frac{\sin t\theta}{\sin\theta} \right)^{d-1-j} \mathrm{d}t \, \mathcal{H}^{d-1-j}(\mathrm{d}n)$$

$$= \frac{1}{|p_L m|} \int_{L^\perp \cap S^{d-1}} \int_0^1 J_{d-j}\phi(n,t) \left(\frac{\sin((1-t)\theta)}{\sin\theta} \right)^{j-1-k} \mathrm{d}t \, \mathcal{H}^{d-1-j}(\mathrm{d}n)$$

$$= \frac{1}{|p_L m|^{j-k}} \int_{\operatorname{im}\phi} \left(\frac{u \cdot p_L m}{|p_L m|} \right)^{j-1-k} \mathcal{H}^{d-j}(\mathrm{d}u).$$

The image if ϕ, $\operatorname{im}\phi$, is a hemisphere with pole m of dimension $d-j$, and a routine calculation yields

$$\int_{\operatorname{im}\phi} \left(u \cdot \frac{p_L m}{|p_L m|} \right)^{j-1-k} \mathcal{H}^{d-j}(\mathrm{d}u) = \frac{1}{2} \int_{S^{d-j}} |u \cdot e|^{j-1-k} \mathcal{H}^{d-j}(\mathrm{d}u) = \frac{\mathcal{O}_{d-1-k}}{\mathcal{O}_{j-1-k}}$$

($e \in S^{d-j}$ is arbitrarily chosen). Substituting into the above integral, we obtain (6.15), and the proof is complete. □

6.3 Absolute Curvature Measures

The curvature measures are signed measures and for certain applications, nonnegative variants of these are needed. The first obvious idea is to consider the total variation measures, $C_k^{\mathrm{var}}(X,\cdot)$, of $C_k(X,\cdot)$ (for sets X for which the curvature measures are defined, of course). It turns out, however, that these measures do not satisfy the Crofton formula. Therefore, other nonnegative variants of curvature measures are considered in integral geometry, see [San74] (where the case of smooth submanifolds is considered). These measures are defined by using the "touching affine subspaces" of given dimension.

Let us assume that reach $X > 0$ (though, the results can be further extended, see Sect. 6.4.4). The total variation of the curvature-direction measures can be expressed, due to Definition 4.28, as integral of the absolute value of the generalized symmetric function of principal curvatures:

$$\widetilde{C}_k^{\mathrm{var}}(X,E)=\frac{1}{\mathcal{O}_{d-1-k}}\int_{E\cap \operatorname{nor} X}|s_{d-1-k}(X,x,n)|\,\mathcal{H}^{d-1}(\mathrm{d}(x,n)), \tag{6.16}$$

for any bounded Borel set $E\subset\mathbb{R}^d\times S^{d-1}$, $k=0,\dots,d-1$.

Remark 6.21

1. Note that the equality $C_k^{\mathrm{var}}(X,A)=\widetilde{C}_k^{\mathrm{var}}(X,A\times S^{d-1})$ is not true in general. Indeed, if $X=S^1$ is the unit circle in $\mathbb{R}^2$ then $C_0^{\mathrm{var}}(S^1,\cdot)=0$, while $\widetilde{C}_0^{\mathrm{var}}(S^1,S^1\times S^1)=2$.
2. The measure $A\mapsto\widetilde{C}_k^{\mathrm{var}}(X,A\times S^{d-1})$ is independent of the chosen embedding space of X (see Lemma 4.36(iii)).

In order to introduce nonnegative variants of curvature measures that do satisfy the Crofton formula, it is natural to define absolute curvature measures by means of the Crofton formula (cf. Theorem 6.2). Recall that μ_j^d denotes the standard invariant measure on the family of j-flats $\mathcal{A}(d,j)$ (see Sect. 2.3).

Definition 6.22 Let X be a set with positive reach in $\mathbb{R}^d$ and $k\in\{0,1,\dots,d-1\}$. The *absolute curvature measure* of X of order k is defined by

$$C_k^{\mathrm{abs}}(X,B)=\gamma(d,k,d-k)^{-1}\int_{\mathcal{A}(d,d-k)}\widetilde{C}_0^{\mathrm{var}}(X\cap F,(B\cap F)\times S^{d-1})\,\mu_{d-k}^d(\mathrm{d}F), \tag{6.17}$$

where B is any bounded Borel subset of $\mathbb{R}^d$.

Clearly, $C_0^{\mathrm{abs}}(X,\cdot)=\widetilde{C}_0^{\mathrm{var}}(X,\cdot\times S^{d-1})$, but this is not true for all $k>0$, as will be seen later.

Immediately from the definition we obtain the analogue of Theorem 6.2:

Theorem 6.23 (Crofton formula for absolute curvature measures) *Let X be a set with positive reach in $\mathbb{R}^d$, $0 \le k \le j \le d$, $B \subset \mathbb{R}^d$ a bounded Borel set. Then*

$$\int_{\mathcal{A}(d,j)} C_k^{\text{abs}}(X \cap F, B \cap F)\, \mu_j^d(\mathrm{d}F) = \gamma(d, d+k-j, j) C_{d+k-j}^{\text{abs}}(X, A).$$

Proof By definition, we have

$$\begin{aligned}
&\int_{\mathcal{A}(d,j)} C_k^{\text{abs}}(X \cap F, B \cap F)\, \mu_j^d(\mathrm{d}F) \\
&= \gamma(j, k, j-k)^{-1} \int_{\mathcal{A}(d,j)} \int_{\mathcal{A}(F,j-k)} C_0^{\text{abs}}(X \cap G, A \cap G)\, \mu_{j-k}^F(\mathrm{d}G)\, \mu_j^d(\mathrm{d}F).
\end{aligned}$$

Since

$$\mu_{j-k}^F(\mathrm{d}G)\, \mu_j^d(\mathrm{d}F)$$

defines a motion invariant measure on $\mathcal{A}_{j-k}^d$, this must be equal to $\mu_{j-k}^d(\mathrm{d}G)$ up to a constant factor (depending only on d, k, j), and we get

$$\begin{aligned}
\int_{\mathcal{A}(d,j)} C_k^{\text{abs}}(X \cap F, B \cap F)\, \mu_j^d(\mathrm{d}F) &= \text{const} \int_{\mathcal{A}_{j-k}} C_0^{\text{abs}}(X \cap G, A \cap G)\, \mu_{j-k}^d(\mathrm{d}G) \\
&= \text{const}\, C_{d+k-j}^{\text{abs}}(X, B),
\end{aligned}$$

using Definition 6.22 again. If X is a convex body then $C_k^{\text{abs}}(X, \cdot) = C_k(X, \cdot)$ and a comparison with Theorem 6.2 yields the correct value of the constant. □

We will derive now explicit representations of absolute curvature measures. To save space, we will use the abbreviated notation $\sum_I \kappa_i$ in place of $\sum_{i \in I} \kappa_i$ (and similarly with products or exterior products). Fix a set X with positive reach and $k \in \{0, 1, \ldots, d-1\}$ and a bounded Borel set $B \subset \mathbb{R}^d$. Applying Theorem 6.20, we obtain for almost all $L \in G(d, d-k)$ and any bounded measurable function g on $\mathbb{R}^d \times S^{d-k-1}$

$$\begin{aligned}
&\int_{L^\perp} \mathbf{1}_B(x) g(x, v) \widetilde{C}_0(X \cap (L+z), \mathrm{d}(x, v))\, \mathcal{L}^k(\mathrm{d}z) \\
&= \mathcal{O}_{d-1-k}^{-1} \int_{\operatorname{nor} X} \frac{\mathbf{1}_B(x) g(x, \pi_L(n))}{|p_L n|^{d-k}} \frac{\sum_{|I|=k} \left(\prod_{I^c} \kappa_i\right) \left[\bigwedge_I a_i, L\right]^2}{\prod_{i=1}^{d-1} \sqrt{1+\kappa_i^2}} \mathcal{H}^{d-1}(\mathrm{d}(x, n)).
\end{aligned}$$

Maximizing the above expression with respect to all measurable functions g with $|g| \leq 1$, we obtain for a generic subspace $L \in G(d, d-k)$

$$\int_{L^\perp} \widetilde{C}_0^{\mathrm{var}}(X \cap (L+z), (B \cap (L+z)) \times S^{d-k-1})\, \mathcal{L}^k(\mathrm{d}z)$$

$$= \mathcal{O}_{d-1-k}^{-1} \int_{\operatorname{nor} X} \frac{\mathbf{1}_B(x)}{|p_L n|^{d-k}} \frac{\left|\sum_{|I|=k} \left(\prod_{I^c} \kappa_i\right) \left[\bigwedge_I a_i, L\right]^2\right|}{\prod_{i=1}^{d-1} \sqrt{1+\kappa_i^2}} \mathcal{H}^{d-1}(\mathrm{d}(x,n)).$$

Taking into account Definition 6.22, we get

$$\gamma(d,k,d-k) C_k^{\mathrm{abs}}(X,B) = \mathcal{O}_{d-1-k}^{-1} \int_{\operatorname{nor} X} \mathbf{1}_B(x) \Psi_k(x,n) \mathcal{H}^{d-1}(\mathrm{d}(x,n))$$

with

$$\Psi_k(x,n) = \int_{G(d,d-k)} \frac{1}{|p_L n|^{d-k}} \frac{\left|\sum_{|I|=k} \left(\prod_{I^c} \kappa_i\right) \left[\bigwedge_I a_i, L\right]^2\right|}{\prod_{i=1}^{d-1} \sqrt{1+\kappa_i^2}} \nu_{d-k}^d(\mathrm{d}L)$$

$$= \frac{1}{2c_{d,d-k}} \int_{G_0(d,d-k)} \frac{\left|\sum_{|I|=k} \left(\prod_{I^c} \kappa_i\right) \left[\bigwedge_I a_i, L(\xi)\right]^2\right|}{|p_{L(\xi)} n|^{d-k} \prod_{i=1}^{d-1} \sqrt{1+\kappa_i^2}} \mathcal{H}^{k(d-k)}(\mathrm{d}\xi)$$

(recall that $c_{d,j} = \mathcal{H}^{j(d-j)}(G(d,j)))$. Denoting $V := L \cap n^\perp \in G(d, d-1-k)$ if $n \not\perp L$ (which is true for ν_{d-k}^d-almost all L), we can write

$$\left[\bigwedge\nolimits_I a_i, L\right] = \left[\bigwedge\nolimits_I a_i \wedge n, V\right] |p_L n| = \left[\bigwedge\nolimits_{I^C} a_i, V^\perp\right] |p_L n|.$$

We will apply now the Coarea formula for the mapping $p : \xi \mapsto \zeta := \frac{\xi \llcorner n^*}{|\xi \llcorner n^*|}$ from $G_0(d, d-k)$ to $G_0^{n^\perp}(d-1, d-1-k)$ (note that $L(\zeta) = V$ if $n \not\perp V$) with Jacobian $J_{(d-1-k)(d-1)} p(\xi) = |p_L n|^{-(d-k-1)}$ (see Lemma 1.72):

$$\Psi_k(x,n) = \frac{1}{2c_{d,d-k}} \int_{G_0^{n^\perp}(d-1,d-1-k)} \mathcal{I}(n, L(\zeta))$$

$$\times \frac{\left|\sum_{|I|=d-1-k} \left(\prod_I \kappa_i\right) \left[\bigwedge_I a_i, L(\zeta)^\perp\right]^2\right|}{\prod_{i=1}^{d-1} \sqrt{1+\kappa_i^2}} \mathcal{H}^{(d-1-k)(d-1)}(\mathrm{d}\zeta)$$

$$= \frac{c_{d-1,d-1-k}}{c_{d,d-k}} \int_{G^{n^\perp}(d-1,d-1-k)} \mathcal{I}(n,V)$$

$$\times \frac{\left|\sum_{|I|=d-1-k} \left(\prod_I \kappa_i\right) \left[\bigwedge_I a_i, V^\perp\right]^2\right|}{\prod_{i=1}^{d-1} \sqrt{1+\kappa_i^2}} \nu_{d-1-k}^{d-1}(\mathrm{d}V),$$

where

$$\mathcal{I}(n,V) = \int_{p^{-1}\{V\}} |p_L n| \, \mathcal{H}^k(\mathrm{d}L).$$

The mapping $L \mapsto \pi_L n$ maps isometrically $p^{-1}\{V\}$ onto S_+^k, the unit hemisphere in $V^\perp$ with pole n, and we obtain after a straightforward calculation

$$\mathcal{I}(n,V) = \int_{S_+^k} |v \cdot n| \, \mathcal{H}^k(\mathrm{d}v) = \frac{\mathcal{O}_{k-1}}{k}.$$

After evaluation of the constants (use (1.12))

$$\frac{c_{d-1,d-1-k}}{c_{d,d-k}} \frac{\mathcal{O}_{k-1}}{k} \gamma(d,k,d-k)^{-1} = \binom{d-1}{k},$$

we arrive at the following expression.

Proposition 6.24 *For a set X with positive reach and $k \in \{0, 1, \ldots, d = 1\}$, we have*

$$C_k^{\mathrm{abs}}(X,B) = \binom{d-1}{k} \mathcal{O}_{d-1-k}^{-1} \int_{\operatorname{nor} X} \mathbf{1}_B(x) \int_{G^{n^\perp}(d-1,d-1-k)}$$

$$\times \frac{\left|\sum_{|I|=d-1-k} \left(\prod_I \kappa_i\right) \left[\bigwedge_I a_i, V^\perp\right]^2\right|}{\prod_{i=1}^{d-1} \sqrt{1+\kappa_i^2}} \nu_{d-1-k}^{d-1}(\mathrm{d}V) \, \mathcal{H}^{d-1}(\mathrm{d}(x,n)),$$

for any bounded Borel set $B \subset \mathbb{R}^d$.

Remark 6.25

1. If X is a convex body then all principal curvatures are nonnegative, the absolute value in the formula can be omitted and we can can integrate (see Example 1.18)

$$\int_{G^{n^\perp}(d-1,d-1-k)} \left[\bigwedge_I a_i, V^\perp\right]^2 \nu_{d-1-k}^{d-1}(\mathrm{d}V) = \binom{d-1}{k}^{-1}.$$

   Thus, we obtain $C_k^{\mathrm{abs}}(X,\cdot) = C_k(X,\cdot)$ in this case, cf. Definition 4.28.

2. Using the inequality $\int |g| \geq |\int g|$ for the inner integral over the Grassmannian, we obtain

$$C_k^{\mathrm{var}}(X,\cdot) \leq C_k^{\mathrm{abs}}(X,\cdot).$$

Example 6.26 Let X be a full-dimensional body in $\mathbb{R}^3$ with sufficiently smooth boundary, and let $\kappa_1(x), \kappa_2(x)$ be the principal curvatures, $x \in \partial X$. Then

$$C_1^{\mathrm{abs}}(X,B) = \frac{1}{\pi}\int_{B\cap\partial X} H_1^{\mathrm{abs}}(X;x)\,\mathcal{H}^1(\mathrm{d}x),$$

where

$$H_1^{\mathrm{abs}}(X;x) = \frac{1}{\pi}\int_0^{\pi}\left|\frac{\kappa_1(x)\cos^2\theta + \kappa_2(x)\sin^2\theta}{2}\right|\mathrm{d}\theta.$$

It is easy to see that $H_1^{\mathrm{abs}}(X;x) \geq \left|\frac{\kappa_1(x)+\kappa_2(x)}{2}\right|$ and a strict inequality holds if $\kappa_1(x)\kappa_2(x) < 0$.

In the sequel, we will derive another representation of absolute curvature measures, namely as measures of "locally colliding planes". Denote

$$G(X,d,k) := \{(x,n,V) : (x,n) \in \operatorname{nor} X, V \in G^{n^\perp}(d-1,k)\}.$$

We call $G(X,d,k)$ *the kth Grassmann bundle of* X. We will need also its "oriented version"

$$G_0(X,d,k) := \{(x,n,\xi) : (x,n) \in \operatorname{nor} X, \xi \in G_0^{n^\perp}(d-1,k)\}.$$

Applying Lemma 1.74, we get the following.

Lemma 6.27 *$G_0(X,d,k)$ (and, hence, also $G(X,d,k)$) is locally p-rectifiable and $\mathcal{H}^p$-measurable with*

$$p = d-1+k(d-1-k).$$

Further, for $\mathcal{H}^p$-almost all $(x,n,\xi) \in G_0(X,d,k)$, $\operatorname{Tan}^p(G_0(X,d,k),(x,n,\xi))$ is is the linear hull of the vectors

$$(0,0,\eta), \quad \eta \in \operatorname{Tan}(G_0^{n^\perp}(d-1,k),\xi),$$

and

$$\left(u,v,n\wedge(\xi \llcorner v^*)\right), \quad (u,v) \in \operatorname{Tan}^{d-1}(\operatorname{nor} X,(x,n)).$$

Definition 6.28 Let the mapping $\phi : G(X, d, k) \to \mathcal{A}(d, k)$ be defined by

$$\phi : (x, n, V) \mapsto (p_{V^\perp} x, V).$$

Its image

$$\mathcal{A}(X, d, k) := \phi(G(X, d, k))$$

will be called the *affine tangent Grassmannian of* X. The measure μ_k^X on $\mathcal{A}(X, d, k)$ is defined as follows. If h is a measurable nonnegative function on $\mathcal{A}(X, d, k)$ then

$$\int_{\mathcal{A}(X,d,k)} h(z, V)\, \mu_k^X(\mathrm{d}(z, V)) = \int_{G(d,k)} \int_{V^\perp} h(z, V)\, \mathcal{H}^{d-k-1}(\mathrm{d}z)\, \nu_k^d(\mathrm{d}V),$$

where we apply the usual representation of affine subspaces in the form $E = z + V$ with a linear subspace V and $z \perp V$. For given $V \in G(d, k)$, we call

$$T_V X := \{z \in V^\perp : (z, V) \in \mathcal{A}(X, d, k)\}$$

the *tangential projection of* X *in direction* V. We also denote

$$\phi_0 : (x, n, \xi) \mapsto (p_{L(\xi)^\perp} x, \xi), \quad (x, n, \xi) \in G_0(X, d, k),$$

and $\mathcal{A}_0(X, d, k) := \phi_0(G_0(X, d, k))$.

Let further $P : (z, \xi) \mapsto \xi$ be the projection defined on $\mathcal{A}_0(X, d, k)$. Applying the Coarea formula (Theorem 1.15) to P, we obtain a relation of μ_k^X to the Hausdorff measure:

$$\begin{aligned} &\int_{\mathcal{A}(X,d,k)} f(z, V)\, \mu_k^X(\mathrm{d}(z, V)) \\ &= (2c_{d,k})^{-1} \int_{\mathcal{A}_0(X,d,k)} f(z, L(\xi)) J_{k(d-k)} P(z, \xi) \mathcal{H}^p(\mathrm{d}(z, \xi)), \end{aligned}$$

where f is any measurable function with bounded support defined on $\mathcal{A}(X, d, k)$.

In the following we express absolute curvature measures as measures over the affine tangent Grassmannian.

Theorem 6.29 *For a set* X *with positive reach,* $k = 0, 1, \ldots, d - 1$ *and for a bounded Borel set* $B \subset \mathbb{R}^d$ *we have,*

$$C_k^{\mathrm{abs}}(X, B) = \beta_{d,k} \int_{\mathcal{A}(X,d,d-1-k)} \sum_{(x,n,V) \in \phi^{-1}\{z,V\}} \mathbf{1}_B(x)\, \mu_{d-1-k}^X(\mathrm{d}(z, V)),$$

with

$$\beta_{d,k} = \frac{1}{2} \binom{d-1}{k} \gamma(d-1, k, d-1-k).$$

Proof Denoting

$$D_k(x,n,V) := \frac{\left|\sum_{|I|=d-1-k} \left(\prod_I \kappa_i\right) \left[\bigwedge_I a_i, V^\perp\right]^2\right|}{\prod_{i=1}^{d-1} \sqrt{1+\kappa_i^2}}, \tag{6.18}$$

we have using Proposition 6.24 and the Coarea formula for the projection $\Pi : (x,n,\xi) \mapsto (x,n)$ defined on $G_0(X,d,d-1-k)$,

$$\begin{aligned}
&C_k^{\mathrm{abs}}(X,B) \\
&= \binom{d-1}{k} \mathcal{O}_{d-1-k}^{-1} \int_{\operatorname{nor} X} \mathbf{1}_B(x) \\
&\quad \times \int_{G^{n^\perp}(d-1,d-1-k)} D_k(x,n,V) \nu_{d-1-k}^{d-1}(\mathrm{d}V)\, \mathcal{H}^{d-1}(\mathrm{d}(x,n)) \\
&= \binom{d-1}{k} \mathcal{O}_{d-1-k}^{-1} (2c_{d-1,d-1-k})^{-1} \\
&\quad \times \int_{G_0(X,d,d-1-k)} \mathbf{1}_B(x) J_{d-1}\Pi(x,n,\xi) D_k(x,n,L(\xi))\, \mathcal{H}^p(\mathrm{d}(x,n,\xi)).
\end{aligned}$$

Using now Lemma 6.30 below and the Area formula for the mapping ϕ_0, we get

$$\begin{aligned}
&C_k^{\mathrm{abs}}(X,B) \\
&= \binom{d-1}{k} \mathcal{O}_{d-1-k}^{-1} (2c_{d-1,d-1-k})^{-1} \\
&\quad \times \int_{G_0(X,d,d-1-k)} \mathbf{1}_B(x) J_p\phi_0(x,n,\xi) J_q P(\phi_0(x,n,\xi))\, \mathcal{H}^p(\mathrm{d}(x,n,\xi)) \\
&= \binom{d-1}{k} \mathcal{O}_{d-1-k}^{-1} (2c_{d-1,d-1-k})^{-1} \\
&\quad \times \int_{\mathcal{A}_0(X,d,d-1-k)} \sum_{(x,n,\xi)\in\phi_0^{-1}\{z,\xi\}} \mathbf{1}_B(x)\, J_q P(\phi_0(z,\xi))\, \mathcal{H}^p(\mathrm{d}(z,\xi)) \\
&= \binom{d-1}{k} \mathcal{O}_{d-1-k}^{-1} \frac{2c_{d,d-1-k}}{2c_{d-1,d-1-k}} \\
&\quad \times \int_{\mathcal{A}(X,d,d-1-k)} \sum_{(x,n,V)\in\phi^{-1}\{z,V\}} \mathbf{1}_B(x)\, \mu_{d-1-k}^X(\mathrm{d}(z,V)),
\end{aligned}$$

and an evaluation of the constant yields the assertion. □

Lemma 6.30 *For $\mathcal{H}^p$-almost all $(x,n,\xi)\in G_0(X,d,k)$,*

$$J_p\phi_0(x,n,\xi)J_qP(\phi_0(x,n,\xi))=J_{d-1}\Pi(x,n,\xi)D_k(x,n,L(\xi)),$$

where $\Pi:(x,n,\xi)\mapsto(x,n)$ is defined on $G_0(X,d,k)$ and $D_k(x,n,V)$ is given by (6.18).

Proof Assume that (x,n) is a regular point of $\operatorname{nor} X$, $a_X(x,n)$ the associated unit simple $(d-1)$-vector, $(x,n,\xi)\in G_0(X,d,k)$, and let V denote the k-subspace associated with ξ. Let $\{w_1\dots,w_{d-1},n\}$ be an orthonormal basis of $\mathbb{R}^d$ such that $w_1,\dots,w_k\in V$ and $w_{k+1},\dots,w_{d-1}\in V^\perp$. Denote

$$\xi_v:=n\wedge(\xi\llcorner v^*),\quad v\in\mathbb{R}^d,$$
$$\eta_{pq}:=w_p\wedge(\xi\llcorner w_q^*),\quad 1\le p\le k,\ k+1\le q\le d-1.$$

Denote $L:=D\phi_0(x,n,\xi)$ for brevity. If $a_X=\bigwedge_{i=1}^{d-1}(u_i,v_i)$ then (u_i,v_i,ξ_{v_i}) and $(0,0,\eta_{pq})$ are tangent vectors to $G_0(X,d,k)$ at (x,n,ξ) by Lemma 6.27, and we have

$$J_p\phi_0(x,n,\xi)=\frac{\left|\bigwedge_{i=1}^{d-1}L\left(u_i,v_i,\xi_{v_i}\right)\wedge\bigwedge_{p=1}^{k}\bigwedge_{q=k+1}^{d-1}L(0,0,\eta_{pq})\right|}{\left|\bigwedge_{i=1}^{d-1}\left(u_i,v_i,\xi_{v_i}\right)\right|\left|\bigwedge_{p=1}^{k}\bigwedge_{q=k+1}^{d-1}(0,0,\eta_{pq})\right|}$$

and

$$J_{d-1}\Pi(x,n,\xi)=\frac{\left|\bigwedge_{i=1}^{d-1}(u_i,v_i)\right|}{\left|\bigwedge_{i=1}^{d-1}\left(u_i,v_i,\eta_{v_i}\right)\right|}.$$

Using Lemma 1.70, we get

$$L(u_i,v_i,\xi_{v_i})=(p_{V^\perp}u_i+(p_Vv_i\cdot p_Vx)n+(n\cdot p_{V^\perp}x)p_Vv_i,\xi_{v_i}),\ i=1,\dots,d-1.$$

Note that $\eta_v=0$ whenever $v\perp V$. We can choose the vectors (u_i,v_i) so that $v_{k+1},\dots,v_{d-1}\perp V$ and that $(p_{V^\perp}u_i)\cdot(p_{V^\perp}u_j)=0$ whenever $1\le i\le k<j\le d-1$. Then $\xi_{v_j}=0$ if $j>k$ and

$$L(u_j,v_j,0)=(p_{V^\perp}u_j,0),\quad j=k+1,\dots,d-1,$$

are orthogonal to $L(u_i,v_i,\xi_{v_i})$, $i=1,\dots,k$. Further, let β_{pq} be the orthogonal projection of $L(0,0,\eta_{pq})$ onto the orthogonal complement to

$$\operatorname{Lin}\{(p_{V^\perp}u_j,0),\ j=k+1,\dots,d-1\}.$$

(Note that the last component of β_{pq} is again η_{pq}.) Then

$$J_p\phi_0(x,n,\xi) = \frac{\left|\bigwedge_{j=k+1}^{d-1}(p_{V^\perp}u_j,0)\right| \left|\bigwedge_{i=1}^{k} L\left(u_i,v_i,\eta_{v_i}\right) \wedge \bigwedge_{p=1}^{k}\bigwedge_{q=k+1}^{d-1}\beta_{pq}\right|}{\left|\bigwedge_{i=1}^{d-1}\left(u_i,v_i,\eta_{v_i}\right)\right| \left|\bigwedge_{p=1}^{k}\bigwedge_{q=k+1}^{d-1}(0,0,\eta_{pq})\right|}$$

Further, $L(u_i,v_i,\xi_{v_i})$, $i=1,\dots,k$, and β_{pq}, $p=1,\dots,k$, $q=k+1,\dots,d-1$, form a basis of the orthogonal complement to the kernel of P in the space $\operatorname{Tan}^p(G_0(X,d,k),(x,n,\xi))$ and, hence,

$$J_qP(\phi_0(x,n,\xi)) = \frac{\left|\bigwedge_{i=1}^{k}\xi_{v_i} \wedge \bigwedge_{p=1}^{k}\bigwedge_{q=k+1}^{d-1}\eta_{pq}\right|}{\left|\bigwedge_{i=1}^{k} L(u_i,v_i,\xi_{v_i}) \wedge \bigwedge_{p=1}^{k}\bigwedge_{q=k+1}^{d-1}\beta_{pq}\right|}.$$

Putting this together, we have (omitting the arguments of the Jacobians)

$$\frac{J_p\phi_0\, J_qP}{J_{d-1}\Pi} = \frac{\left|\bigwedge_{i=1}^{k}\xi_{v_i}\right| \left|\bigwedge_{j=k+1}^{d-1}p_{V^\perp}u_j\right|}{\left|\bigwedge_{i=1}^{d}(u_i,v_i)\right|}.$$

Since the mapping $\xi_v \mapsto p_V v$ is an isometry (see Lemma 1.30) and $a_X(x,n)$ is a unit multivector, we can write

$$\frac{J_p\phi_0\, J_qP}{J_{d-1}\Pi} = \left|\bigwedge_{i=1}^{k} p_V v_i\right| \left|\bigwedge_{j=k+1}^{d-1} p_{V^\perp}u_j\right| = |\langle a_X(x,n),\psi\rangle|$$

with the $(d-1)$-form ψ given by

$$\left\langle \bigwedge_{i=1}^{d-1}(u_i,v_i),\psi \right\rangle = \sum_{|I|=k} \left\langle \bigwedge_{i\in I} p_V v_i, w_1^*\wedge\cdots\wedge w_k^* \right\rangle \left\langle \bigwedge_{i\in I^c} p_{V^\perp}u_i, w_{k+1}^*\wedge\cdots\wedge w_{d-1}^* \right\rangle.$$

Writing a_X in the form given in Proposition 4.23, we obtain that the last expression equals $B_k(x,n,V)$, and the proof is finished. □

6.4 Bibliographical Notes

1. The first version of a translative intersection formula for curvature measures was presented by Schneider and Weil in [SW86] in the setting of convex bodies. The proof was done by approximation with polytopes and the mixed functionals were not given explicitly.

2. The intersection formula (Theorem 6.10) can be extended to more than two bodies. These can be found in [Wei90] for the case of convex bodies and in [Rat96] for sets with positive reach. An explicit integral representation of mixed curvature measures of more than two bodies with positive reach can be found in [Hug99] for convex bodies, and in [HR18] for sets with positive reach.
3. The translative Crofton formula (Theorem 6.20) and some related results were proved in [Rat99].
4. Absolute curvature measures were treated by Santaló in [San76], in the setting of smooth submanifolds. The Crofton formula was proved by Baddeley in [Bad80]. An extension to sets with positive reach was carried out in [Zäh89], see also [RZ90a]. Absolute curvature measures for unions of sets with positive reach were introduced in [Rat02].
5. The total absolute curvature measures

$$\mathbf{C}_j^{\mathrm{abs}}(X) = C_j^{\mathrm{abs}}(X, \mathbb{R}^d),$$

$j = 0, \ldots, d-1$, are related to the total measure of tangent planes as follows.

Theorem 6.31 *Let X be a set with positive reach and compact boundary, and let the following full-dimensionality condition be satisfied:*

$$\mathcal{H}^{d-1}\{x \in \partial X : \exists n \in S^{d-1}, (x, n) \in \operatorname{nor} X, (x, -n) \in \operatorname{nor} X\} = 0. \qquad (6.19)$$

Then

$$\mathbf{C}_k^{\mathrm{abs}}(X) = \mu_k^X(\mathcal{A}(X, d, d-1-k)).$$

This results follows from Theorem 6.29 and from the following lemma which can be found in [RZ02, Lemma 1].

Lemma 6.32 *For $\mathcal{H}^p$ almost all $(z, V) \in (X, d, k)$, ϕ^{-1} is either a singleton, or a pair of points of the type (x, n, V), $(x, -n, V)$.*

6. Curvature measures can be extended to tensor-valued valuations. An extension of the Principal kinematic formula to this setting was obtained by Hug and Schneider [HS08].
7. Rotational integral geometry concerns integrals over the special orthogonal group $\mathrm{SO}(d)$ only (not over translations). This theory was developed by Jensen in [Jen98]. The following rotational integral formula for curvature measures was proved in [JR08]:

Theorem 6.33 *Let $X \subset \mathbb{R}^d$ be a set with positive reach such that $0 \notin \partial X$ and assume that for ν_j^d-almost all $L \in G(d, j)$, there is no point $(x, n) \in \operatorname{nor} X$ with $x \in L$ and $n \perp L$. Then, for all $0 \leq k < j < d$,*

$$\int_{G(d,j)} \mathbf{C}_k(X \cap L)\, \nu_j^d(\mathrm{d}L) = \mathcal{O}_{j-1-k}^{-1} \int_{\operatorname{nor} X} \frac{1}{|x|^{d-j}}$$

$$\times \sum_{|I|=j-1-k} Q_j(x, n, A_{I^c}) \frac{\prod_{i \in I} \kappa_i(x, n)}{\prod_{i=1}^{d-1} \sqrt{1 + \kappa_i^2(x, n)}} \mathcal{H}^{d-1}(\mathrm{d}(x, n)),$$

provided that the integral on the right side exists. Here A_{I^c} denotes the linear subspace spanned by the principal direction $a_i(x, n)$, $i \notin I$, and the weight function Q_j is given by

$$Q_j(x, n, A_{I^c}) = \int_{G(d,j;x)} \frac{[L, A_{I^c}]^2}{|p_L n|^{j-k}}\, \nu_{j;1}^d(\mathrm{d}L),$$

with $G(d, j; x)$ denoting the submanifold of $G(d, j)$ of j-subspaces containing x, and $\nu_{j;1}^d$ is its invariant measure.

An overview on rotational integral formulas can be found in [JK17].

Chapter 7
Approximation of Curvatures

7.1 Approximation by Parallel Sets

Recall that the curvature measures $C_k(X_r, \cdot)$ of the r-parallel sets to a set X with positive reach converge vaguely to those of X itself (see Corollary 4.35). This stability result motivates a natural question whether curvature measures of more general sets can be introduced through approximation with parallel sets. This will indeed be the case, as it will be clear in Chap. 9. However, not only parallel sets may be used for approximation. Classically, approximations by polyhedral (piecewise linear) sets are frequently used in differential geometry, or approximation by smooth sets in different ways.

Here we will use flat convergence of normal cycles as an appropriate tool for convergence of the curvature-direction measures and their total values by means of the Lipschitz-Killing curvature forms φ_k. Note that, in general, convergence with respect to the Hausdorff distance d_H is not sufficient.

Recall that **(F)** lim means convergence of $(d-1)$-currents in $\mathbb{R}^d \times \mathbb{R}^d$ with respect to the flat norm

$$\mathbf{F}(T) = \sup\big\{T(\varphi) : \varphi \in \mathcal{D}^{d-1}(\mathbb{R}^d \times \mathbb{R}^d),\ \sup_{(x,n)\in\mathbb{R}^d\times\mathbb{R}^d} ||\varphi(x,n)|| \le 1, \\ \sup_{(x,n)\in\mathbb{R}^d\times\mathbb{R}^d} ||d\varphi(x,n)|| \le 1\big\}$$

and $\mathbf{M}$ is the mass norm of the currents (cf. Sect. 1.3.2).

Proposition 7.1 *Suppose that $X, X^{\varepsilon} \in \mathcal{U}_{\mathcal{PR}}$, $X, X^{\varepsilon} \subset K$, $\varepsilon > 0$, for some compact $K \subset \mathbb{R}^d$, then for $k = 0, 1, \dots, d-1$,*

$$\mathbf{(F)} \lim_{\varepsilon\to 0} N_{X^{\varepsilon}} = N_X \ \text{ implies } \ \lim_{\varepsilon\to 0} \mathbf{C}_k(X^{\varepsilon}) = \mathbf{C}_k(X)\,.$$

J. Rataj, M. Zähle, *Curvature Measures of Singular Sets*, Springer Monographs in Mathematics, https://doi.org/10.1007/978-3-030-18183-3_7

If additionally

$$\sup_{\varepsilon} \mathbf{M}(N_{X^\varepsilon}) < \infty\,,$$

then we get

$$\mathrm{w} - \lim_{\varepsilon\to 0} \widetilde{C}_k(X^\varepsilon, \cdot) = \widetilde{C}_k(X, \cdot)\,.$$

Proof For any smooth functions f and h on $\mathbb{R}^d \times \mathbb{R}^d$ such that h has compact support and is equal to 1 on $K \times S^{d-1}$ we get $hf\varphi_k \in \mathcal{D}^{d-1}(\mathbb{R}^d \times \mathbb{R}^d)$ and for $k = 0, 1, \ldots, d-1$,

$$c_k(f) := \max\Big(\sup_{(x,n)\in\mathbb{R}^d\times\mathbb{R}^d} ||hf\varphi_k(x,n)||, \sup_{(x,n)\in\mathbb{R}^d\times\mathbb{R}^d} ||d(hf\varphi_k)(x,n)||\Big) < \infty\,.$$

Here we have used that h, f, φ_k as well as the corresponding derivatives are smooth and therefore bounded on $\operatorname{spt} h$. Then we infer

$$\left|N_{X^\varepsilon}(f\varphi_k) - N_X(f\varphi_k)\right| = \left|N_{X^\varepsilon}(hf\varphi_k) - N_X(hf\varphi_k)\right|$$

and the right-hand side does not exceed $c_k(f)\,\mathbf{F}(N_{X^\varepsilon} - N_X)$ in view of the definition of the flat norm $\mathbf{F}$. Consequently,

$$\lim_{\varepsilon\to 0} \left|N_{X^\varepsilon}(f\varphi_k) - N_X(f\varphi_k)\right| = 0\,.$$

Choosing $f \equiv 1$ we obtain the first assertion, since $C_k(\cdot) = N_{(\cdot)}(\varphi_k)$.
Let now g be any continuous function on $\mathbb{R}^d \times S^{d-1}$. It has a continuous extension f to $\mathbb{R}^d \times \mathbb{R}^d$. Furthermore, f can be approximated by smooth functions f_j, $j = 1, 2, \ldots$, uniformly on compact sets. Then we get

$$\begin{aligned}\left|N_{X^\varepsilon}(f\varphi_k) - N_X(f\varphi_k)\right| &\le \left|N_{X^\varepsilon}((f - f_j)\varphi_k)\right| + \left|N_X((f - f_j)\varphi_k)\right| \\ &\quad + \left|N_{X^\varepsilon}(f_j\varphi_k) - N_X(f_j\varphi_k)\right|.\end{aligned}$$

Recall that the norm of the form φ_k is bounded on the compact set $K \times S^{d-1}$ containing $\operatorname{spt} N_{X^\varepsilon}$ and $\operatorname{spt} N_X$. Then the first two summands tend to 0 as $j \to \infty$ uniformly in ε. For the first one we have used the assumption that the mass norms of the N_{X_i} are uniformly bounded. The third summand goes to 0 as $\varepsilon \to 0$ for every j by the above arguments for smooth functions. Hence, we obtain

$$\lim_{\varepsilon\to 0} \left|N_{X^\varepsilon}(f\varphi_k) - N_X(f\varphi_k)\right| = 0\,.$$

Using that $\widetilde{C}_k(\cdot, g) = N_{(\cdot)}(g\varphi_k) = N_{(\cdot)}(f\varphi_k)$ for g and f as before this yields

$$\lim_{i\to\infty} (\widetilde{C}_k(X_i, g) - \widetilde{C}_k(X, g)) = 0$$

for all continuous g, i.e., the second assertion. □

This will be applied below to several approximation problems.

Turning back to the problem of approximation by parallel sets we first show that the reach of a set is upper semicontinuous with respect to the Hausdorff distance (see (2.3)).

Lemma 7.2 *Let X^ε, X be subsets of $\mathbb{R}^d$, $\varepsilon \in (0, \varepsilon_0)$, such that* reach $X^\varepsilon \geq r$ *for all $\varepsilon \in (0, \varepsilon_0)$ and $X^\varepsilon \overset{d_H}{\to} X$. Then* reach $X \geq r$.

Proof Assume, for the contrary, that reach $X < r$, and let z be a point with $s := \operatorname{dist}(z, X) < r$ and with two different nearest points $x, y \in X$, $|x - z| = |y - z| = s$. For given $\varepsilon > 0$, denote $s_\varepsilon := \operatorname{dist}(z, X^\varepsilon)$ and let $w_\varepsilon \in X^\varepsilon$ be a point with $|w_\varepsilon - z| = s_\varepsilon$. Let $\varepsilon_i \to 0$ be a sequence for which $w := \lim_{i\to\infty} w_{\varepsilon_i} \in X$ exists. We can assume that $w \neq x$ (otherwise, we interchange x and y). Then choose a sequence of points $x_i \in X^{\varepsilon_i}$ such that $\lim_{i\to\infty} x_i = x$. Since reach $X^\varepsilon \geq r$, we have by Corollary 4.6

$$(x_i - w_{\varepsilon_i}) \cdot (z - w_{\varepsilon_i}) \leq \frac{|x_i - w_{\varepsilon_i}|^2 |z - w_{\varepsilon_i}|}{2r}.$$

Letting $i \to \infty$, we get

$$(x - w) \cdot (z - w) \leq \frac{|x - w|^2 |z - w|}{2r}.$$

In the isosceles triangle zxw, the cosine of the angle at vertex w equals $|w - x|/(2s)$. Thus we have

$$(x - w) \cdot (z - w) = \frac{|x - w|^2 |z - w|}{2s}.$$

This, however, contradicts the last displayed inequality, since $s < r$. □

A basic tool for approximations is the following theorem saying that Hausdorff convergence of sets with uniformly positive reach implies already flat convergence of normal cycles.

Theorem 7.3 *For $\varepsilon \in (0, \varepsilon_0)$ let X^ε, X be compact subsets of $\mathbb{R}^d$ such that $X^\varepsilon \overset{d_H}{\to} X$ and* $\inf_{0<\varepsilon<\varepsilon_0}$ reach $X^\varepsilon > 0$. *Then we get*

$$\sup_{0<\varepsilon<\varepsilon_0} \mathbf{M}(N_{X^\varepsilon}) < \infty \quad \text{and} \quad (\mathbf{F}) \lim_{\varepsilon\to 0} N_{X^\varepsilon} = N_X .$$

For the proof, we will use the following technical lemma which will also be applied to the case of polyhedral approximation of curvatures.

Lemma 7.4 *For* $0 < \varepsilon < \varepsilon_0$ *let* S^ε *and* T^ε *be integer-multiplicity k-rectifiable currents without boundary (cycles), whose supports are compact, and* $f^\varepsilon : \operatorname{spt} T^\varepsilon \to \operatorname{spt} S^\varepsilon$ *be Lipschitz mappings with the properties*

(i) $\sup\{|f^\varepsilon(x) - x| : x \in \operatorname{spt} T^\varepsilon\} \to 0,\ \varepsilon \to 0,$
(ii) $(f^\varepsilon)_\# T^\varepsilon = S^\varepsilon,\ 0 < \varepsilon < \varepsilon_0,$
(iii) $\sup_{0<\varepsilon<\varepsilon_0} \operatorname{Lip}(f^\varepsilon) < \infty,$
(iv) $\sup_{0<\varepsilon<\varepsilon_0} \mathbf{M}(T^\varepsilon) < \infty\ .$

Then we get

$$\sup_{0<\varepsilon<\varepsilon_0} \mathbf{M}(S^\varepsilon) < \infty \quad \text{and} \quad (\mathbf{F}) \lim_{\varepsilon\to 0} (S^\varepsilon - T^\varepsilon) = 0\,.$$

Proof Assumption (ii) and the definition of push forward and mass norm of a current imply

$$\mathbf{M}(S^\varepsilon) = \mathbf{M}((f^\varepsilon)_\# T^\varepsilon) \le \operatorname{Lip}(f^\varepsilon)^k \mathbf{M}(T^\varepsilon)\,,$$

which yields the first assertion in view of (iii) and (iv).
For the second assertion we can apply Lemma 1.61 together with (ii) to the currents $T := T^\varepsilon$. The corresponding mappings there are $f := f^\varepsilon$ and the identity $g := \mathrm{id}$. Since S^ε and T^ε have no boundaries, we infer for the flat norm

$$\mathbf{F}(S^\varepsilon - T^\varepsilon) = \mathbf{F}((f^\varepsilon)_\# T^\varepsilon - T^\varepsilon) \le (\max\{1, \operatorname{Lip}(f^\varepsilon)\})^k \int |f^\varepsilon - \mathrm{id}|\, \mathrm{d}\|T^\varepsilon\|\,,$$

and the last expressions tends to 0 as $\varepsilon \to 0$ by (i), (iii) and (iv). □

Proof (of Theorem 7.3*)* Our aim is to apply Lemma 7.4 to the currents $S^\varepsilon := N_{X^\varepsilon}$ and $T^\varepsilon := N_X$ and suitable mappings f^ε.
Denote

$$r := \inf_{0<\varepsilon<\varepsilon_0} \operatorname{reach} X^\varepsilon\,.$$

We have then reach $X \ge r$ by Lemma 7.2. In order to construct the f^ε consider the mapping

$$h : (x, n) \mapsto x + \frac{r}{3} n, \quad (x, n) \in \operatorname{nor} X,$$

which is clearly Lipschitz and maps nor X bijectively onto $\partial X_{r/3}$. Note also that the image of h, $\partial X_{r/3} \subseteq (X^\varepsilon)_{r/2} \setminus (X^\varepsilon)_{r/4}$ when ε is small enough, by the definition of the Hausdorff distance. Further, let g^ε be the restriction of the mapping

$$\widetilde{\Pi}_{X^\varepsilon} : y \mapsto \left(\Pi_{X^\varepsilon} y, \frac{y - \Pi_{X^\varepsilon} y}{|y - \Pi_{X^\varepsilon} y|} \right)$$

(see Sect. 4.5) onto $\partial(X^\varepsilon)_{r/3}$; g^ε is an injective Lipschitz mapping from $\partial X_{r/3}$ to nor X^ε. Since the domain of g^ε is contained in $(X^\varepsilon)_{r/2} \setminus (X^\varepsilon)_{r/4}$ for sufficiently small ε, the Lipschitz constant of g^ε is bounded independently of ε, see Lemma 4.21.

We will show now that g^ε is onto nor X^ε. Let $(z, m) \in$ nor X^ε be given; we will find a point $y \in \partial X_{r/3}$ with $\widetilde{\Pi}_{X^\varepsilon} y = (z, m)$. Equivalently, we have to show that $y = z + tm$ for some $t > 0$. Assume that ε is small enough so that $d_H(X, X^\varepsilon) < \frac{r}{6}$. Then, by the triangular inequality, $\operatorname{dist}(z + \frac{r}{6} m, X) < \frac{r}{3}$, whereas $\operatorname{dist}(z + \frac{r}{2} m, X) > \frac{r}{3}$. Since the distance is continuous, there must be a $t \in (\frac{r}{3}, \frac{r}{2})$ with the desired property and, hence, g^ε is onto.

In this way we obtain bijective mappings

$$f^\varepsilon := g^\varepsilon \circ h : \text{nor}\, X \to \text{nor}\, X^\varepsilon$$

with uniformly bounded Lipschitz constants (fulfilling thus (iii) in Lemma 7.4). We will show now that

$$|f^\varepsilon(x, n) - (x, n)| \to 0, \quad \varepsilon \to 0, \tag{7.1}$$

uniformly in (x, n). This yields condition (i) in Lemma 7.4. Choose $(x, n) \in$ nor X and denote $y = f(x, n) = x + \frac{r}{3} n$ and $(z, m) = g^\varepsilon(y)$. By the assumptions, there exists a point $x' \in X \cap B(z, \varepsilon)$. By Corollary 4.6,

$$(x' - x) \cdot n \le \frac{|x' - x|^2}{2r}.$$

Hence,

$$\begin{aligned}
|x' - y|^2 &= |(x' - x) - (y - x)|^2 = |(x' - x) - \frac{r}{3} n|^2 \\
&= |x' - x|^2 + \frac{r^2}{9} - \frac{2r}{3}(x' - x) \cdot n \\
&\ge |x' - x|^2 + \frac{r^2}{9} - \frac{2r}{3} \frac{|x' - x|^2}{2r} \\
&= \frac{2}{3}|x' - x|^2 + \frac{r^2}{9}.
\end{aligned}$$

On the other hand, $|z-y| = \operatorname{dist}(y, X^\varepsilon) \le \operatorname{dist}(y, X) + \varepsilon = \frac{r}{3} + \varepsilon$, thus $|x'-y| \le \frac{r}{3} + 2\varepsilon$. Putting these two estimates together, we infer

$$\frac{2}{3}|x'-x|^2 + \frac{r^2}{9} \le \left(\frac{r}{3} + 2\varepsilon\right)^2 = \frac{r^2}{9} + \frac{4r\varepsilon}{3} + 4\varepsilon^2,$$

whence $|x'-x|^2 \le 2r\varepsilon + 6\varepsilon^2$ and

$$|z-x| \le \sqrt{2r\varepsilon + 6\varepsilon^2} + \varepsilon.$$

Thus $z - x \to 0$. Since

$$|m-n| = \left| \frac{y-x}{|y-z|} - \frac{y-z}{|y-z|} \right| \le 2\frac{|z-x|}{|y-x|} = \frac{6}{r}|z-x|,$$

also $m - n \to 0$ uniformly as $\varepsilon \to 0$. Hence, (7.1) is fulfilled.

It remains to verify condition (ii) from Lemma 7.4. Using the area formula for currents (Theorem 1.49), we get

$$(f^\varepsilon)_\# N_X = (\mathcal{H}^{d-1} \llcorner \operatorname{nor} X^\varepsilon) \wedge \xi$$

with

$$\xi(f^\varepsilon(x,n)) = \frac{(\bigwedge_{d-1} \operatorname{ap} Df^\varepsilon(x,n)) a_X(x,n)}{|(\bigwedge_{d-1} \operatorname{ap} Df^\varepsilon(x,n)) a_X(x,n)|}$$

and $a_X(x,n) = a_1 \wedge \cdots \wedge a_{d-1}$, $a_i = (1+\kappa_i^2)^{-1/2}(b_i, \kappa_i b_i)$ and κ_i, b_i are the principal curvatures and directions of X at $(x,n) \in \operatorname{nor} X$, $i = 1, \ldots, d-1$, such that $\{b_1, \ldots, b_{d-1}, n\}$ is a positively oriented orthonormal basis. Similarly, we have

$$N_{X^\varepsilon} = (\mathcal{H}^{d-1} \llcorner \operatorname{nor} X^\varepsilon) \wedge a_{X^\varepsilon},$$

where $a_{X^\varepsilon}(y,m) = a_1^\varepsilon \wedge \cdots \wedge a_{d-1}^\varepsilon$, $a_i^\varepsilon = (1+(\kappa_i^\varepsilon)^2)^{-1/2}(b_i^\varepsilon, \kappa_i^\varepsilon b_i)$ and $\kappa_i^\varepsilon, b_i^\varepsilon$ are the principal curvatures and directions of X^ε at $(y,m) = f^\varepsilon(x,n) \in \operatorname{nor} X^\varepsilon$, $i = 1, \ldots, d-1$, such that $\{b_1^\varepsilon, \ldots, b_{d-1}^\varepsilon, m\}$ is a positively oriented orthonormal basis.

Thus, in order to verify (ii) from Lemma 7.4, we have to show that $\xi = a_{X^\varepsilon}$ $\mathcal{H}^{d-1}$-almost everywhere on $\operatorname{nor} X^\varepsilon$. Since both ξ and a_{X^ε} are unit $(d-1)$-vectors associated with the same subspace, $\operatorname{Tan}^{d-1}(\operatorname{nor} X^\varepsilon, (y,m))$, it will be enough to check that their signs agree, which is equivalent to:

$$(\textstyle\bigwedge_{d-1} \operatorname{ap} Df^\varepsilon(x,n))\, a_X(x,n) \bullet a_{X^\varepsilon}(f^\varepsilon(x,n)) > 0 \tag{7.2}$$

for $\mathcal{H}^{d-1}$-almost all $(x, n) \in \operatorname{nor} X$. Note that, by the construction of f^ε, we have

$$\operatorname{ap} Df^\varepsilon(x, n)(u, v) = \operatorname{ap} Dg^\varepsilon(x + \tfrac{r}{3}n)(u + \tfrac{r}{3}v)$$

for any $u, v \in \mathbb{R}^d$, hence,

$$\operatorname{ap} Df^\varepsilon(x, n)a_i = \operatorname{ap} Dg^\varepsilon(x + \tfrac{r}{3}n)(s_i b_i)$$

with $s_i = (1 + \frac{r}{3}\kappa_i)/\sqrt{1+\kappa_i^2}$, $i = 1, \dots, d-1$. Note that, since $\kappa_i \geq -(\operatorname{reach} X)^{-1} \geq -r^{-1}$ by Proposition 4.23, we get $s_i > 0$, $i = 1, \dots, d-1$.

Further, using (4.5) and the relations from Corollary 4.26, we get

$$\operatorname{ap} D\widetilde{\Pi}_{X^\varepsilon}(y + tm)(b_j^\varepsilon) = s_j^{\varepsilon,t} a_j^\varepsilon, \quad \operatorname{ap} D\widetilde{\Pi}_{X^\varepsilon}(y + tm)(m) = 0,$$

whenever $0 < t < r$, where $s_j^{\varepsilon,t} = \sqrt{1 + (\kappa_j^\varepsilon)^2}/(1 + t\kappa_j^\varepsilon)$, $j = 1, \dots, d-1$. Again, $s_j^{\varepsilon,t} > 0$ since $\kappa_j^\varepsilon \geq -r^{-1}$. Thus, expressing

$$b_i = \sum_{j=1}^{d-1} (b_i \cdot b_j^\varepsilon) b_j^\varepsilon + (b_i \cdot m)m,$$

we obtain

$$\operatorname{ap} Df^\varepsilon(x, n)a_i = \sum_{j=1}^{d-1} s_i s_j^{\varepsilon,t} (b_i \cdot b_j^\varepsilon) a_j^\varepsilon, \quad i = 1, \dots, d-1, \tag{7.3}$$

and

$$\begin{aligned}
&(\textstyle\bigwedge_{d-1} \operatorname{ap} Df^\varepsilon(x, n))\, a_X(x, n) \bullet a_{X^\varepsilon}(f^\varepsilon(x, n)) \\
&= \det\left(s_i s_j^{\varepsilon,t} (b_i \cdot b_j^\varepsilon)\right)_{i,j=1}^{d-1} \\
&= (s_1 \cdots s_{d-1}(b_1 \wedge \cdots \wedge b_{d-1})) \bullet \left(s_1^{\varepsilon,t} \cdots s_{d-1}^{\varepsilon,t}(b_1^\varepsilon \wedge \cdots \wedge b_{d-1}^\varepsilon)\right) \\
&= s_1 s_1^{\varepsilon,t} \cdots s_{d-1} s_{d-1}^{\varepsilon,t} (n \cdot m) > 0,
\end{aligned}$$

since $s_i, s_i^{\varepsilon,t} > 0$ and m, n are close unit vectors. This shows (7.2) and the proof is complete. □

Below we will use also the following special version for approximation with parallel sets.

Lemma 7.5 *Assume that $X \subset \mathbb{R}^d$ has positive reach and compact boundary. Then*

$$(\mathbf{F}) \lim_{\varepsilon\to 0} N_{X_\varepsilon} = N_X \text{ and } \lim_{\varepsilon\to 0} \mathbf{M}(N_{X_\varepsilon}) = \mathbf{M}(N_X).$$

Proof The arguments are as in the proof of Theorem 7.3, but more simple. Here we consider for $\varepsilon <$ reach X the bijections $f^\varepsilon : \operatorname{nor} X \to \operatorname{nor} X_\varepsilon$ with $f^\varepsilon(x,n) := (x+\varepsilon n, n)$ and apply Lemma 7.4 to the currents $S^\varepsilon := N_{X_\varepsilon}$ and $T^\varepsilon := N_X$ in order to get the flat convergence. For the mass norms we obtain

$$\mathbf{M}(N_{X_\varepsilon}) = \int_{\operatorname{nor} X} \left| \bigwedge_{d-1} \operatorname{ap} Df^\varepsilon(x,n) a_X(x,n) \right| \mathcal{H}^{d-1}(d(x,n)),$$

where $a_X = a_1 \wedge \cdots \wedge a_{d-1}$, $a_i = (1+(\kappa_i)^2)^{-1/2}(b_i, \kappa_i b_i)$ and κ_i, b_i are the principal curvatures and directions of X at $(x,n) \in \operatorname{nor} X$, $i = 1, \ldots, d-1$. (Recall the notions from Sect. 4.4.) Since $\bigwedge_{d-1} \operatorname{ap} Df^\varepsilon(x,n) a_X(x,n) = a_1^\varepsilon \wedge \cdots \wedge a_{d-1}^\varepsilon$ with $a_i^\varepsilon = (1+(\kappa_i)^2)^{-1/2}((1+\varepsilon\kappa_i)b_i, \kappa_i b_i)$, we infer that the above integrand goes to $|a_X(x,n)|$ as $\varepsilon \to 0$ uniformly in $(x,n) \in \operatorname{nor} X$. Since $|a_X(x,n)| = 1$, the above integral converges to $\mathcal{H}^{d-1}(\operatorname{nor} X) = \mathbf{M}(N_X)$. □

We will apply now Theorem 7.3 together with Proposition 7.1 in order to obtain a continuity result for parallel sets to compact $\mathcal{U}_{\mathcal{PR}}$-sets.

Theorem 7.6 *Suppose that $X = \bigcup_{i=1}^m X^i$ is a $\mathcal{U}_{\mathcal{PR}}$-representation of a compact set X and $\varepsilon_j \to 0$ is a sequence such that*

$$\liminf_{j\to\infty} \operatorname{reach} \bigcap_{i\in I} (X^i)_{\varepsilon_j} > 0$$

for all $I \subset \{1, \ldots, m\}$. Then we get

$$\limsup_{j\to\infty} \mathbf{M}(N_{X_{\varepsilon_j}}) < \infty, \quad (\mathbf{F}) \lim_{j\to\infty} N_{X_{\varepsilon_j}} = N_X$$

and

$$\mathrm{w} - \lim_{j\to\infty} \widetilde{C}_k(X_{\varepsilon_j}, \cdot) = \widetilde{C}_k(X, \cdot), \quad j = 0, 1, \ldots, d-1.$$

Consequently, if a compact set X admits a regular $\mathcal{U}_{\mathcal{PR}}$-representation in the sense of Definition 5.20, then

$$\limsup_{\varepsilon\to 0} \mathbf{M}(N_{X_\varepsilon}) < \infty, \quad (\mathbf{F}) \lim_{\varepsilon\to 0} N_{X_\varepsilon} = N_X$$

and

$$\mathrm{w}-\lim_{\varepsilon\to 0}\widetilde{C}_k(X_\varepsilon,\cdot)=\widetilde{C}_k(X,\cdot),\quad k=0,1,\dots,d-1.$$

Proof For the first part we will use the $\mathcal{U}_{\mathcal{PR}}$-representations $X_{\varepsilon_j}=\bigcup_{i=1}^m(X^i)_{\varepsilon_j}$ for $\varepsilon_0:=0$ and for sufficiently large j. The additivity of the unit normal cycles yields

$$N_{X_{\varepsilon_j}}=\sum_{I\subset\{1,\dots,m\}}(-1)^{|I|-1}N_{\bigcap_{i\in I}(X^i)_{\varepsilon_j}},$$

where $|I|$ denotes the cardinality of I. Since $\bigcap_{i\in I}(X^i)_{\varepsilon_j}\searrow\bigcap_{i\in I}X^i$, we have

$$d_H\left(\bigcap_{i\in I}(X^i)_{\varepsilon_j},\bigcap_{i\in I}X^i\right)\to 0,\quad j\to\infty,$$

and, hence,

$$\text{(F)}\lim_{j\to\infty}N_{\bigcap_{i\in I}(X^i)_{\varepsilon_j}}=N_{\bigcap_{i\in I}X^i}$$

and $\limsup_{j\to\infty}\mathbf{M}(N_{\bigcap_{i\in I}(X^i)_{\varepsilon_j}})<\infty$ for any I by Theorem 7.3. Since the above sum is finite, the first two assertions of the first statement follow. Weak convergence of the curvature-direction measures then follows according to Proposition 7.1.

The second statement is a consequence, since in view of Proposition 5.25 for a regular $\mathcal{U}_{\mathcal{PR}}$-representation any sequence $\varepsilon_j\to 0$ fulfills the conditions of the theorem. □

The following regularity condition on the above sequences ε_j will be used in the next section for polytopal approximations.

Definition 7.7 We say that a sequence $\varepsilon_j\to 0$ is *regular* with respect to a $\mathcal{U}_{\mathcal{PR}}$-representation $X=\bigcup_i X^i$ if for any finite index set $I\subset\mathbb{N}$:

(i) $\liminf_{j\to\infty}\operatorname{reach}\bigcap_{i\in I}(X^i)_{\varepsilon_j}>0$,
(ii) for all sufficiently large j, $X_{\varepsilon_j}=\bigcup_i(X^i)_{\varepsilon_j}$ is a regular $\mathcal{U}_{\mathcal{PR}}$-representation (cf. Definition 5.20).

Remark 7.8

(i) For a regular $\mathcal{U}_{\mathcal{PR}}$-representation of a compact set any sequence $\varepsilon_j\to 0$ is regular due to Proposition 5.25.
(ii) The latter remains valid for any finite union of convex bodies.
(iii) We do not know whether any $\mathcal{U}_{\mathcal{PR}}$-representation has a regular sequence $\varepsilon_j\to 0$.

7.2 Polytopal Approximation

In the next chapter, we will characterize curvature measures like in classical geometry by its natural properties. To this aim we need an appropriate approximation with polytopes.

Recall that a simplicial d-polytope is the union $P = \bigcup \Sigma$ of a simplicial d-complex Σ and that $\mathcal{F}_k(P)$ denotes the set of all k-faces of P (Sect. 2.5). Clearly, any d-polytope P belongs to $\mathcal{U}_{\mathcal{PR}}$ and, hence, its normal cycle N_P and curvature measures $C_k(P, \cdot)$ $(k = 0, \ldots, d)$ are defined. It is not difficult to see that

$$\operatorname{nor} P \subset \bigcup_{k=0}^{d-1} \bigcup_{\mu \in \mathcal{F}_k(P)} \mu \times \Gamma_\mu, \tag{7.4}$$

where Γ_μ is the spherical convex hull of all outer unit normal vectors $\nu(\mu')$ to facets μ' of P such that μ is a face of μ'. Clearly, Γ_μ is contained in the unit $(d-1-k)$-sphere S^{d-1-k} of $\mu^\perp$, the orthogonal complement of the affine hull of μ.

Note that, though the notion of a face of P depends on the simplicial d-complex Σ, the unit normal bundle nor P is independent of the choice of Σ and is uniquely determined by $P = \bigcup \Sigma$.

As mentioned in the previous section, convergence in the Hausdorff metric together with flat convergence of the normal cycles with uniformly bounded mass norms implies weak convergence of the curvature-direction measures. Therefore we will consider appropriate flat approximations of the normal cycles of unions of sets with positive reach by those of d-polytopes.

An immediate consequence of Lemma 7.4 is the following, where the currents S^ε are replaced by the normal cycle of the $\mathcal{U}_{\mathcal{PR}}$-set and the T^ε by those of the polytopes.

Proposition 7.9 *Suppose that $X \in \mathcal{U}_{\mathcal{PR}}$ is compact and P^ε, $0 < \varepsilon < \varepsilon_0$, are compact d-polytopes such that*

(i) $\sup_{0<\varepsilon<\varepsilon_0} \mathbf{M}(N_{P^\varepsilon}) < \infty$,

and assume that there exist Lipschitz mappings

$$f^\varepsilon : \operatorname{nor} P^\varepsilon \to \mathbb{R}^d \times S^{d-1}$$

such that:

(ii) $\sup_{0<\varepsilon<\varepsilon_0} \operatorname{Lip} f^\varepsilon < \infty$,
(iii) $\lim_{\varepsilon \to 0+} \sup\{|f^\varepsilon(x,n) - (x,n)| : (x,n) \in \operatorname{nor} P^\varepsilon\} = 0$,
(iv) $(f^\varepsilon)_\# N_{P^\varepsilon} = N_X$, $0 < \varepsilon < \varepsilon_0$.

Then we have $\lim_{\varepsilon\to 0} d_H(P^\varepsilon, X) = 0$ *and*

$$(\mathbf{F}) \lim_{\varepsilon\to 0} N_{P^\varepsilon} = N_X .$$

This enables us to prove the *Polytopal approximation theorem* for normal cycles.

Theorem 7.10 *Suppose that* $X \in \mathcal{U}_{\mathcal{PR}}$ *is compact and admits a* $\mathcal{U}_{\mathcal{PR}}$*-representation with a regular sequence* $\varepsilon_j \to 0$ *(see Definition 7.7). Then there exist d-polytopes* $P^1, P^2, \dots$ *such that*

$$\lim_{j\to\infty} d_H(P^j, X) = 0 \ \textit{and}\ (\mathbf{F}) \lim_{j\to\infty} N_{P^j} = N_X .$$

In particular,

$$\lim_{j\to\infty} \mathbf{C}_k(P^j) = \mathbf{C}_k(X).$$

If X *is the parallel set to a set* Y *of positive reach, i.e.,* $X = Y_\delta$*, at distance* $0 < \delta <$ reach Y*, then the approximating d-polytopes* P^j *can be found in such a way that, moreover,*

$$\sup_j \mathbf{M}(N_{P^j}) < \infty,$$

hence,

$$\mathrm{w} - \lim_{j\to\infty} \widetilde{C}_k(P^j, \cdot) = \widetilde{C}_k(X, \cdot) \quad k = 0, 1, \dots, d-1.$$

In order to show this we first prove a lemma on approximation of parallel sets to sets with positive reach by simplicial d-polytopes.

Lemma 7.11 *Let* $Y \subset \mathbb{R}^d$ *be compact with positive reach and* $0 < \delta <$ reach Y*. Then, there exist* $\varepsilon_0, \theta > 0$ *and simplicial d-polytopes* $P^\varepsilon = \bigcup \Sigma^\varepsilon \subset \mathbb{R}^d$ *such that for all* $0 < \varepsilon < \varepsilon_0$*:*

(i) *all edges of* Σ^ε *have lengths at least* ε*,*
(ii) *all simplices of* Σ^ε *have circumradii at most* $\frac{5}{2}\varepsilon$*,*
(iii) $\Theta(\sigma) \geq \theta$, $\sigma \in \Sigma^\varepsilon$*,*
(iv) *all vertices of* P^ε *lie in* ∂Y_δ*,*
(v) $Y_{\delta-3\varepsilon} \subset P^\varepsilon \subset Y_{\delta+3\varepsilon}$.

Proof Let an

$$0 < \varepsilon < \varepsilon_0 := \frac{1}{3}\min\{\delta, \text{reach}\, Y - \delta\}$$

be given. Let N^ε be an $(\varepsilon, 1)$-net in ∂Y_δ (i.e., $|y - x| \geq \varepsilon$ whenever x, y are two different points from N^ε, and $\partial Y_\delta \subset \bigcup_{x \in N^\varepsilon} B(x, \varepsilon)$, see Sect. 2.5). Let $m \in \mathbb{N}$ be such that $Y_{2\delta} \subset [-m, m]^d$, consider Y_δ and N^ε to be embedded into the torus (cube with periodic boundary) $\mathbb{T}^d := \mathbb{R}^d / m\mathbb{Z}^d$, and let $M^\varepsilon \subset \operatorname{int} Y_{\delta-2\varepsilon} \cup (\mathbb{T}^d \setminus Y_{\delta+2\varepsilon})$ be an $(\varepsilon, 1)$-net of $\operatorname{int} Y_{\delta-2\varepsilon} \cup (\mathbb{T}^d \setminus Y_{\delta+2\varepsilon})$. From the maximality, we get by an elementary consideration

$$\mathbb{T}^d \subset \bigcup_{x \in N^\varepsilon \cup M^\varepsilon} B(x, 2\varepsilon),$$

hence, $M^\varepsilon \cup N^\varepsilon$ is an $(\varepsilon, 2)$-net in $\mathbb{T}^d$ (or, equivalently, an $m\mathbb{Z}^d$-periodic $(\varepsilon, 2)$-net in $\mathbb{R}^d$). By Theorem 2.8, there exists a $\delta > 0$ (independent of ε) and a weighted Delaunay d-complex Σ_0^ε of $N^\varepsilon \cup M^\varepsilon$ with weight function $w \in [0, \varepsilon/2)$ and such that

$$\Theta(\sigma) \geq \theta, \quad \sigma \in \Sigma_0^\varepsilon.$$

Let $\Sigma^\varepsilon \subset \Sigma_0^\varepsilon$ consist of those $\sigma \in \Sigma_0^\varepsilon$ with all vertices lying in Y_δ. We shall verify that the simplicial d-polytopes $P^\varepsilon = \bigcup \Sigma^\varepsilon$ satisfy the assertion of the lemma.

Assertions (i) and (iii) follow easily from the construction, (ii) is a consequence of (2.9). We shall verify now that all vertices of Σ^ε are from N^ε (and, hence, lie in ∂Y_δ). Assume, for the contrary, that this is not the case, i.e., there is a point $x \in M^\varepsilon$ which is a vertex of P^ε. Then, clearly, $x \in \operatorname{int} Y_{\delta-2\varepsilon}$, and x must be a vertex of some simplex $\sigma \in \Sigma_0^\varepsilon$ which has another vertex, say y, outside of Y_δ (since, otherwise, x would not be a boundary point of P^ε). Hence, $y \in \mathbb{R}^d \setminus Y_{\delta+2\varepsilon}$. Let $B(c_\sigma, \hat{R}_\sigma)$ be the orthoball of σ (see Sect. 2.5) which contains no point of N^ε in its interior. One of the line segments $[c_\sigma, x]$ and $[c_\sigma, y]$ hits ∂Y_δ in a point s which must have distance greater than 2ε from both x and y. Hence, $B(s, \varepsilon) \subset \operatorname{int} B(c_\sigma, \hat{R}_\rho)$ (since the distance of both x and y from the orthoball $B(c_\sigma, \hat{R}_\rho)$ is less or equal to $\frac{1}{2}\varepsilon$) and, thus, $B(s, \varepsilon) \cap N^\varepsilon = \emptyset$, which contradicts that N^ε is an $(\varepsilon, 1)$-net in ∂Y_δ, and (iv) is proved.

In order to verify (v), take a point $x \in P^\varepsilon$ and let $\sigma \in \Sigma^\varepsilon$ be such that $x \in \sigma$. Then, there must be at least one vertex y of σ with $|x - y| \leq \frac{5}{2}\varepsilon$ (by (ii)) and since $d_Y(y) \leq \delta$ and d_Y is 1-Lipschitz, we get $d_Y(x) \leq \delta + \frac{5}{2}\varepsilon$. Analogously we can prove that any point $z \in \mathbb{R}^d \setminus P^\varepsilon$ fulfills $d_Y(z) \geq \delta - \frac{5}{2}\varepsilon$, proving (v). □

Recall that $\nu(\mu)$ denotes the unit normal vector to a facet μ of a d-polytope.

Proposition 7.12 *In the situation of Lemma* 7.11, *there exist constants* $Q, L > 0$ *such that*

(vi) $|\nu(\mu) - \nu(\mu')| < Q\varepsilon$ *whenever* μ, μ' *are two adjacent facets of* P^ε,
(vii) $|\mathcal{H}^{d-1}(\partial P^\varepsilon) - \mathcal{H}^{d-1}(\partial Y_\delta)| \leq L\varepsilon$.

Proof Consider the mapping

$$G^{\varepsilon} := \Pi_{\partial Y_\delta}|\partial P^{\varepsilon} : \partial P^{\varepsilon} \to \partial Y_\delta$$

(note that $\partial P^{\varepsilon} \subset \mathrm{Unp}\,(\partial Y_\delta)$ by Lemma 7.11(v) if ε is small enough, hence, G^{ε} is well defined). Let $x \in \partial P^{\varepsilon}$, let $\mu = \mathrm{conv}\,\{x_0, \dots, x_{d-1}\}$ be a facet of P^{ε} containing x and let $u \in S^{d-1} \cap (\mathrm{aff}\,\mu - x)$. Let further $\sigma = \mathrm{conv}\,\{x_0, \dots, x_d\} \in \Sigma^{\varepsilon}$ be the d-simplex containing μ. We can write u in the form

$$u = \sum_{i=1}^{d-1} \alpha_i (x_i - x_0).$$

Since $u \in \alpha_i(x_i - x_0) + \mathrm{Lin}\,(\mu_i)$, $i = 1, \dots, d-1$ (recall that $\mu_i = \mathrm{conv}\,\{x_j : 0 \le j \le d-1,\ j \ne i\}$), we get

$$1 = |u| \ge |\alpha_i|\,\mathrm{dist}\,(x_i, \mathrm{aff}(\mu_i)) \ge |\alpha_i| h(\mu) \ge |\alpha_i| h(\sigma),$$

hence,

$$|\alpha_i| \le \frac{1}{h(\sigma)} = \frac{1}{d\rho(\sigma)\Theta(\sigma)} \le \frac{1}{d\theta\varepsilon}, \qquad i = 1, \dots, d-1, \tag{7.5}$$

by Lemma 7.11(iii). Denote for brevity $y = G^{\varepsilon}(x)$ and $\nu(y) = \nu_{Y_\delta}(y)$ (the unit outer normal vector to Y_δ at y). Since

$$\mathrm{reach}\,(\partial Y_\delta) \ge r_0 := \min\{\delta, \mathrm{reach}\, Y - \delta\},$$

we have by Corollary 4.6

$$|\nu(y) \cdot (x - x_i)| \le \frac{|y - x_i|^2}{2r_0}, \qquad i = 0, \dots, d-1.$$

Thus,

$$\begin{aligned}
|\nu(y) \cdot u| &= \left| \sum_{i=1}^{d-1} \alpha_i (\nu(y) \cdot (x_i - x_0)) \right| \\
&= \left| \sum_{i=1}^{d-1} \alpha_i (\nu(y) \cdot (x_i - y) + \nu(y) \cdot (y - x_0)) \right| \\
&\le \sum_{i=1}^{d-1} |\alpha_i| (|\nu(y) \cdot (x_i - y)| + |\nu(y) \cdot (y - x_0)|) \\
&\le \sum_{i=1}^{d-1} |\alpha_i| \left(\frac{|x_i - y|^2}{2r_0} + \frac{|y - x_0|^2}{2r_0} \right).
\end{aligned}$$

Using further the estimate $|y - x_i| \le |y - x| + |x - x_i| \le 3\varepsilon + 5\varepsilon$ (which follows from (i) and (iii)) and (7.5), we get

$$|\nu(y) \cdot u| \le \frac{d-1}{d}\frac{64}{\theta r_0}\varepsilon =: q\varepsilon. \tag{7.6}$$

Since this is true for any unit vector $u \perp \nu(\mu)$, we infer

$$|\nu(\mu) \cdot \nu(y)| \ge \sqrt{1 - q^2\varepsilon^2}. \tag{7.7}$$

We distinguish now two possibilities. If $\nu(\mu) \cdot \nu(y) > 0$ then

$$|\nu(\mu) - \nu(y)|^2 = 2(1 - \nu(\mu) \cdot \nu(y)) \le 1 - \sqrt{1 - q^2\varepsilon^2} \le q^2\varepsilon^2 \tag{7.8}$$

if $\varepsilon < q^{-1}$. If, on the other hand, $\nu(\mu) \cdot \nu(y) \le 0$ then

$$|\nu(\mu) + \nu(y)|^2 = 2(1 + \nu(\mu) \cdot \nu(y)) \le 1 - \sqrt{1 - q^2\varepsilon^2} \le q^2\varepsilon^2 \tag{7.9}$$

if $\varepsilon < q^{-1}$. We will show that (7.9) is impossible. Since reach $Y^\delta \ge r_0$, we have again by Corollary 4.6

$$\nu(y) \cdot (y - x_d) \ge -\frac{|y - x_d|^2}{2r_0}.$$

Therefore,

$$\nu(y)\cdot(x_0 - x_d) = \nu(y)\cdot(x_0 - y) + \nu(y)\cdot(y - x_d) \ge -\frac{|x_0 - y|^2 + |y - x_d|^2}{2r_0} \ge -\frac{64\varepsilon^2}{r_0}.$$

On the other hand,

$$\nu(\mu) \cdot (x_0 - x_d) \ge h(\sigma) = d\rho(\sigma)\Theta(\sigma) \ge d\theta\varepsilon.$$

Thus, $(\nu(\mu) + \nu(y)) \cdot (x_0 - x_d) \ge \frac{d\theta}{2}\varepsilon$ if ε is small enough, hence, $|\nu(\mu) + \nu(y)| \ge \frac{d\theta}{2}|x_0 - x_d|^{-1}\varepsilon \ge \frac{d\theta}{10}$. This excludes (7.9) for sufficiently small $\varepsilon > 0$. If now μ, μ' are two intersecting facets, $x \in \mu \cap \mu'$ and $y = G^\varepsilon(x)$, (7.8) implies that

$$|\nu(\mu) - \nu(\mu')| \le |\nu(\mu) - \nu(y)| + |\nu(y) - \nu(\mu')| \le 2q\varepsilon,$$

which proves (vi).

We claim that G^ε must be a bijection. Indeed, if there would be two different points $x, x' \in \partial P^\varepsilon$ and $y := G^\varepsilon(x) = G^\varepsilon(x')$ then both x, x' would lie on a ray starting from y with direction $\nu(y)$. But then, either $\nu(\mu)$, or $\nu(\mu')$ would form an obtuse angle with $\nu(y)$, which would contradict (7.8).

Now we have from the area formula

$$\mathcal{H}^{d-1}(\partial Y_\delta) = \int_{\partial P^\varepsilon} J_{d-1}G^\varepsilon(x)\,\mathcal{H}^{d-1}(\mathrm{d}x),$$

hence,

$$\mathcal{H}^{d-1}(\partial Y_\delta) - \mathcal{H}^{d-1}(\partial P^\varepsilon) = \int_{\partial P^\varepsilon} (J_{d-1}G^\varepsilon(x) - 1)\,\mathcal{H}^{d-1}(\mathrm{d}x). \tag{7.10}$$

By (iii) and Lemma 4.7, $\mathrm{Lip}\,(G^\varepsilon) \le r_0/(r_0 - 3\varepsilon) \le 2$ and, hence, $\|\mathrm{ap}\,DG^\varepsilon(x)\| \le 2$ if $\varepsilon \le r_0/6$. We can express G^ε in the form

$$G^\varepsilon(x) = x \pm d(x)\nu(G^\varepsilon(x)),$$

where $d(x) := d_{\partial Y_\delta}(x)$ and $\nu(y) = \nu_{Y_\delta}(y)$. Recall that d is 1-Lipschitz, $\mathrm{grad}\,d(x) = \pm\nu(g(x))$ (Lemma 4.2) and $\|D\nu(y)\| \le r_0^{-1}$ (4.4). Differentiating, we get

$$\mathrm{ap}\,DG^\varepsilon(x) = \mathrm{id} \pm Dd(x)\nu(G^\varepsilon(x)) \pm d(x)D\nu(G^\varepsilon(x)) \circ \mathrm{ap}\,DG^\varepsilon(x),$$

and for any $u \in \mathrm{Tan}(P^\varepsilon, x)$

$$\begin{aligned} |\mathrm{ap}\,DG^\varepsilon(x)u - u| &\le |Dd(x)u| + |d(x)|\|D\nu(y)\|\|\mathrm{ap}\,DG^\varepsilon\||u| \\ &\le |\nu(y)\cdot u| + |d(x)|r_0^{-1}2 \\ &\le q\varepsilon + 3\varepsilon r_0^{-1}2, \end{aligned}$$

where we have used (7.6) and (v) in the last line. Thus, $|J_{d-1}G^\varepsilon(x) - 1| \le L'\varepsilon$ for some $L' > 0$ and for all sufficiently small $\varepsilon > 0$, and (7.10) implies

$$|\mathcal{H}^{d-1}(\partial Y_\delta) - \mathcal{H}^{d-1}(\partial P^\varepsilon)| \le L'\varepsilon\mathcal{H}^{d-1}(\partial P^\varepsilon).$$

Since this surely implies the uniform boundedness of $\mathcal{H}^{d-1}(\partial P^\varepsilon)$ in ε, (vii) follows. □

Proof (of Theorem 7.10) Let $\mathcal{PR}^+$ denote the family of all compact sets $X \subset \mathbb{R}^d$ which are parallel sets $X = Y_\delta$ of a set with positive reach and $0 < \delta < \mathrm{reach}\,Y$. Let a compact set $X \in \mathcal{U}_{\mathcal{PR}}$ and a regular sequence (ε_j) w.r.t. a $\mathcal{U}_{\mathcal{PR}}$-representation of X be given. We first show that there exists a sequence $X^j \in \mathcal{PR}^+$ such that

$$\lim_{j\to\infty} d_H(X^j, X) = 0, \quad (\mathbf{F})\lim_{j\to\infty} N_{X^j} = N_X \text{ and } \sup_j \mathbf{M}(N_{X^j}) < \infty. \tag{7.11}$$

Theorem 7.6 guaranties that (7.11) is true if we replace X^j with $Z^j := X_{\underset{\sim}{\varepsilon_j}}$. The sets Z^j need not have positive reach, but the closures to their complements, $\widetilde{Z^j}$, do, and we have $N_{\widetilde{Z^j}} = \rho_{\#} N_{Z^j}$, see Theorem 5.26. Applying Lemma 7.5 to the sets $\widetilde{Z^j}$, and using the reflection principle (Theorem 5.26) for the parallel sets $(\widetilde{Z^j})_{\delta_j}$ with some small $\delta_j > 0$, we obtain that (7.11) is true with X^j being the closure of the complement to $(\widetilde{Z^j})_{\delta_j}$, and it is easy to see that these sets belong to $\mathcal{PR}^+$.

In order to prove the theorem, it will thus be sufficient to show that for any compact set $X \in \mathcal{PR}^+$ there exists a sequence (P^j) of d-polytopes such that

$$\lim_{j\to\infty} d_H(P^j, X) = 0, \quad \textbf{(F)} \lim_{j\to\infty} N_{P^j} = N_X, \text{ and } \sup_j \mathbf{M}(N_{P^j}) < \infty.$$

Recall that X is a compact $C^{1,1}$-domain in $\mathbb{R}^d$ (Corollary 4.22) and let $\nu_X : \partial X \to S^{d-1}$ be its (Lipschitz) unit outer normal vector field. The unit normal bundle nor X is a Lipschitz submanifold of $\mathbb{R}^d \times \mathbb{R}^d$ with orienting $(d-1)$-vector field a_X and

$$N_X = (\mathcal{H}^{d-1} \llcorner \operatorname{nor} X)) \wedge a_X$$

(cf. Sect. 4).

In view of positive reach the components of the manifold ∂X, and thus of nor X, have distances greater than some positive constant. Therefore we can reduce the problem to the case when nor X is connected. In this situation the Constancy theorem for Lipschitz manifolds (Theorem 1.53) implies the following: If the current

$$T := (\mathcal{H}^{d-1} \llcorner \operatorname{nor} X) \wedge \phi\, a_X$$

for some integrable function ϕ is a cycle, i.e., $\partial T = 0$, then ϕ is constant, hence

$$T = \phi N_X\,.$$

Below T will be specified for our purposes.

So let $X = Y_\delta$ be such that $0 < \delta < \operatorname{reach} Y$, let $\varepsilon_0, \theta > 0$ and $P^\varepsilon : 0 < \varepsilon < \varepsilon_0$ be the approximating d-polytopes from Lemma 7.11 and let $Q, L > 0$ be the constants from Proposition 7.12; assume (without loss of generality) that

$$\varepsilon_0 < \tfrac{1}{4} \min\{\delta, \operatorname{reach} Y - \delta\}.$$

We will first verify condition (i) from Proposition 7.9. Let Σ^ε be locally finite simplicial d-complexes such that $P^\varepsilon = \bigcup \Sigma^\varepsilon$ and $\Theta(\sigma) \geq \theta$, $\sigma \in \Sigma^\varepsilon$. Let Σ_x^ε denote the set of simplices $\sigma \in \Sigma^\varepsilon$ containing a point $x \in \partial P^\varepsilon$. Using Remark 2.7(ii), we have $\mathcal{L}^d(\sigma) \geq \theta^{d-1}\varepsilon^d$. Since all edge lengths are at most 5ε, we get $\bigcup \Sigma_x^\varepsilon \subset B(x, 5\varepsilon)$. Consequently, the number of simplices in Σ_x^ε fulfills

$$\#\Sigma_x^\varepsilon \cdot \theta^{d-1}\varepsilon^d \leq \sum_{\sigma \in \Sigma_x^\varepsilon} \mathcal{L}^d(\sigma) = \mathcal{L}^d(\bigcup \Sigma_x^\varepsilon) \leq \omega_d (5\varepsilon)^d,$$

hence,

$$\#\Sigma_x^\varepsilon \le 5^d \omega_d \theta^{1-d}. \tag{7.12}$$

It follows, taking into account the definition and additivity of the index function i_{P^ε} (see Theorem 5.6), that it is uniformly bounded in ε. Therefore in order to verify (i) it suffices to show that

$$\sup_\varepsilon \mathcal{H}^{d-1}(\operatorname{nor} P^\varepsilon) < \infty.$$

Let $\Sigma_j^\varepsilon := \bigcup \mathcal{F}_j(\Sigma^\varepsilon)$ be the union of all j-faces of Σ^ε (j-skeleton), $j = 0, \dots, d-1$. Since $\Sigma_{d-1}^\varepsilon \subset \partial P^\varepsilon$ and $\mathcal{H}^{d-1}(\partial P^\varepsilon \setminus \Sigma_{d-1}^\varepsilon) = 0$, we get from Proposition 7.12(vii) that $\mathcal{H}^{d-1}(\Sigma_{d-1}^\varepsilon)$ is bounded from above by a constant $C > 0$ (independent of ε). As any facet F of Σ^ε has surface area at least

$$\mathcal{H}^{d-1}(F) \ge \theta^{d-2}\varepsilon^{d-1}$$

by the fatness condition, we get

$$\#\mathcal{F}_{d-1}(\Sigma^\varepsilon)\theta^{d-2}\varepsilon^{d-1} \le \mathcal{H}^{d-1}(\Sigma_{d-1}^\varepsilon) \le C,$$

which implies

$$\#\mathcal{F}_{d-1}(\Sigma^\varepsilon) \le C\theta^{2-d}\varepsilon^{-(d-1)}.$$

Further, for $0 \le j \le d-1$, each facet F has $\binom{d-1}{j}$ j-faces and each j-face has j-dimensional measure at most $c_j\varepsilon^j$ for some constant $c_j > 0$ (since the edge lengths are bounded by 5ε). Consequently,

$$\mathcal{H}^j(\Sigma_j^\varepsilon) \le C\theta^{2-d}\varepsilon^{-(d-1)} \cdot \binom{d-1}{j} c_j \varepsilon^j = \beta_j \varepsilon^{-d-1-j}$$

for some constants $\beta_j > 0$. On the other hand, using the property (vi) from Proposition 7.12, we see that for any $0 \le j \le d-1$ and for any face $F \in \mathcal{F}_j(\Sigma^\varepsilon)$, the set Γ_F has diameter less or equal to $Q\varepsilon$, hence,

$$\mathcal{H}^{d-1-j}(\Gamma_F) \le \omega_{d-1-j} Q^{d-1-j}\varepsilon^{d-1-j}.$$

Taking into account (7.4) we obtain

$$\mathcal{H}^{d-1}(\operatorname{nor} P^\varepsilon) \le \sum_{j=0}^{d-1} \beta_j Q^{d-1-j}\omega_{d-1-j},$$

proving (i).

Note that, since $X = Y_\delta$, we have

$$Y_r \setminus Y \subset \operatorname{Unp}(\partial X)$$

whenever $0 < r < \operatorname{reach} Y$. Further, for $0 < \varepsilon < \varepsilon_0$ we define the mappings $f^\varepsilon : \operatorname{nor} P^\varepsilon \to \mathbb{R}^d \times S^{d-1}$ by

$$f^\varepsilon : (x, n) \mapsto (\Pi_{\partial X}(x), \nu_X(\Pi_{\partial X}(x))).$$

We get from Lemma 7.11(v) that

$$\partial P^\varepsilon \subset Y_{(\operatorname{reach} Y+\delta)/2} \setminus Y_{\delta/2}$$

if $0 < \varepsilon < \varepsilon_0$, and the mapping $\Pi_{\partial X}$ is Lipschitz on $Y_{(\operatorname{reach} Y+\delta)/2} \setminus Y_{\delta/2}$ by Lemma 4.7. Since clearly ν_X is Lipschitz on $\partial X = \partial Y_\delta$, we get that the mappings f^ε are uniformly Lipschitz in ε, which is condition (ii) of Proposition 7.9.

Condition (iii) of Proposition 7.9 follows immediately from the construction since we have $|x - \Pi_{\partial X}(x)| < q\varepsilon$ whenever $x \in \partial P^\varepsilon$, by Lemma 7.11(v), and ν_X is continuous on ∂X. It remains to show that

$$(f^\varepsilon)_\# N_{P^\varepsilon} = N_X .$$

Recall that $N_{P^\varepsilon} = (\mathcal{H}^{d-1} \llcorner \operatorname{nor} P^\varepsilon) \wedge i_{P^\varepsilon} a_{P^\varepsilon}$ (for the index function i_{P^ε} and the orienting $(d-1)$-vector field a_{P^ε}) is a cycle. From this we infer $\partial((f^\varepsilon)_\# N_{P^\varepsilon}) = (f^\varepsilon)_\# \partial N_{P^\varepsilon} = 0$, i.e.,

$$T^\varepsilon := (f^\varepsilon)_\# N_{P^\varepsilon}$$

is also a cycle. Furthermore, the area formula for rectifiable currents yields

$$T^\varepsilon = (\mathcal{H}^{d-1} \llcorner \operatorname{nor} X) \wedge \eta^\varepsilon,$$

with

$$\begin{aligned} \eta^\varepsilon(y, m) &= \sum_{(x,n)\in(f^\varepsilon)^{-1}\{(y,m)\}} i_{P^\varepsilon}(x, n) \frac{(\bigwedge_{d-1} \operatorname{ap} Df^\varepsilon(x, n)) a_{P^\varepsilon}(x, n)}{|(\bigwedge_{d-1} \operatorname{ap} Df^\varepsilon(x, n)) a_{P^\varepsilon}(x, n)|} \\ &= \phi^\varepsilon(y, m) a_X(y, m) \end{aligned}$$

where

$$\phi^\varepsilon(y, m) := \sum_{(x,n)\in(f^\varepsilon)^{-1}\{(y,m)\}} i_{P^\varepsilon}(x, n) \operatorname{sgn}(f^\varepsilon; x, n)$$

and $\operatorname{sgn}(f^{\varepsilon};x,n)$ is the corresponding sign of orientation of the $(d-1)$-vector associated with $\operatorname{Tan}(\operatorname{nor}X, f^{\varepsilon}(x,n))$, i.e., $\operatorname{sgn}(f^{\varepsilon};x,n)$ equals 1 or -1 if $\frac{(\bigwedge_{d-1}\operatorname{ap} Df^{\varepsilon}(x,n))a_{P^{\varepsilon}}(x,n)}{|(\bigwedge_{d-1}\operatorname{ap} Df^{\varepsilon}(x,n))a_{P^{\varepsilon}}(x,n)|}$ agrees with $a_X(y,m)$ or $-a_X(y,m)$, respectively.

According to the arguments at the beginning of the proof the function ϕ^{ε} must be constant. Therefore it suffices to show that $\phi^{\varepsilon}(y,m)=1$ for a set of (y,m) with positive $\mathcal{H}^{d-1}$-measure. Then we obtain

$$T^{\varepsilon}=N_X\,.$$

To this aim we consider an open subset Δ of some facet μ of P^{ε} with outer unit normal n, such that the mapping f^{ε} on $\Delta\times\{n\}$ is bijective. For $x\in\Delta$ we have $i_{P^{\varepsilon}}(x,n)=1$ and $a_{P^{\varepsilon}}(x,n)=a_1\wedge\ldots\wedge a_{d-1}$ with $a_i=(b_i,0)$ for an orthonormal basis $b_1,\ldots,b_{d-1}$ in the (constant) tangent space to P^{ε} at x such that $\langle b_1\wedge\ldots\wedge b_{d-1}\wedge n,\Omega_d\rangle=1$. Then we obtain for these pairs (x,n), $\phi^{\varepsilon}(y,m)=\operatorname{sgn}(f^{\varepsilon};x,n)$ is the sign of the scalar product

$$\big((\textstyle\bigwedge_{d-1}\operatorname{ap} Df^{\varepsilon}(x,n))\,a_{P^{\varepsilon}}(x,n)\big)\cdot a_X(y,m)$$

and, applying the same procedure as in the proof of Theorem 7.3, we get (7.3) again, now with $s_i=0$, $i=1,\ldots,d-1$, and the sign will be positive. At the same time the area theorem implies that the set $f^{\varepsilon}(\Delta\times\{n\})\subset\operatorname{nor}X$ has positive Hausdorff measure.

Thus, all conditions of Proposition 7.9 are fulfilled and we get the desired convergence. □

7.3 Bibliographical Notes

1. A more direct proof of a consequence of Theorem 7.3 was given by Federer [Fed59, §5.9]. He showed, under the same assumptions, the vague (in his notation weak) convergence of curvature measures using the Steiner formula.
2. Approximation of $C^{1,1}$ surfaces by polyhedral ones and convergence of the associated curvature measures was shown by Cheeger, Müller and Schrader in [CMS84]. They triangulate the smooth surfaces and refine these triangulations in order to get "fat triangulations", where the approximation works. Fu [Fu93] obtained the corresponding approximation result for the associated normal cycles. Approximation of the normal cycles of C^2-submanifolds of $\mathbb{R}^d$ by polyhedra is also treated in [Mor08, §20.3].
3. Cohen-Steiner and Morvan [CSM06] proved a stability result for normal cycles, which yields for a smooth compact domain $X\subset\mathbb{R}^d$ and a compact d-polytope $P\subset\mathbb{R}^d$ such that ∂P lies within the distance reach ∂X from ∂X,

$$\mathbf{F}(N_X-N_P)\le(d_H(X,P)+\alpha(X,P))\,C_X(\mathbf{M}(N_P)+\mathbf{M}(\partial N_P)),$$

where $\alpha(X, P)$ is the supremum of angular differences $|\angle(\nu_X(\Pi_{\partial X}(y)), v)|$ over all $(y, v) \in \operatorname{nor} P$, and C_X is a constant depending on the second fundamental form of X. In fact, the result is proved in a significantly higher generality, in the Riemannian setting and for a much more general set P admitting a normal cycle, roughly in the sense of Definition 9.5.

4. The result of Fu & Scott [FS13] implies that if $X \subset \mathbb{R}^3$ is a parallel set to a compact set with positive reach ($X \in \mathcal{PR}^+$) then there exist approximating 3-polytopes P^j such that $P^j \stackrel{d_H}{\to} X$, (**F**) $\lim_{j\to\infty} N_{P^j} = N_X$, and

$$\sup_j \mathbf{M}(N_{P^j}) \leq C\, \mathbf{M}(N_X),$$

with a constant C (independent of X). Using this in the proof of Theorem 7.10 we can get approximating polytopes P^j to a compact $\mathcal{U}_{\mathcal{PR}}$ set X in $\mathbb{R}^3$ with a regular sequence (ε_j) (in particular, to a compact set with positive reach) such that, in addition, $\sup_j \mathbf{M}(N_{P^j}) < \infty$. This implies, in view of Proposition 7.1, that

$$\mathrm{w} - \lim_{j\to\infty} \widetilde{C}_k(P^j, \cdot) = \widetilde{C}_k(X, \cdot) \quad k = 0, 1, \ldots, d-1.$$

An extension to higher dimension remains open.

Chapter 8
Characterization Theorems

8.1 Characterization of Lipschitz-Killing Curvatures

In the classical case of convex geometry the Lipschitz-Killing curvatures and related measures can be characterized as certain *basic Euclidean invariants*, which underlines their geometric importance. Recall that a well-known result of Hadwiger (Theorem 2.1) states that any motion invariant continuous valuation on the space of compact convex sets is a linear combination of Minkowski's quermasssintegrals. This was localized by Schneider to the case of their curvature measures and surface area measures, see Theorems 2.2 and 2.3. The aim of the present chapter is to extend such characterizations to compact $\mathcal{U}_{\mathcal{PR}}$-sets. Here the main idea is to combine continuity with respect to the Hausdorff distance with flat convergence of the associated normal cycles. This enables us to reduce everything to the special case of polytopes via approximation. Moreover, in this way we can show that functionals or measures with such properties on polytopes have unique extensions to compact $\mathcal{U}_{\mathcal{PR}}$-sets.

First recall some classical notions:

Definition 8.1 A mapping Ψ defined on a space $\mathcal{S}$ of subsets of $\mathbb{R}^d$ and with values in a vector space is said to be a *valuation* or *additive* if

$$\Psi(A \cup B) = \Psi(A) + \Psi(B) - \Psi(A \cap B), \text{ provided } A, B, A \cap B, A \cup B \in \mathcal{S},$$

and $\Psi(\emptyset) = 0$ whenever $\emptyset \in \mathcal{S}$.

By induction we infer the so-called *inclusion-exclusion principle* for a valuation Ψ:

$$\Psi\Big(\bigcup_{i=1}^{n} A_i\Big) = \sum_{k=1}^{n}(-1)^{k-1} \sum_{1\leq i_1<\ldots<i_k\leq n} \Psi(A_{i_1} \cap \ldots \cap A_{i_k}). \tag{8.1}$$

provided all sets under consideration are elements of $\mathcal{S}$.

J. Rataj, M. Zähle, *Curvature Measures of Singular Sets*, Springer Monographs in Mathematics, https://doi.org/10.1007/978-3-030-18183-3_8

Definition 8.2 A family $\mathcal{S}$ of subsets of $\mathbb{R}^d$ is said to be *intersection stable* if $\emptyset \in \mathcal{S}$ and $A \cap B \in \mathcal{S}$ whenever $A, B \in \mathcal{S}$. If $\mathcal{S}$ is intersection stable we denote

$$\mathcal{L}(\mathcal{S}) := \{A_1 \cup \cdots \cup A_n : A_1, \ldots, A_n \in \mathcal{S},\ n \in \mathbb{N}\}$$

the *lattice* generated by $\mathcal{S}$. For a general (not necessarily intersection stable) family $\mathcal{S}$ we use the same notation for the *conditional lattice* generated by $\mathcal{S}$:

$$\begin{aligned}\mathcal{L}(\mathcal{S}) := \ &\{A_1 \cup \cdots \cup A_n : A_{i_1} \cap \cdots \cap A_{i_k} \in \mathcal{S},\\ &\qquad 1 \le i_1 < \cdots < i_k \le n,\ 1 \le k \le n\}.\end{aligned}$$

If the family $\mathcal{S}$ is invariant under Euclidean motions and spatial scaling the following notions are relevant.

Definition 8.3 A mapping $\Psi : \mathcal{S} \to \mathbb{R}$ is called
motion invariant if $\Psi \circ g = \Psi$ for all Euclidean motions g,
homogeneous of degree k if $\Psi(\lambda(\cdot)) = \lambda^k \Psi(\cdot)$, $\lambda > 0$.

Here we are interested in the following special cases for $\mathcal{S}$:

$\mathcal{P}$	–	convex polytopes (convex hulls of finite sets)
$\mathcal{C}$	–	convex compact sets
$\mathcal{L}(\mathcal{P})$	–	polytopes
$\mathcal{R} = \mathcal{L}(\mathcal{C})$	–	convex ring
$\mathcal{PR}^c$	–	compact sets of positive reach
$\mathcal{U}_{\mathcal{PR}}^c = \mathcal{L}(\mathcal{PR}^c)$	–	compact $\mathcal{U}_{\mathcal{PR}}$-sets.

The first auxiliary result from convex geometry is an extension theorem for polytopes.

Proposition 8.4 *Any valuation* Ψ *on* $\mathcal{P}$ *has a unique additive extension to* $\mathcal{L}(\mathcal{P})$ *according to the inclusion-exclusion principle.*

Proof If a (nonconvex) polytope $P \in \mathcal{L}(\mathcal{P})$ has a representation $P = P_1 \cup \cdots \cup P_n$ with $P_1, \ldots, P_n \in \mathcal{P}$, we clearly wish to define the extension using the inclusion-exclusion formula (8.1), i.e.,

$$\Psi(P) := \sum_I (-1)^{|I|-1} \Psi\left(\bigcup_{i \in I} P_i\right),$$

where the summation is taken over all index sets $I \subset \{1, \ldots, n\}$ and $|I|$ denotes the cardinality of I. We have to check that this value does not depend on the particular representation of P. Do do this, let $P = Q_1 \cup \cdots \cup Q_m$ be another representation with $Q_1, \dots . Q_m \in \mathcal{P}$, and denote

$$\widetilde{\Psi}(P) := \sum_J (-1)^{|J|-1} \Psi\left(\bigcup_{j \in J} Q_j\right)$$

(here, of course, the summation is taken over all subsets $J \subset \{1, \dots, m\}$). For any $I \subset \{1, \dots, n\}$, we can write

$$\bigcap_{i \in I} P_i = \bigcup_{j=1}^{m} \bigcap_{i \in I} (P_i \cap Q_j)$$

and the inclusion-exclusion formula yields

$$\Psi\left(\bigcap_{i \in I} P_i\right) = \sum_J (-1)^{|J|-1} \Psi\left(\bigcap_{i \in I} \bigcap_{j \in J} (P_i \cap Q_j)\right),$$

hence,

$$\Psi(P) = \sum_I \sum_J (-1)^{|I|+|J|} \Psi\left(\bigcap_{i \in I} \bigcap_{j \in J} (P_i \cap Q_j)\right).$$

Due to the symmetry of the last expression, we would obtain the same when expressing $\widetilde{\Psi}(P)$. Thus the extension is defined uniquely and it is additive, again due to the inclusion-exclusion formula. □

A main tool for our purposes is the following classical version of characterizing the intrinsic volumes. We formulate it in the notation of the Lipschitz-Killing curvature functionals $\mathbf{C}_k$ including the volume functional $\mathbf{C}_d$. Here *continuity with respect to the Hausdorff distance* is an important assumption. It should be noted that the Hausdorff distance is a metric only on $\mathcal{C}' = \mathcal{C} \setminus \{\emptyset\}$ (or $\mathcal{P}' = \mathcal{P} \setminus \{\emptyset\}$); we consider the topology on $\mathcal{C}$ (or $\mathcal{P}$) with $\emptyset$ being an isolated point. Hence, in the sequel the corresponding *continuity condition concerns only nonempty sets*.

Proposition 8.5

(i) *Any motion invariant continuous valuation Ψ on $\mathcal{C}$ has the unique representation*

$$\Psi = \sum_{k=0}^{d} c_k \, \mathbf{C}_k$$

for some real constants c_k. For $\Psi \geq 0$ we have $c_k \geq 0$, $k = 0, \dots, d$.

(ii) *If, additionally, Ψ is homogeneous of degree k, then $\Psi = c_k \, \mathbf{C}_k$.*

(iii) *Ψ has a unique additive extension to $\mathcal{L}(\mathcal{P})$ with the same representation as in* (i) *and* (ii).

Proof

(i) is Hadwiger's characterization Theorem 2.1 for the case of $\mathcal{C}'$. By definition, $\Psi(\emptyset) = 0$, which yields the desired equality for the empty set.
(ii) is an easy and well-known consequence, using the homogeneity properties of the Lipschitz-Killing curvatures (see Proposition 4.47).
(iii) follows from (i) together with Proposition 8.4 and additivity of the $\mathbf{C}_k$.

□

In order to extend this to more general sets we first introduce the essential notions of *polytopal continuity* and *flat continuity*. Recall Definition 1.58 for flat convergence of currents.

Definition 8.6 Suppose that $\mathcal{P} \subset \mathcal{S} \subset \mathcal{U}_{\mathcal{PR}}^c$. Then a valuation Ψ on $\mathcal{S}$ is said to be

(i) *F-continuous at $X \in \mathcal{S}$*, if for any sequence $X_n \in \mathcal{S}$ converging to X in the Hausdorff distance

$$(\mathbf{F}) \lim_{n\to\infty} N_{X_n} = N_X\,, \quad \text{implies} \quad \lim_{n\to\infty} \Psi(X_n) = \Psi(X)\,,$$

(ii) *P-continuous at $X \in \mathcal{S}$*, if (i) is fulfilled for polytopes $X_n = P_n \in \mathcal{L}(\mathcal{P})$ (and the additive extension of Ψ to $\mathcal{L}(\mathcal{P})$ if necessary).

Remark 8.7

1. In the case $\mathcal{S} := \mathcal{C}$ the notion of F-convergence agrees with convergence in the Hausdorff distance. (Cf. Theorem 7.3 which also works without ε-parametrization of the sets.) Therefore our approach is consistent with that from classical convex geometry. However, it is easy to see that continuity with respect to the Hausdorff distance does not fit to the problem already in the case of non-convex polytopes.
2. By Proposition 7.1 the curvature functionals $\mathbf{C}_k$, $k = 0, \ldots, d-1$, are F-continuous in the general case. For $k = d$ this follows from continuity of the Lebesgue measure with respect to the Hausdorff distance.

Basing on the above notions the approximation results from the previous chapter enable us to reduce the characterization of motion invariant F-continuous valuations on $\mathcal{S} := \mathcal{U}_{\mathcal{PR}}^c$ to the classical case of polytopes.

Theorem 8.8 (Characterization of Lipschitz-Killing curvatures) *Let Ψ be a motion invariant valuation on $\mathcal{C}$, which is continuous with respect to the Hausdorff distance. Then it admits a unique extension to a motion invariant valuation on $\mathcal{U}_{\mathcal{PR}}^c$ being P-continuous at all nonempty $X \in \mathcal{PR}^c$. It has the representation*

$$\Psi(X) = \sum_{k=0}^{d} c_k\, \mathbf{C}_k(X), \qquad X \in \mathcal{U}_{\mathcal{PR}}^c,$$

for some uniquely determined real constants c_k and the Lipschitz-Killing curvatures $\mathbf{C}_k$. Therefore Ψ is F-continuous. $\Psi(P) \geq 0$, $P \in \mathcal{P}$, implies $c_k \geq 0$ for all k. If Ψ is homogeneous of degree k, we get $\Psi(X) = c_k \, \mathbf{C}_k(X)$.

Proof Applying Proposition 8.5 we first obtain the unique extension to general polytopes

$$\widetilde{\Psi}(P) = \sum_{k=0}^{d} c_k \, \mathbf{C}_k(P) \,, \quad P \in \mathcal{L}(\mathcal{P}) \,,$$

where the constants c_k fulfill the desired properties. Furthermore, the curvature functionals $\mathbf{C}_k$ are F-continuous motion invariant valuations on $\mathcal{U}^c_{\mathcal{PR}}$. Therefore the functional

$$\Psi(X) := \sum_{k=0}^{d} c_k \, \mathbf{C}_k(X) \,, \quad X \in \mathcal{U}^c_{\mathcal{PR}} \,,$$

possesses the same properties, and it is an extension of Ψ.

Let now $\widetilde{\Psi}$ be any extension with the desired properties and $X \in \mathcal{PR}^c \setminus \emptyset$. According to Theorem 7.10 there exists a sequence of polytopes P_n converging to X in the Hausdorff metric, such that

$$\text{(F)} \lim_{n \to \infty} N_{P_n} = N_X \,.$$

Since $\widetilde{\Psi}$ is assumed to be $\mathcal{P}$-continuous on $\mathcal{PR}^c$, we get

$$\lim_{n \to \infty} \widetilde{\Psi}(P_n) = \widetilde{\Psi}(X) \,.$$

Furthermore, by the above arguments,

$$\widetilde{\Psi}(P_n) = \sum_{k=0}^{d} c_k \, \mathbf{C}_k(P_n) \,.$$

The summands on the right-hand side converge to $c_k \, \mathbf{C}_k(X)$ as $n \to \infty$ by F-continuity of the $\mathbf{C}_k$. Hence,

$$\widetilde{\Psi}(X) = \sum_{k=0}^{d} c_k \, \mathbf{C}_k(X) = \widetilde{\Psi}(X) \,, \quad X \in \mathcal{PR}^c \,.$$

The corresponding equality for all $X \in \mathcal{U}^c_{\mathcal{PR}}$ is a consequence of the inclusion-exclusion principle for $\widetilde{\Psi}$ and Ψ. Hence, such an extension is unique. □

An immediate consequence of Theorem 8.8 is the following.

Corollary 8.9 *Let Ψ be a motion invariant valuation on $\mathcal{S}$ and suppose that one of the following additional three conditions is fulfilled:*

(i) $\mathcal{S} = \mathcal{C}$ *and Ψ is continuous with respect to the Hausdorff distance.*
(ii) $\mathcal{S} = \mathcal{PR}^c$ *and Ψ is P-continuous (for the unique additive extension of Ψ from $\mathcal{P}$ to $\mathcal{L}(\mathcal{P})$).*
(iii) $\mathcal{S} = \mathcal{U}^c_{\mathcal{PR}}$ *and Ψ is F-continuous.*

Then then one obtains the unique representation

$$\Psi = \sum_{k=0}^{d} c_k \, \mathbf{C}_k$$

with constants c_k as above.

It is an open problem, whether in (ii) the F-continuity on $\mathcal{PR}^c$ is sufficient.

8.2 Characterization of Associated Measures

The Characterization Theorem 8.8 for the (total) Lipschitz-Killing curvatures can be localized. Here we derive three measure versions. The first one concerns the basic curvature-direction measures $\widetilde{C}_k(X, \cdot)$ on $\mathbb{R}^d \times S^{d-1}$ and the other two their marginal measures on $\mathbb{R}^d$ and on S^{d-1}, respectively. In the first case the proof is more direct.

Denote the vector spaces of *finite signed measures* on a metric space $\mathcal{Y}$ by $\mathcal{M}(\mathcal{Y})$.

The notion of $\mathcal{M}(\mathcal{Y})$*-valued valuation* Ψ on some space $\mathcal{S} \subset \mathcal{U}^c_{\mathcal{PR}}$ is analogous to the real-valued case in Definition 8.1. We use the notation $\Psi(X, \cdot)$. In the corresponding notions of continuity we additionally suppose that the mass norms $\mathbf{M}$ of the approximating sequences are uniformly bounded.

Definition 8.10 Suppose that $\mathcal{P} \subset \mathcal{S} \subset \mathcal{U}^c_{\mathcal{PR}}$. Then a valuation $\Psi : \mathcal{S} \to \mathcal{M}(\mathcal{Y})$ is said to be

(i) *F-continuous at $X \in \mathcal{S}$*, if for any sequence $X_n \in \mathcal{S}$ converging to X in the Hausdorff distance, such that $\sup_n \mathbf{M}(N_{X_n}) < \infty$ and

$$\textbf{(F)} \lim_{n\to\infty} N_{X_n} = N_X \,, \quad \text{we get w} - \lim_{n\to\infty} \Psi(X_n, \cdot) = \Psi(X, \cdot) \,,$$

(ii) *P-continuous at $X \in \mathcal{S}$*, if (i) is fulfilled for polytopes $X_n = P_n \in \mathcal{L}(\mathcal{P})$ (and the additive extension of Ψ to $\mathcal{L}(\mathcal{P})$ if necessary).

Again the empty set is considered as an isolated point, so that the notions of continuity concern only nonempty sets from $\mathcal{S}$.

We are interested in the cases $\mathcal{Y} := \mathbb{R}^d \times S^{d-1}$, $\mathcal{Y} := \mathbb{R}^d$ and $\mathcal{Y} := S^{d-1}$.

Finally, as in previous chapters the action of the group $\mathcal{G}_d$ of Euclidean motions is transferred to $\mathbb{R}^d \times S^{d-1}$ and to S^{d-1} by means of the orthogonal component g^o of $g \in \mathcal{G}_d$. For brevity we use here the same symbols:

$$g(x,n) := (gx, g^o n),\ (x,n) \in \mathbb{R}^d \times S^{d-1}, \text{ and } gn := g^o n, \ , \ n \in S^{d-1}.$$

Similarly, we write for spatial scaling $\lambda(x,n) := (\lambda x, n)$, $(x,n) \in \mathbb{R}^d \times S^{d-1}$, $\lambda > 0$.

Definition 8.11 A mapping $\Psi : \mathcal{S} \to \mathcal{M}(\mathcal{Y})$ is

(i) *motion covariant* if $\Psi(gX, g(\cdot)) = \Psi(X, \cdot)$, $g \in \mathcal{G}_d$, $X \in \mathcal{S}$,
(ii) *homogeneous of degree* k if $\Psi(\lambda X, \lambda(\cdot)) = \lambda^k \Psi(X, \cdot)$, $\lambda > 0$, $X \in \mathcal{S}$, for $\mathcal{Y} = \mathbb{R}^d \times S^{d-1}$ or $\mathcal{Y} = \mathbb{R}^d$,
(iii) *locally determined* if for all Borel sets $B \subset \mathcal{Y}$ and $X, Y \in \mathcal{S}$,

$$N_X \llcorner \mathbf{1}_{\widetilde{B}} = N_Y \llcorner \mathbf{1}_{\widetilde{B}} \text{ implies } \Psi(X, B) = \Psi(Y, B),$$

where $\widetilde{B} := B$ for $\mathcal{Y} = \mathbb{R}^d \times S^{d-1}$, $\widetilde{B} := B \times S^{d-1}$ for $\mathcal{Y} = \mathbb{R}^d$, and $\widetilde{B} := \mathbb{R}^d \times B$ for $\mathcal{Y} = S^{d-1}$.

We first consider the version $\mathcal{Y} := \mathbb{R}^d \times S^{d-1}$ concerning base points and normal directions, i.e., the *curvature-direction measures* $\widetilde{C}_k(X, \cdot)$, $k = 0, \ldots, d-1$. For the special case of convex polytopes, where $\mathcal{S} = \mathcal{P}$, the well-known characterizations of Lebesgue measure and spherical Lebesgue measure can be used and no additivity or continuity assumption is needed.

Lemma 8.12 (Characterization of curvature-direction measures for polytopes)

(i) *Let* $\Psi : \mathcal{P} \to \mathcal{M}(\mathbb{R}^d \times S^{d-1})$ *be motion covariant and locally determined. Then it has the unique representation*

$$\Psi(P, \cdot) = \sum_{k=0}^{d-1} c_k \widetilde{C}_k(P, \cdot)$$

for some real constants c_k. *If* Ψ *is homogeneous of degree* k, *then* $\Psi(P, \cdot) = c_k \widetilde{C}_k(P, \cdot)$.
(ii) Ψ *has a unique extension to a motion covariant locally determined valuation on* $\mathcal{L}(\mathcal{P})$. *It has the same representation as in* (i), hence it is $\mathcal{P}$-continuous.

Proof (i) In order to show that $\Psi(\emptyset, \cdot) = 0$ we use that the marginal measure $\Psi(\emptyset, (\cdot) \times S^{d-1})$ on $\mathbb{R}^d$ is translation invariant and hence, a constant multiple of Lebesgue measure. Since it is finite, the constant must be 0, which implies the above assertion.

For arbitrary $P \in \mathcal{P}$ the locality of Ψ yields that $\operatorname{spt} \Psi(P, \cdot) \subset \operatorname{nor} P$. Fix now an affine subspace $L \in \mathcal{A}(d,k)$. Then for Borel sets $A \subset L$, bounded, and $B \subset$

$S^{d-1} \cap L^{\perp}$ the values $\Psi(P, A \times B)$ do not depend on P such that $A \times B \subset \operatorname{nor} P$, again by the locality of Ψ. Denote $\mu_L(A \times B) := \Psi(P, A \times B)$ for such P. From the motion invariance of Ψ we obtain

$$\mu_L(A \times B) = c_L \mathcal{L}^k(A)\mathcal{H}^{d-1-k}(B),$$

for some $c_L \in \mathbb{R}$, and that $c_L =: c_k$ is independent of $L \in \mathcal{A}(d,k)$. For a k-face F of the convex polytope P denote the set of unit vectors from the normal cone of P at arbitrary points of $\operatorname{rel\,int} F$ by $\operatorname{nor}(P, F)$. Then we infer $\Psi(P, A \times B) = c_k \mathcal{L}^d(A)\mathcal{H}^{d-1-k}(B)$ if $A \subset \operatorname{rel\,int} F$ and $B \subset \operatorname{nor}(P, F)$. Moreover,

$$\operatorname{nor} P = \bigcup_{k=0}^{d-1} \bigcup_{F \in \mathcal{F}_k(P)} \operatorname{rel\,int} F \times \operatorname{nor}(P, F)$$

is a disjoint union and consequently,

$$\begin{aligned}\Psi(P, A \times B) &= \sum_{k=0}^{d-1} c_k \sum_{F \in \mathcal{F}_k(P)} \mathcal{L}^k(F \cap A)\mathcal{H}^{d-1-k}(\operatorname{nor}(P, F) \cap B) \\ &= \sum_{k=0}^{d-1} c_k \widetilde{C}_k(P, A \times B)\end{aligned}$$

for all Borel sets $A \subset \mathbb{R}^d$ and $B \subset S^{d-1}$. Since such product sets form an intersection stable generator of the Borel σ-algebra on $\mathbb{R}^d \times S^{d-1}$ we get

$$\Psi(P, \cdot) = \sum_{k=0}^{d-1} c_k \widetilde{C}_k(P, \cdot).$$

If Ψ is homogeneous of degree k we infer $c_j = 0$ for $j \neq k$ because of the scaling properties of the $\widetilde{C}_j$.

(ii) The right-hand side of the above representation is defined for all polytopes, which provides an extension to a P-continuous motion covariant locally determined valuation on $\mathcal{L}(\mathcal{P})$. Uniqueness follows then from the convex case by the inclusion-exclusion principle applied to an arbitrary extension with these properties. □

Using the approximation results from the last section the general case for $\mathcal{U}^c_{\mathcal{PR}}$-sets is a consequence. Recall that $\mathcal{PR}^+$ denotes the set of $X \subset \mathbb{R}^d$ such that $X = Y_\delta$ for some $Y \in \mathcal{PR}^c$ and $0 < \delta < \operatorname{reach} Y$.

Theorem 8.13 (Characterization of curvature-direction measures) *Let* Ψ : $\mathcal{P} \to \mathcal{M}(\mathbb{R}^d \times S^{d-1})$ *be motion covariant and locally determined. Then it admits a unique extension to a motion covariant valuation on* $\mathcal{U}^c_{\mathcal{PR}}$ *being* P*-continuous on*

$\mathcal{PR}^+$ and F-continuous on $\mathcal{PR}^c$. It has the representation

$$\Psi(X,\cdot) = \sum_{k=0}^{d-1} c_k\, \widetilde{C}_k(X,\cdot), \qquad X \in \mathcal{U}^c_{\mathcal{PR}},$$

for some uniquely determined real constants c_k and the curvature-direction measures $\widetilde{C}_k(X,\cdot)$. Therefore Ψ is F-continuous on $\mathcal{U}^c_{\mathcal{PR}}$.

If Ψ is homogeneous of degree k, we get $\Psi(X,\cdot) = c_k\, \widetilde{C}_k(X,\cdot)$ for all X.

Proof Here the arguments for measures are similar as in the proof of Theorem 8.8 for functionals. Lemma 8.12 provides the corresponding extension

$$\Psi(P,\cdot) = \sum_{k=0}^{d-1} c_k\, \widetilde{C}_k(P,\cdot)$$

for general polytopes $P \in \mathcal{L}(\mathcal{P})$. According to Theorem 7.10 for any $X \in \mathcal{PR}^+$ there exists a sequence $P^j \in \mathcal{L}(\mathcal{P})$ converging to X in the Hausdorff distance such that $\sup_j \mathbf{M}(N_{P^j}) < \infty$, (F) $\lim_{j\to\infty} N_{P^j} = N_X$ and $\mathrm{w}-\lim_{j\to\infty} \widetilde{C}_k(P^j,\cdot) = \widetilde{C}_k(X,\cdot)$, $k = 0,\dots d-1$. By P-continuity of Ψ we infer $\mathrm{w}-\lim_{j\to\infty} \Psi(P^j,\cdot) = \Psi(X,\cdot)$. This yields the required extension for $X \in \mathcal{PR}^+$.

Next one uses the approximation of arbitrary compact $\mathcal{PR}$-sets by parallel sets in the sense of Lemma 7.5 together with F-continuity in order to obtain the corresponding extension of $\Psi$to $\mathcal{PR}^c$. The latter together with the inclusion-exclusion principle resulting from additivity leads to the general version for $\mathcal{U}^c_{\mathcal{PR}}$. □

Remark 8.14

1. The above proof shows that the condition on F-continuity of the extension of Ψ to $X \in \mathcal{PR}^c$ could be replaced by the version where the approximating sets X_n in the definition are parallel sets of of X with small distances. F-continuity is in this case a consequence.
2. For the total values $\Psi(\mathbb{R}^d \times S^{d-1})$ the results are consistent with those from the previous section.

As in Corollary 8.9 the last theorem implies the following.

Corollary 8.15 *Let $\Psi : \mathcal{S} \to \mathcal{M}(\mathbb{R}^d \times S^{d-1})$ be a motion covariant locally determined valuation on $\mathcal{S}$ and suppose that one of the following three additional conditions is fulfilled:*

(i) *$\mathcal{S} = \mathcal{C}$ and Ψ is weakly continuous with respect to the Hausdorff distance.*
(ii) *$\mathcal{S} = \mathcal{PR}^c$, Ψ is F-continuous, and it is P-continuous at all sets from $\mathcal{PR}^+$ (for the unique additive extension of Ψ from $\mathcal{P}$ to $\mathcal{L}(\mathcal{P})$).*
(iii) *$\mathcal{S} = \mathcal{U}^c_{\mathcal{PR}}$ and Ψ is F-continuous.*

Then Ψ admits the unique representation

$$\Psi(X, \cdot) = \sum_{k=0}^{d-1} c_k\, \widetilde{C}_k(X, \cdot)\,, \quad X \in \mathcal{S}\,,$$

with constants c_k as above.

Next we consider the marginal *Lipschitz-Killing curvature measures*

$$C_k(X, \cdot) = \widetilde{C}_k(X, (\cdot) \times S^{d-1})\,, \quad k = 0, \ldots, d-1\,,$$

and include again the Lebesgue measure, restricted to X, in form of $C_d(X, \cdot)$.

Recall that for convex polytopes weak continuity with respect to the Hausdorff distance agrees with F-continuity.

Theorem 8.16 (Characterization of curvature measures) *Let $\Psi : \mathcal{P} \to \mathcal{M}(\mathbb{R}^d)$ be a motion covariant locally determined valuation, which is weakly continuous with respect to the Hausdorff distance. Suppose additionally, that $\Psi(P, \cdot) \geq 0$, $P \in \mathcal{P}$. Then it admits a unique extension to a valuation on $\mathcal{U}^c_{\mathcal{PR}}$ being P-continuous on $\mathcal{PR}^+$ and F-continuous on $\mathcal{PR}^c$. It has the representation*

$$\Psi(X, \cdot) = \sum_{k=0}^{d} c_k\, C_k(X, \cdot), \qquad X \in \mathcal{U}^c_{\mathcal{PR}},$$

for some uniquely determined constants $c_k \geq 0$. Therefore Ψ is an $\mathcal{M}(\mathbb{R}^d)$-valued F-continuous motion covariant locally determined valuation on $\mathcal{U}^c_{\mathcal{PR}}$.

If Ψ is homogeneous of degree k, we get $\Psi(X, \cdot) = c_k\, C_k(X, \cdot)$ for all X.

Proof Theorem 2.2 provides the unique representation

$$\Psi(P, \cdot) = \sum_{k=0}^{d} c_k\, C_k(P, \cdot)$$

for nonempty convex polytopes P. (For the empty set this is trivial, since both sides of the equation vanish.) Its proof is more involved than that of Lemma 8.12, see [Sch78, 6.1]. Note that the notion of locality used there is equivalent to that from Definition 8.11 (iii). For arbitrary polytopes the equality follows by additive extension on both sides.

The remaining part concerning approximations and additive extension is as in the proof of Theorem 8.13. □

Finally, the *direction versions*, i.e., the marginal measures with respect to the normal components, are denoted by

$$S_k(X, \cdot) := \widetilde{C}_k(X, \mathbb{R}^d \times (\cdot))\,, \quad k = 0, \ldots, d-1\,.$$

Recall that in the convex case they agree with the *k-th order surface area measures*.

Theorem 8.17 (Characterization of direction measures) *Let $\Psi : \mathcal{C} \to \mathcal{M}(\mathbb{R}^d)$ be a motion covariant locally determined valuation, which is weakly continuous with respect to the Hausdorff distance. Then it admits a unique extension to a valuation on $\mathcal{U}^c_{\mathcal{PR}}$ being being P-continuous on $\mathcal{PR}^+$ and F-continuous on $\mathcal{PR}^c$. It has the representation*

$$\Psi(X, \cdot) = \sum_{k=0}^{d-1} c_k \, S_k(X, \cdot), \quad X \in \mathcal{U}^c_{\mathcal{PR}},$$

for some uniquely determined real constants. Therefore Ψ is an $\mathcal{M}(S^{d-1})$-valued F-continuous motion covariant locally determined valuation.

Proof Here the representation for convex bodies is given by Theorem 2.3 (see Schneider [Sch75]). The remaining arguments are as above. □

Remark 8.18

1. As in Corollary 8.15 the corresponding characterizations of the curvature measures C_k or the direction measures S_k on the spaces $\mathcal{C}$, $\mathcal{PR}^c$ and $\mathcal{U}^c_{\mathcal{PR}}$ are consequences of Theorems 8.16 and 8.17, respectively.
2. Theorem 8.16 is formulated under the additional condition of positivity of Ψ on $\mathcal{P}$. The question, whether arbitrary linear combinations of the Lipschitz-Killing curvature measures can be characterized in such a way, may be reduced to the following classical problem, which is still unsolved: Is any simple rotation invariant real valuation on spherical polytopes P a multiple of the corresponding $(d-1)$-volume $\mathcal{H}^{d-1}(P)$? (*Simple* means here vanishing on lower-dimensional polytopes. See proof of Theorem 6.1 in Schneider [Sch78].) For the case of non-negative functionals a proof of this relationship was given [Sch78, 6.2].
3. All results of this chapter concerning $\mathcal{U}^c_{\mathcal{PR}}$-sets can be extended to more general classes of sets from the next chapter with associated normal cycles, provided they admit polytopal approximations as in Theorem 7.10. Therefore the Lipschitz-Killing curvatures and the curvature-direction measures as well as both their marginal variants can be considered as *complete systems of Euclidean invariants, which are F-continuous (locally determined measure-valued) valuations.* (For the case of the curvature measures the above positivity is still assumed.)

8.3 Bibliographical Notes

1. Hadwiger's characterization Theorem 2.1 is a special case of Corollary 8.9. It has been used in the auxiliary Proposition 8.5 (i).
2. The characterization of the direction measures S_k on the space $\mathcal{C}$ in the sense of Remark 8.18(i) can be found in Schneider [Sch75].

3. The corresponding version for the curvature measures C_k on $\mathcal{C}$ was shown in Schneider [Sch78].
4. Federer [Fed59, 5.17] already posed the problem of a suitable characterization of the curvature measures for $\mathcal{PR}$-sets, where a corresponding notion of continuity was not yet developed.
5. The unique additive extendability of a valuation on $\mathcal{P}$ (Proposition 8.4) was shown by Volland [Vol57]. Later, Groemer [Gro78] proved that any continuous valuation on $\mathcal{C}$ has a unique extension to a valuation on the convex ring $\mathcal{L}(\mathcal{C})$. See also [Sch14, §6.2] for an overview and further references.
6. The characterization of the curvature-direction measures for convex polytopes (cf. Lemma 8.12 (i)) was given in Glasauer [Gla97].

Chapter 9
Extensions of Curvature Measures to Larger Set Classes

So far we have introduced curvature measures for sets with positive reach and their locally finite unions (such that any finite intersection has positive reach). This setting is still not satisfactory since it does not encompass some natural set classes as closures of complements to convex bodies, or boundaries of convex bodies, though these sets should apparently admit a natural definition of curvatures.

The aim of this chapter is to show how the technique of normal cycles and the compactness theorem for currents can be used to extend curvature measures in a unique way to more general classes, covering the examples mentioned above. Different extensions appeared up to now in the literature and they are usually technically demanding. We aim rather to present the method than to give an extension as wide as possible (which seems to be an open question up to now).

The basic idea is as follows. We approximate a given set $X \subset \mathbb{R}^d$ (and we will restrict ourselves to compact sets in this chapter) by appropriate "nice" sets X^i (smooth sets, sets with positive reach etc.) for which the normal cycles N_{X^i} are defined. If one can show that the normal cycles are bounded as currents in the mass norm

$$\sup_i \mathbf{M}(N_{X^i}) < \infty, \tag{9.1}$$

then the Compactness theorem for currents (Theorem 1.57) implies that there exists a subsequence $N_{X^{i_k}}$ converging to an integral current T in the flat seminorm. Moreover, T retains some other important geometric properties of the normal cycles N_{X^i}, namely it is a cycle ($\partial T = 0$) and it is Legendrian ($T \llcorner \alpha = 0$, cf. Proposition 4.40). Of course, if this limit would change from subsequence to subsequence, we could not expect it to be a good candidate for a normal cycle of X. Nevertheless, there is a uniqueness theorem due to J. Fu [Fu89b] (cf. Theorem 9.4 below with another proof) saying that there is at most one integral cycle with the Legendrian property and with another property linking to topological properties of the set (Euler-Poincaré characteristic of intersection with halfspaces). This makes it

J. Rataj, M. Zähle, *Curvature Measures of Singular Sets*, Springer Monographs in Mathematics, https://doi.org/10.1007/978-3-030-18183-3_9

possible to introduce normal cycles (and, hence, also curvature measures, through the relation given in Theorem 4.42) in a unique way, and guaranteeing even a local version of the Gauss-Bonnet formula.

9.1 Legendrian Cycles

Definition 9.1 A *Legendrian cycle* in $\mathbb{R}^d$ is a locally integral $(d-1)$-current $T \in \mathbf{I}^{\text{loc}}_{d-1}(\mathbb{R}^d \times \mathbb{R}^d)$ with the following properties:

$$\operatorname{spt} T \subset \mathbb{R}^d \times S^{d-1}, \tag{9.2}$$

$$\partial T = 0 \quad (T \text{ is a cycle}), \tag{9.3}$$

$$T \llcorner \alpha = 0 \quad (T \text{ is Legendrian}), \tag{9.4}$$

where α is the *contact* 1*-form* in $\mathbb{R}^{2d}$ acting as

$$\langle (u, v), \alpha(x, n)\rangle = u \cdot n, \qquad u, v, x, n \in \mathbb{R}^d.$$

Any locally integral current $T \in \mathbf{I}^{\text{loc}}_{d-1}(\mathbb{R}^d \times \mathbb{R}^d)$ is integer-multiplicity locally $(d-1)$-rectifiable, hence, it has an integral representation of the form

$$T = (\mathcal{H}^{d-1} \llcorner W_T)\, i_T \wedge a_T, \tag{9.5}$$

with a $\mathcal{H}^{d-1}$-measurable and locally $(d-1)$-rectifiable set $W_T \subset \mathbb{R}^d \times S^{d-1}$, a unit simple measurable $(d-1)$-vector field a_T on W_T (prescribing an orientation of W_T) and an integer-valued $\mathcal{H}^{d-1}$-integrable function i_T on W_T (index function), cf. Definitions 1.55 and 1.43.

The exterior derivative of the contact form α is the *symplectic form* $\omega := d\alpha$ which is a constant 2-form in $\mathbb{R}^d \times \mathbb{R}^d$ acting as

$$\langle (u, v) \wedge (u', v'), \omega\rangle = u \cdot v' - v \cdot u'. \tag{9.6}$$

The standard description using the notation $dx_i = (e_i, 0)^*$, $dy_i = (0, e_i)^*$, $i = 1, \dots, d$, is

$$\omega = \sum_{i=1}^{d} dx_i \wedge dy_i.$$

A Legendrian cycle T clearly satisfies

$$T \llcorner \omega = T \llcorner d\alpha = \partial(T \llcorner \alpha) = 0, \tag{9.7}$$

and this is called *Lagrangian property*. Using the representation (9.5) of T, we get that

$$T \llcorner \omega = \mu \wedge \xi$$

is representable by integration, with

$$\mathrm{d}\mu = i_T \, |a_T \llcorner \omega| \, \mathrm{d}(\mathcal{H}^{d-1} \llcorner W_T)$$

and

$$\xi = \frac{a_T \llcorner \omega}{|a_T \llcorner \omega|}$$

if $|a_T \llcorner \omega| > 0$, and $\xi = 0$ otherwise. Thus, (9.7) and Proposition 1.36 imply that $\mu = 0$, which means that

$$a_T(x, n) \llcorner \omega = 0, \quad \|T\| - \text{a.a.} \tag{9.8}$$

Similarly we obtain from the Legendrian property $T \llcorner \alpha = 0$ that

$$a_T(x, n) \llcorner \alpha = 0, \quad \|T\| - \text{a.a.} \tag{9.9}$$

Recall that π_0, π_1 denote the two component projections in $\mathbb{R}^d \times \mathbb{R}^d$, i.e.,

$$\pi_0(x, n) = x, \quad \pi_1(x, n) = n.$$

For a general Legendrian cycle we can find a representation of its approximate tangent spaces as in the case of the normal cycle of a set with positive reach (Proposition 4.23).

Theorem 9.2 *Let T be a Legendrian cycle in $\mathbb{R}^d$. Then, for $\|T\|$-almost all (x, n), $\mathrm{Tan}^{d-1}(W_T, (x, n))$ is a $(d-1)$-dimensional linear subspace of $\mathbb{R}^{2d}$ and there exists a positively oriented orthonormal basis*

$$\{b_1(x, n), \dots, b_{d-1}(x, n), n\}$$

of $\mathbb{R}^d$ and numbers $\kappa_1(x, n), \dots, \kappa_{d-1}(x, n) \in (-\infty, \infty]$ such that the vectors

$$a_i(x, n) := \left(\frac{1}{\sqrt{1 + \kappa_i^2(x, n)}} b_i(x, n), \frac{\kappa_i(x, n)}{\sqrt{1 + \kappa_i^2(x, n)}} b_i(x, n) \right), \quad i = 1, \dots, d-1,$$

form an orthonormal basis of $\mathrm{Tan}^{d-1}(W_T, (x, n))$. (We set $\frac{1}{\sqrt{1+\infty^2}} = 0$ and $\frac{\infty}{\sqrt{1+\infty^2}} = 1$.) The numbers $\kappa_i(x, n)$ are uniquely determined, up to the order, and *the subspace spanned by the vectors $b_j(x,n)$ belonging to a fixed value among the $\kappa_i(x,n)$ $(1 \le i \le d-1)$ is uniquely determined.*

In analogy to sets with positive reach, we will call the numbers $\kappa_i(x,n)$ *principal curvatures* and vectors $b_i(x,n)$ *principal directions* of T at (x,n).

Proof By rectifiability, for $\|T\|$-almost all (x,n), $T_{x,n} := \operatorname{Tan}^{d-1}(W_T,(x,n))$ is a $(d-1)$-dimensional subspace of $\mathbb{R}^d$ associated with $a_T(x,n)$ and

$$a_T(x,n) \llcorner \omega = a_T(x,n) \llcorner \alpha = 0$$

by (9.8) and (9.9). Note that the last equality implies that $T_{x,n} \subset n^\perp \times n^\perp$.

Fix such an $(x,n) \in W_T$ and an $\varepsilon > 0$, and consider the linear mapping

$$L_\varepsilon : T_{x,n} \to n^\perp, \quad (u,v) \mapsto u + \varepsilon v.$$

Decomposing $T_{x,n} = \ker L_\varepsilon \oplus (\ker L_\varepsilon)^\perp$, if $p := \dim\ker L_\varepsilon > 0$ we can find and orthonormal basis of $\ker L_\varepsilon$ of the form

$$\left(\frac{\varepsilon}{\sqrt{1+\varepsilon^2}} b_i, -\frac{1}{\sqrt{1+\varepsilon^2}} b_i\right), \quad 1 \le i \le p,$$

hence, we can put $\kappa_1 = \cdots = \kappa_p = -1/\varepsilon$.

Any vector $(u,v) \in (\ker L_\varepsilon)^\perp$ satisfies

$$0 = (u,v)\cdot(\varepsilon b_i, -b_i) = \varepsilon(u\cdot b_i) - (v \cdot b_i), \quad i = 1,\dots,p.$$

From the Lagrangian property $a_T \llcorner \omega = 0$ and (9.6), we get additionally

$$0 = \langle (u,v)\wedge(\varepsilon b_i, -b_i), \omega\rangle = -(u\cdot b_i) - \varepsilon(v\cdot b_i), \quad i = 1,\dots,p.$$

These two linear equations imply that $u \cdot b_i = v \cdot b_i = 0$, $i = 1,\dots,p$. Thus, denoting by V the ($(d-p-1)$-dimensional) orthogonal complement to $\operatorname{Lin}\{b_1,\dots,b_p,n\}$, we get that $(\ker L_\varepsilon)^\perp \subset V \times V$. Hence, the linear mapping

$$L := \pi_1 \circ (L_\varepsilon | (\ker L_\varepsilon)^\perp)^{-1}$$

acts from V to V and it is self-adjoint. This follows again from the Lagrangian property since for any $(u,v), (u',v') \in (\ker L_\varepsilon)^\perp$,

$$v\cdot(u'+\varepsilon v') - (u+\varepsilon v)\cdot v' = u'\cdot v - u \cdot v' = \langle (u',v')\wedge(u,v), \omega\rangle = 0.$$

Thus, we can choose an orthonormal basis $\{b_i = u_i + \varepsilon v_i,\ i = p+1, \dots, d-1\}$ of eigenvectors of L and the corresponding real eigenvalues λ_i satisfy $v_i = \lambda_i(u_i + \varepsilon v_i)$, $i = 1, \dots, d-1$, hence

$$((1-\lambda_i\varepsilon)b_i, \lambda_i b_i), \quad i = p+1, \dots, d-1,$$

are basis vectors of $(\ker L_\varepsilon)^\perp$. Setting $\kappa_i = \lambda_i/(1-\varepsilon\lambda_i)$ (which is defined as ∞ if $\lambda_i\varepsilon = 1$), we get the assertion on existence. The uniqueness (and, thus, independence of ε) can be proved exactly as in Lemma 4.24. □

If we orient the carrier W_T by the unit $(d-1)$-vectorfield

$$a_T(x,n) := a_1(x,n) \wedge \cdots \wedge a_{d-1}(x,n),$$

there must exist an integrable integer-valued index function i_T on W_T such that (9.5) holds. Let $\lambda(x,n)$ be the number of negative principal curvatures an (x,n) (defined $\mathcal{H}^{d-1}$-almost everywhere on W_T) and set

$$\iota_T(x,n) := (-1)^{\lambda(x,n)} i_T(x,n). \tag{9.10}$$

This particular form of the index function in (9.5) will often be used in the sequel.

For the uniqueness theorem we will use the restriction of a Legendrian current T to the *Gauss curvature form* φ_0 (see Definition 4.31)

$$\langle a_1 \wedge \cdots \wedge a_{d-1}, \varphi_0(x,n)\rangle = (d\omega_d)^{-1}\langle \pi_1(a_1) \wedge \cdots \wedge \pi_1(a_{d-1}) \wedge n, \Omega_d\rangle,$$

$a_1, \dots, a_{d-1} \in \mathbb{R}^d \times \mathbb{R}^d$. We can also describe φ_0 in a shorter way by using the operation of interior multiplication of multivectors and multicovectors (see Sect. 1.2.1):

$$\varphi_0(x,n) = (d\omega_d)^{-1}(\pi_1)^\#(n \lrcorner \Omega_d), \quad (x,n) \in \mathbb{R}^d \times \mathbb{R}^d.$$

To any unit vector u, the Hodge star operator assigns the unit simple $(d-1)$-vector $\star u$ associated with $\mathrm{Tan}(S^{d-1}, u)$ (see Definition 1.29). Let us denote this mapping by

$$\sigma : u \mapsto \star u$$

which orients the unit sphere.

Recall also that $(\pi_1)_\# T$ is the push-forward of the current T by π_1 (see (1.27)). It can be described as follows.

Theorem 9.3 *For any compactly supported Legendrian cycle T we have*

(i) $(\pi_1)_\# T = T(\varphi_0)(\mathcal{H} \llcorner S^{d-1}) \wedge \sigma$,
(ii) $T(\varphi_0) = \sum_{x\in\mathbb{R}^d} \iota_T(x,n)$ *for almost all* $n \in S^{d-1}$, *and*
(iii) *for any* $\|T\|$*-integrable function* g *on* $\mathbb{R}^d \times S^{d-1}$,

$$d\omega_d T(g\varphi_0) = \int_{S^{d-1}} \sum_{x\in\mathbb{R}^d} g(x,n)\iota_T(x,n)\,\mathcal{H}^{d-1}(\mathrm{d}x).$$

Proof Since T is a cycle, we get $\partial(\pi_1)_\# T = (\pi_1)_\#\partial T = 0$, hence, $(\pi_1)_\# T$ is a cycle as well. The Constancy theorem (Theorem 1.53) applied to the manifold S^{d-1} oriented by σ yields

$$(\pi_1)_\# T = c(\mathcal{H} \llcorner S^{d-1}) \wedge \sigma$$

with a constant $c \in \mathbb{Z}$. Further, by the definition of φ_0 we have for any smooth function h on S^{d-1}

$$\begin{aligned} d\omega_d T((h\circ\pi_1)\varphi_0) &= T((h\circ\pi_1)\pi_1^\#(\mathrm{id} \lrcorner \Omega_d)) \\ &= ((\pi_1)_\# T)(h(\mathrm{id} \lrcorner \Omega_d)) \\ &= c\int_{S^{d-1}} h(n)\langle\sigma(n), n \lrcorner \Omega_d\rangle\,\mathcal{H}^{d-1}(\mathrm{d}n) \\ &= c\int_{S^{d-1}} h\,\mathrm{d}\mathcal{H}^{d-1}. \end{aligned} \tag{9.11}$$

Setting $h \equiv 1$, we get $c = T(\varphi_0)$ and (i) is proved.

In order to show (ii), note that, by Theorem 9.2, the Jacobian of $\pi_1|W_T$ equals $\mathcal{H}^{d-1}$- almost everywhere

$$J_{d-1}(\pi_1|W_T)(x,n) = \left|\prod_{i=1}^{d-1} \frac{\kappa_i(x,n)}{\sqrt{1+\kappa_i(x,n)^2}}\right|.$$

On the other hand, we have

$$\begin{aligned} d\omega_d\langle a_T(x,n), \varphi_0(x,n)\rangle &= \langle \pi_1 a_1(x,n)\wedge\cdots\wedge\pi_1 a_{d-1}(x,n), n \lrcorner \Omega_d\rangle \\ &= \prod_{i=1}^{d-1}\frac{\kappa_i(x,n)}{\sqrt{1+\kappa_i(x,n)^2}}\langle b_1(x,n)\wedge\cdots\wedge b_{d-1}(x,n)\wedge n, \Omega_d\rangle \\ &= (-1)^{\lambda(x,n)} J_{d-1}(\pi_1|W_T)(x,n). \end{aligned}$$

Now, the Area formula (Theorem 1.49) applied to $\pi_1|W_T$ and a $\|T\|$-integrable function g implies

$$
\begin{aligned}
d\omega_d T(g\varphi_0) &= \int_{W_T} g(x,n) i_T(x,n) \langle a_T(x,n), \varphi_0(n)\rangle\, \mathcal{H}^{d-1}(\mathrm{d}(x,n)) \\
&= \int_{W_T} g(x,n) i_T(x,n) \prod_{i=1}^{d-1} \frac{\kappa_i(x,n)}{\sqrt{1+\kappa_i(x,n)^2}} \mathcal{H}^{d-1}(\mathrm{d}(x,n)) \qquad (9.12) \\
&= \int_{W_T} g(x,n) \iota_T(x,n) J_{d-1}(\pi_1|W_T)(x,n) \mathcal{H}^{d-1}(\mathrm{d}(x,n)) \\
&= \int_{S^{d-1}} \sum_{x\in\mathbb{R}^d} g(x,n) \iota_T(x,n)\, \mathcal{H}^{d-1}(\mathrm{d}n),
\end{aligned}
$$

which proves (iii). Comparing the last formula with (9.11), we get

$$
\int_{S^{d-1}} g(n) \left(T(\varphi_0) - \sum_{x\in\mathbb{R}^d} \iota_T(x,n) \right) \mathcal{H}^{d-1}(dn) = 0
$$

for any bounded measurable function g on S^{d-1}, and this implies (ii). □

Now we can formulate and prove the uniqueness theorem for Legendrian cycles.

Theorem 9.4 *Let T be a compactly supported Legendrian cycle such that $T \llcorner \varphi_0 = 0$. Then $T = 0$.*

Proof Using (9.12), we get for any smooth function g on $\mathbb{R}^d \times S^{d-1}$

$$
(T \llcorner \varphi_0)(g) = T(g\varphi_0) = (d\omega_d)^{-1} \int_{W_T} \iota_T\, g\, s_{d-1}\, \mathrm{d}\mathcal{H}^{d-1},
$$

where

$$
s_{d-1}(x,n) = \prod_{i=1}^{d-1} \frac{\kappa_i(x,n)}{\sqrt{1+\kappa_i(x,n)^2}}
$$

("generalized Gauss curvature" defined $\mathcal{H}^{d-1}$-almost everywhere on W_T). The assumption $T \llcorner \varphi_0 = 0$ thus implies $s_{d-1} = 0$ almost everywhere on W_T. In particular, denoting $f = \pi_1|W_T$, we have rank $Df < d-1$ almost everywhere.

Assume, for the contrary, that $T \neq 0$, and let

$$
m := \operatorname{ess\,sup} \operatorname{rank} Df < d-1.
$$

We can then assume, without loss of generality (removing a subset of $(d-1)$-dimensional measure zero), that at all $(x, n) \in W_T$, $\operatorname{Tan}^{d-1}(W_T, (x, n))$ has the basis given in Theorem 9.2 with principal curvatures $\kappa_i(x, n) = 0$ whenever $i > m$. (In the case $d = 3$ and $m = 1$ it might happen that, in order to preserve the prescribed orientation, we have to choose $\kappa_1 = 0$ instead of $\kappa_2 = 0$, and the role of the indices must be exchanged in the whole proof.) Further, the set

$$Y := \{n \in S^{d-1} : \mathcal{H}^{d-1-m}(f^{-1}\{n\}) > 0\}$$

is countably $\mathcal{H}^m$-rectifiable (see Theorem 1.24) and, applying the Area formula (Theorem 1.21), we get for $\mathcal{H}^{d-1}$-almost all $(x, n) \in f^{-1}(Y)$,

$$\text{either ap } J_m f(x, n) = 0 \text{ or } \operatorname{Tan}^m(f(W_T), n) = \operatorname{Lin}\{b_1(x, n), \dots, b_m(x, n)\}.$$

Note that if $(x, n) \in f^{-1}(Y)$ and $\operatorname{rank} Df(x, n) = m$ then the subspaces $U(n) := \operatorname{Lin}\{b_i(x, n) : 1 \le i \le m\}$ and $V(n) := \operatorname{Lin}\{b_i(x, n) : m + 1 \le i \le d - 1\}$ do not depend on x for which $(x, n) \in f^{-1}(Y)$, due to the property above.

Let p be the orthogonal projection from $\mathbb{R}^d$ into an m-dimensional subspace L_m of $\mathbb{R}^d$, and let Ω_m denote the volume form in L_m. Due to the choice of m, we have

$$T \llcorner (p \circ \pi_1)^\# \Omega_m \neq 0$$

for some fixed subspace L_m. Slicing T by $p \circ \pi_1$, we get for the masses $\mathbf{M}$ of the currents (see (1.34))

$$0 \neq \mathbf{M}(T \llcorner (p \circ \pi_1)^\# \Omega_m) = \int_{L_m} \mathbf{M}\langle T, p \circ \pi_1, y\rangle \, \mathcal{H}^m(\mathrm{d}y).$$

Hence, there exists a measurable subset Z of L_m of positive measure such that for $y \in Z$:

(i) $\langle T, p \circ \pi_1, y\rangle \neq 0$ is a cycle (cf. Proposition 1.66);
(ii) (see Theorem 1.65)

$$\langle T, p \circ \pi_1, y\rangle = (\mathcal{H}^{d-1-m} \llcorner (f^{-1}(p^{-1}\{y\}))) \wedge \zeta(x, n),$$

where for $(x, n) \in W_m \cap f^{-1}(p^{-1}\{y\})$,

$$\begin{aligned} \zeta(x, n) &= \iota_T(x, n) a_T(x, n) \llcorner (p \circ \pi_1)^\# \Omega_m / \text{ap } J_m(p \circ f)(x, n) \\ &= \iota_T(x, n)(b_{m+1}(x, n), 0) \wedge \cdots \wedge (b_{d-1}(x, n), 0). \end{aligned}$$

Take a point $y \in Y$, let R be an indecomposable component (cf. Definition 1.67) of $\langle T, p \circ \pi_1, y\rangle$, and take a point $(x, n) \in \operatorname{spt} R$. Applying Theorem 1.69 for the projection π_1, we get

$$\operatorname{spt} R \subset f^{-1}\{n\}.$$

It follows that $n \in Y$ (otherwise, $R = 0$). Denoting by q the orthogonal projection of $\mathbb{R}^d \times \mathbb{R}^d$ to $(V(n) \times \{0\})^\perp$, we have from the description above

$$\zeta \llcorner q = 0 \quad \mathcal{H}^{d-1-m} - a.e. \text{ on } f^{-1}\{n\},$$

which implies that $q|\operatorname{spt} R$ is constant (apply Theorem 1.69 again, this time for the projection $q \circ \pi_1$). Hence, $\operatorname{spt} R$ is contained in the affine $(d - 1 - m)$-subspace $(x, n) + V(n) \times \{0\}$. The Constancy theorem yields that R must be a multiple of the Lebesgue measure. Since T (and, hence, also R) is compactly supported, we get $R = 0$, a contradiction. □

9.2 Normal Cycles

Let us come back to the approximation idea described at the beginning of this chapter. In order to get uniqueness for the Legendrian cycles obtained as flat limits of normal cycles of approximating sequences, we have to assure that all the limit currents have the same restriction to the Gauss curvature form φ_0. It is natural to relate this quantity to geometric properties of the set X.

Assume for a moment that X is a compact domain with C^2 (or even only $C^{1,1}$) boundary and let

$$N_X := (\mathcal{H}^{d-1} \llcorner \operatorname{nor} X) \wedge a_X$$

be its normal cycle. Then, clearly the index function satisfies

$$\iota_X(x, n) := \iota_{N_X}(x, n) = (-1)^{\lambda_X(x)}$$

for $\mathcal{H}^{d-1}$-almost all $(x, n) \in \operatorname{nor} X$. Consider the height function

$$h_v : x \mapsto v \cdot x, \quad x \in \mathbb{R}^d,$$

whose restriction to X is a Morse function for $\mathcal{H}^{d-1}$-almost all $v \in S^{d-1}$ (see Sect. 3.4). Moreover, x is a critical point of $h_v|X$ if and only if $(x, -v) \in \operatorname{nor} X$ and, due to (9.10), Corollary 3.10 yields the formula for the Euler characteristic

$$\chi(X) = \sum_{x \in \mathbb{R}^d} \iota_X(x, -v) \quad \text{for almost all } v \in S^{d-1}.$$

Given $v \in S^{d-1}$ and $t \in \mathbb{R}$, let us denote by

$$H_{v,t} := \{y \in \mathbb{R}^d : y \cdot v \leq t\}$$

the closed halfspace with outer normal vector v and distance t of the boundary hyperplane from the origin. Corollary 3.10 yields

$$\chi(X \cap H_{v,t}) = \sum_{x: x \cdot v \leq t} \iota_X(x, -v) \quad \text{for a.a. } (v, t) \in S^{d-1} \times \mathbb{R}. \tag{9.13}$$

(Here and in the sequel, we write "for a.a. $(v, t) \in S^{d-1} \times \mathbb{R}$" to shorten "for $\mathcal{H}^{d-1}$-almost all $v \in S^{d-1}$ and $\mathcal{L}^1$-almost all $t \in \mathbb{R}$".)

Equation (9.13) can be taken as a principal relation connecting a Legendrian cycle with a compact set. In order to express the right-hand side in its "current form", recall that the slice of a Legendrian cycle T by the projection π_1

$$\langle T, \pi_1, -v \rangle \in \mathbf{I}_0(\mathbb{R}^d \times \mathbb{R}^d), \quad v \in S^{d-1},$$

is an integral 0-current at $\mathcal{H}^{d-1}$-almost all $v \in S^{d-1}$, see Definition 1.63. Applying Theorem 1.65, we get

$$\langle T, \pi_1, -v \rangle = \left(\mathcal{H}^0 \llcorner (\pi_1)^{-1}\{-v\}\right) \wedge \iota_T,$$

in other words, for any $\|T\|$-integrable function g on $\mathbb{R}^d \times S^{d-1}$,

$$\langle T, \pi_1, -v \rangle(g) = \sum_{x \in \mathbb{R}^d} \iota_T(x, -v) g(x, -v) \quad \text{for a.a. } v \in S^{d-1}.$$

Applying the formula with the indicator function of $H_{v,t}$, we get

$$\langle T, \pi_1, -v \rangle(\mathbf{1}_{H_{v,t} \times S^{d-1}}) = \sum_{x: x \cdot v \leq t} \iota_T(x, -v) \quad \text{for a.a. } (v, t) \in S^{d-1} \times \mathbb{R}. \tag{9.14}$$

Using the mapping

$$p : (x, v) \mapsto -x \cdot v, \quad (x, v) \in \mathbb{R}^d \times S^{d-1},$$

we can rewrite the left-hand side as $p_{\#}\langle T, \pi_1, -v \rangle(\mathbf{1}_{(-\infty,t]})$. Motivated by (9.13) and (9.14), we can define normal cycles for rather general sets.

Definition 9.5 We say that a compact set $X \subset \mathbb{R}^d$ admits a *normal cycle* if there exists a compactly supported Legendrian cycle T such that

$$p_{\#}\langle T, \pi_1, -v \rangle(\mathbf{1}_{(-\infty,t]}) = \chi(X \cap H_{v,t}) \text{ for a.a. } (v, t) \in S^{d-1} \times \mathbb{R}. \tag{9.15}$$

In such a case, we write $N_X := T$, call N_X the *normal cycle of* X, and we define the curvature(-direction) measures of X of orders $k = 0, \dots, d-1$ as follows (cf. Theorem 4.42):

$$\begin{aligned}
\widetilde{C}_k(X, E) &:= (N_X \llcorner E)(\varphi_k), \quad E \subset \mathbb{R}^d \times S^{d-1} \text{ Borel}, \\
C_k(X, B) &:= \widetilde{C}_k(X, B \times S^{d-1}), \quad B \subset \mathbb{R}^d \text{ Borel}, \\
\mathbf{C}_k(X) &:= N_X(\varphi_k).
\end{aligned}$$

Proposition 9.6 *Any compact set $X \subset \mathbb{R}^d$ admits at most one normal cycle.*

Proof Let T_1 and T_2 be two normal cycles of a compact set $X \subset \mathbb{R}^d$. Then $T := T_1 - T_2$ is again a Legendrian cycle; we will show that $T = 0$.

By (9.15), we have

$$p_\#\langle T, \pi_1, -v\rangle(\mathbf{1}_{(-\infty,t]}) = 0 \text{ for a.a. } (v, t) \in S^{d-1} \times \mathbb{R},$$

thus (as any Radon measure on $\mathbb{R}$ is determined by its distribution function)

$$p_\#\langle T, \pi_1, -v\rangle = 0 \text{ for a.a. } v \in S^{d-1}.$$

This can be written equivalently as

$$\langle \Lambda_\# T, \pi, v\rangle = 0 \text{ for a.a. } v \in S^{d-1}$$

with the mappings

$$\Lambda : (x, v) \mapsto (v, x \cdot v), \quad (x, v) \in \mathbb{R}^d \times S^{d-1}, \tag{9.16}$$

and $\pi : (v, t) \mapsto v$, $(v, t) \in S^{d-1} \times \mathbb{R}$ (note that $\pi_1 = \pi \circ \Lambda$). Integrating, and applying the coarea formula for currents, we obtain

$$\begin{aligned}
0 &= \int_{S^{d-1}} \langle \Lambda_\# T, \pi, v\rangle(g)\, \mathcal{H}^{d-1}(dv) \\
&= (\Lambda_\# T) \llcorner \pi^\#(\mathrm{id} \lrcorner \Omega_d)(g) \\
&= \Lambda_\#(T \llcorner \Lambda^\# \pi^\#(\mathrm{id} \lrcorner \Omega_d))(g) \\
&= \Lambda_\#(T \llcorner \varphi_0)
\end{aligned}$$

for any smooth function g on S^{d-1}. Since Λ is injective $\mathcal{H}^{d-1}$-almost everywhere on spt $(T \llcorner \varphi_0)$ by Lemma 9.7 below, we get also $T \llcorner \varphi_0 = 0$, and the assertion follows by Theorem 9.4. □

Lemma 9.7 *Let T be a Legendrian cycle with representation* (9.5). *Then,*

$$\mathcal{H}^{d-1}(\{n \in S^{d-1} : \exists x \neq y,\ (x, n), (y, n) \in W_T,\ (x-y)\cdot n = 0\}) = 0.$$

Proof Consider the quadratic surface

$$Q := \{(x, m, y, n) \in \mathbb{R}^{4d} : m = n,\ (x-y)\cdot n = 0\} \subset \mathbb{R}^{4d}$$

and note that $(x-y)\cdot v = (x-y)\cdot v' = 0$ whenever $(x, m, y, n) \in Q$ and $(u, v, u', v') \in \operatorname{Tan}(Q, (x, m, y, n))$. Denote

$$N := \Pi((W_T \times W_T) \cap Q),$$

where $\Pi : (x, m, y, n) \mapsto (x, m)$. We have to show that $\mathcal{H}^{d-1}(\pi_1(N)) = 0$. To this end, we will show that the Jacobian $J_{d-1}(\pi_1|N)$ vanishes $\mathcal{H}^{d-1}$-almost everywhere on N. If $(x, n, y, n) \in (W_T \times W_T) \cap Q$ and $(u, v) \in \operatorname{Tan}(N, (x, n))$ then it follows from the above described tangent property of Q that $(x-y)\cdot v = 0$. But then, $\dim \pi_1(\operatorname{Tan}(N, (x, n))) < d-1$ and, hence, $J_{d-1}(\pi_1|N)(x, n) = 0$ whenever it exists. □

Next we will show that the above definition is consistent with the former approach for special set classes.

Proposition 9.8 *If X is compact and* $\operatorname{reach} X > 0$ *(or $X \in \mathcal{U}_{PR}$) then X admits a normal cycle N_X which agrees with that from Definition* 4.38 *(or Definition* 5.9*, respectively).*

Proof Let X be a compact set with $\operatorname{reach} X > 0$ and let N_X be its normal cycle from Definition 4.38. We will show that N_X satisfies (9.15). First, note that at almost all $(x, v) \in \operatorname{nor} X$,

$$\iota_X(x, v) := \iota_{N_X}(x, v) = (-1)^{\lambda_X(x,v)}.$$

Consider the parallel sets X_r which are compact $C^{1,1}$-domains for $r > 0$ and Corollary 3.10 implies that

$$\chi(X_r \cap H_{v,t}) = \sum_{y \in H_{v,t} \cap \nu_{X_r}^{-1}(-v)} \lambda_{X_r}(y)$$

for almost all (v, t) and all $r > 0$. We let now r tend to 0 on both sides. If X and $H_{v,t}$ do not touch (which is the case for almost all (v, t)) we get using Proposition 4.17 that $\operatorname{reach}(X \cap H_{v,t}) \geq \eta \operatorname{reach} X$ with

$$\eta = \eta(v, t) := \inf\{|u + v| : (x, u) \in \operatorname{nor} X,\ x \cdot v = t\} > 0.$$

Since $\operatorname{nor} X$ is closed (see Lemma 4.11), the function η is lower semicontinuous in t and we infer that $\liminf_{r\to 0_+} \operatorname{reach}(X_r \cap H_{v,t}) > 0$. Since clearly $X_r \cap H_{v,t} \to X \cap H_{v,t}$ in the Hausdorff distance, we get from Theorem 7.3 and Proposition 7.1 that

$$\chi(X_r \cap H_{v,t}) = \mathbf{C}_0(X_r \cap H_{v,t}) \to \mathbf{C}_0(X \cap H_{v,t}) = \chi(X \cap H_{v,t}), \quad r \to 0_+.$$

For the right hand side, note that $y \in \nu_{X_r}^{-1}(-v)$ if and only if $(y + rv, -v) \in \operatorname{nor} X$. If X does not touch $H_{v,t}$, again from the closedness of $\operatorname{nor} X$ we get that there exists a $\delta > 0$ such that there is no $(x, -v) \in \operatorname{nor} X$ with $|x \cdot v - t| < \delta$. But then we get for any $0 < r < \delta$

$$\sum_{y\in H_{v,t}\cap\nu_{X_r}^{-1}(-v)} \lambda_{X_r}(y) = \sum_{x:x\cdot v\le t} (-1)^{\lambda_X(x,-v)} = \sum_{x:x\cdot v\le t} \iota_X(x,-v),$$

hence,

$$\chi(X \cap H_{v,t}) = \sum_{x:x\cdot v\le t} \iota_X(x,-v) = p_\#\langle N_X, \pi_1, -v\rangle(\mathbf{1}_{(-\infty,t]}),$$

for almost all (v, t), proving thus (9.15).

If $X \in \mathcal{U}_{\mathcal{PR}}$ we choose a $\mathcal{U}_{\mathcal{PR}}$ representation $X = \bigcup_i X^i$ and apply the additivity for both N_X (Theorem 5.11) and $\chi(X \cap H_{v,t})$. □

Example 9.9 Let X be convex compact with nonempty interior and contained in the interior of a ball B. Then, both $B \setminus \operatorname{int} X$ and ∂X admit normal cycles. In particular,

$$N_{B\setminus \operatorname{int} X} = \rho_\# N_X + N_B \quad \text{and} \quad N_{\partial X} = N_X + \rho_\# N_X,$$

where $\rho : (x, n) \mapsto (x, -n)$.

Example 9.10 Assume that $\operatorname{reach} X > 0$, $\mathbb{R}^d \setminus X$ is bounded and that

$$(x, n) \in \operatorname{nor} X \text{ implies } (x, -n) \notin \operatorname{nor} X. \tag{9.17}$$

(Note that this is a kind of "full-dimensionality" condition saying that there is no pair of outer normals at a point lying in opposite directions.) Then $\mathbb{R}^d \setminus \operatorname{int} X$ admits a normal cycle, in particular,

$$N_{\mathbb{R}^d\setminus \operatorname{int} X} = \rho_\# N_X.$$

Remark 9.11 Condition (9.17) cannot be avoided in Example 9.10. Indeed, consider an infinite compact totally disconnected set $K \subset [0, 1]$ (embedded in the x-axis) and set

$$X := \{(x, y) \in \mathbb{R}^2 : x^2 + y^2 \ge 4 \text{ or } \operatorname{dist}(x, K)^2 \le |y|\}.$$

then reach $X > 0$, X^C is bounded but $\overline{X^C}$ has infinitely many components, hence it cannot admit a normal cycle (cf. [RZ17, Example 7.12] for more details).

Lemma 9.12 (Motion covariance of normal cycles) *If a compact set $X \subset \mathbb{R}^d$ admits a normal cycle then also gX admits a normal cycle for any Euclidean motion $g \in \mathcal{G}_d$, and the normal cycles are related by*

$$N_{gX} = \tilde{g}_\# N_X,$$

where $\tilde{g} : (x, n) \mapsto (gx, g_0 n)$, g_0 being the linear component of g.

Proof Assume that N_X is the normal cycle of a compact set X and let $g \in \mathcal{G}_d$ be a Euclidean motion of the form $g(x) = b + g_0(x)$, $x \in \mathbb{R}^d$, with a linear isometry g_0. We will show that the current $T := \tilde{g}_\# N_X$ is a normal cycle of gX. Clearly, T is a Legendrian cycle which satisfies (i) from Definition 9.5. It remains to verify condition (ii), i.e.,

$$p_\# \langle T, \pi_1, -v\rangle(\mathbf{1}_{(-\infty,t]}) = \chi(gX \cap H_{v,t}) \quad \text{for a.a. } (v, t).$$

Note that, since χ is motion invariant,

$$\chi(gX \cap H_{v,t}) = \chi(g(X \cap g^{-1} H_{v,t})) = \chi(X \cap H_{\tilde{v},\tilde{t}}),$$

where $\tilde{v} := g_0(v)$ and $\tilde{t} := t - b \cdot v$. Further,

$$\begin{aligned} p_\# \langle T, \pi_1, -v\rangle(\mathbf{1}_{(-\infty,t]}) &= p_\# \tilde{g}_\# \langle N_X, \pi_1 \circ \tilde{g}, -v\rangle(\mathbf{1}_{(-\infty,t]}) \\ &= p_\# \tilde{g}_\# \langle N_X, \pi_1, -\tilde{v}\rangle(\mathbf{1}_{(-\infty,t]}) \\ &= p_\# \langle N_X, \pi_1, -\tilde{v}\rangle(\mathbf{1}_{(-\infty,\tilde{t}]}) \\ &= \chi(X \cap H_{\tilde{v},\tilde{t}}), \end{aligned}$$

and the equality follows. □

Remark 9.13 It is not known whether the normal cycle from Definition 9.5 is locally determined as that of a set with positive reach. In particular, the following question seems to be important: Assume that $X \subset \mathbb{R}^d$ is compact and that for any $x \in X$ there exists $\delta > 0$ and a compact set $Y \subset \mathbb{R}^d$ admitting normal cycle N_Y and such that $X \cap B(x, \delta) = Y \cap B(x, \delta)$. Does it follow that X admits a normal cycle, N_X, and that, for x, δ, Y as above, $N_X \llcorner \mathbf{1}_{B(x,\delta)} = N_Y \llcorner \mathbf{1}_{B(x,\delta)}$?

9.3 Lipschitz Domains

In the sequel, we will limit ourselves to the case when X is a closed Lipschitz domain in $\mathbb{R}^d$, i.e., X coincides locally with the subgraph of a Lipschitz function defined on a hyperplane. A general compact Lipschitz domain does not admit a normal cycle, but under additional assumptions the normal cycle exists, see Theorem 9.22. At the end we also prove a Principal kinematic formula for a class of closed Lipschitz domains.

Definition 9.14

(i) Let $L > 0$, $v \in S^{d-1}$ and an open set $U \subset \mathbb{R}^d$ be given. A closed set $X \subset \mathbb{R}^d$ is an (L, v)-*Lipschitz subgraph in* U if there exists an L-Lipschitz function g defined on the orthogonal complement $v^\perp$ such that

$$X \cap U = \{x \in \mathbb{R}^d : x \cdot v \le g(p_{v^\perp} x)\} \cap U.$$

In the case $U = \mathbb{R}^d$, we call X simply an (L, v)-*Lipschitz subgraph.*

(ii) A closed set $X \subset \mathbb{R}^d$ is a *closed Lipschitz domain* if for any $x \in X$ there exist an $L > 0$, unit vector v, and neighbourhood U of x (in $\mathbb{R}^d$) such that X is an (L, v)-Lipschitz subgraph in U. The unit vector v with the above properties will be called *admissible for* X *at* x, or *admissible for* X *on* U.

Remark 9.15 If X is a closed Lipschitz domain and x, v, g, U are as in the above definition then

$$\phi : x \mapsto p_{v^\perp} x + (x \cdot v - g(p_{v^\perp} x))v, \quad x \in U,$$

is a bi-Lipschitz chart of X in the usual sense. Abusing the terminology, we will call the triple (v, g, U) a (Lipschitz) chart of X, and a collection of charts whose domains U cover the whole X will be called an atlas of X.

Lemma 9.16

(i) *If* X *is an* (L, v)-*Lipschitz subgraph in an open set* U *and* $x \in \partial X$ *then the cone*

$$V_{L,v} := \left\{ u \in \mathbb{R}^d : u \cdot v \ge (\sin \gamma)|u| \right\},$$

with $\gamma :=$ arctan L *satisfies*

$$(x - V_{L,v}) \cap U \subset X \text{ and } \operatorname{int}(x + V_{L,v}) \cap U \subset \mathbb{R}^d \setminus X.$$

(ii) *If* X *is an* (L, v_1)- *and* (L, v_2)-*Lipschitz subgraph in* U *and* $v \in S^{d-1}$ *is a spherical convex combination of* v_1 *and* v_2 *then* X *is an* (L, v)-*Lipschitz subgraph in* U.

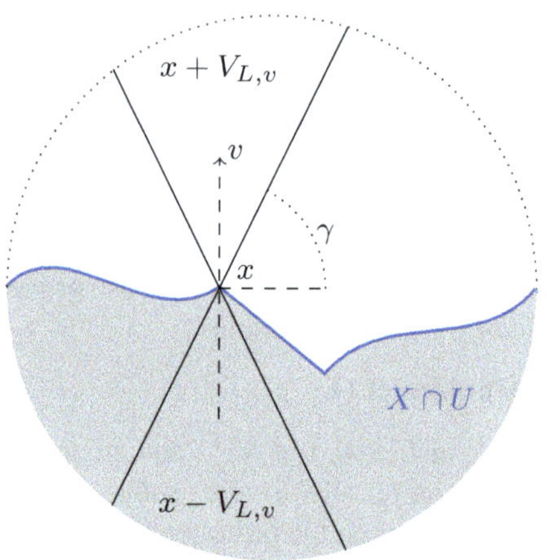

Fig. 9.1 X is an (L, v)-Lipschitz subgraph in U and $V_{L,v}$ is the cone from Lemma 9.16

Proof Assertion (i) follows from the L-Lipschitz property using standard methods. In order to show (ii), note first that (a) implies (in the case $U = \mathbb{R}^d$) that also $x - V_{L,v} \subset X$ whenever $x \in X$. This implies that if X is an (L, v_1)- and (L, v_2)-Lipschitz subgraph then $x - V_{L,v_1} - V_{L,v_2} \subset X$ if $x \in \partial X$, and since $V_{L,v} \subset V_{L,v_1} + V_{L,v_2}$ if v is a spherical convex combination of v_1 and v_2, assertion (ii) follows using (i) again (Fig. 9.1). □

The *Clarke tangent cone* of X at x is defined as

$$C(X, x) = \liminf_{x_i \to x, x_i \in X} \operatorname{Tan}(X, x_i)$$

(i.e., $u \in C(X, x)$ if and only if whenever $x_i \to x$, $x_i \in X$, there exist tangent vectors $u_i \in \operatorname{Tan}(X, x_i)$ with $u_i \to u$). (The definition given in [Cla83, §2.4] is equivalent, see [RW04, Theorem 6.26].) The *Clarke normal cone* $N(X, x)$ of X at x is defined as the polar cone

$$N(X, x) = C(X, x)^o.$$

The *Clarke normal bundle* of X is the sphere bundle

$$\operatorname{nor}^{\mathrm{Cl}} X = \{(x, n) : x \in \partial X, n \in N(X, x) \cap S^{d-1}\}.$$

We will need some properties of these cones.

Lemma 9.17 *If $X \subset \mathbb{R}^d$ is a closed Lipschitz domain and $x \in \mathbb{R}^d$ then*

(i) *both $C(X, x)$ and $N(X, x)$ are closed convex cones,*
(ii) *$u \in \operatorname{int} C(X, x)$ if and only if there exists $\varepsilon > 0$ such that $[y, y + \varepsilon v] \subset X$ whenever $y \in X \cap B(x, \varepsilon)$ and $v \in B(u, \varepsilon)$,*

(iii) *if* $x_i \to x$, $v_i \in N(X, x_i)$ *and* $v_i \to v$ *then* $v \in N(X, x)$ *(hence,* $\operatorname{nor}^{\mathrm{Cl}} X$ *is closed),*

(iv) $v \in S^{d-1}$ *is admissible for* X *at* x *if and only if* $-v \in \operatorname{int} C(X, x)$,

(v) *if* X *is an* (L, v)*-Lipschitz subgraph in an open set* U *then* $-V_{L,v} \subset C(X, x)$ *whenever* $x \in X \cap U$,

(vi) *if* $v \in S^{d-1}$ *is admissible for* X *at* x *then* $v \notin C(X, x)$ *(hence,* $C(X, x)$ *is contained in a halfspace),*

(vii) *if* $x \in \partial X$, $v \in S^{d-1}$ *and* $B(x - \varepsilon v, \varepsilon) \subset X$ *for some* $\varepsilon > 0$ *then* $v \in N(X, x)$,

(viii) *if* $\operatorname{reach}(X, x) > 0$ *then* $C(X, x) = \operatorname{Tan}(X, x)$ *and* $N(X, x) = \operatorname{Nor}(X, x)$.

Proof The closedness and convexity of the cone $C(X, x)$ follow from the properties of the subgradient of f at x, see [Cla83, Proposition 2.1.2]. The normal cone $N(X, x)$ is always closed and convex as a polar cone. Hence, (i) is proved.

Properties (ii) and (iii) are known results due to Rockafellar, see [Cla83, §2.5.8].

Properties (iv), (v) and (vi) follow from (ii) and Lemma 9.16 (i).

Let the assumptions of (vii) hold. If there would be some $u \in \operatorname{int} C(X, x)$ with $u \cdot v > 0$ then we would obtain using (ii) that x is an interior point of X, a contradiction. Thus $u \cdot v \le 0$ whenever $u \in C(X, x)$ and, hence, $v \in N(X, x)$.

If $\operatorname{reach}(X, x) > 0$ we know that the set-valued function $y \mapsto \operatorname{Nor}(X, y) \cap S^{d-1}$ is upper semicontinuous at x (i.e., if $x_i \in X$, $x_i \to x$, $n_i \in \operatorname{Nor}(X, x_i) \cap S^{d-1}$ and $n_i \to n$ then $n \in \operatorname{Nor}(X, x) \cap S^{d-1}$, cf. Lemma 4.11). Since $\operatorname{Tan}(X, x) = \operatorname{Nor}(X, x)^o$, it follows that $y \mapsto \operatorname{Tan}(X, y) \cap S^{d-1}$ is lower semicontinuous at x, which implies that $\operatorname{Tan}(X, x) = C(X, x)$, and the equality $\operatorname{Nor}(X, x) = N(X, x)$ follows immediately. □

We say that two closed Lipschitz domains $X, Y \subset \mathbb{R}^d$ *touch* if there exists $(x, n) \in \operatorname{nor}^{\mathrm{Cl}} X$ such that $(x, -n) \in \operatorname{nor}^{\mathrm{Cl}} Y$.

Lemma 9.18 *If* X, Y *are closed Lipschitz domains that do not touch then* $X \cap Y$ *and* $X \cup Y$ *are closed Lipschitz domains as well.*

Proof Take $x \in X \cap Y$ and observe that $\operatorname{int} C(X, x) \cap \operatorname{int} C(Y, x) \neq \emptyset$. (Indeed, if the interiors would not intersect then the two convex cones could be separated by a hyperplane and each of the two unit vectors perpendicular to this hyperplane would belong to the Clarke normal bundle of one of the sets X, Y, which would violate the non-touching assumption at x.) Thus, by Lemma 9.17 (iv), there exists a unit vector n admissible for both X and Y at x and if g, h are the Lipschitz functions on $n^\perp$ that determine X, Y, respectively, locally at x, then $\min\{g, h\}$, $\max\{g, h\}$ determine $X \cap Y$, $X \cup Y$, respectively, locally at x. □

In the following we consider "interior parallel sets" ($\varepsilon > 0$)

$$X_{-\varepsilon} := \{x \in X : d_{\partial X}(x) \ge \varepsilon\}.$$

Lemma 9.19 *If $X \subset \mathbb{R}^d$ is a closed Lipschitz domain and $K \subset \mathbb{R}^d$ compact then there exist $\eta, \varepsilon_0 > 0$ such that for all $0 < \varepsilon < \varepsilon_0$ and $x \in K \cap X_{-\varepsilon}$,*

$$\operatorname{reach}(X_{-\varepsilon}, x) \geq \eta(\varepsilon + r),$$

where $r := \inf_{y \in K \cap X} \operatorname{reach}(X, y)$.

Proof If $X \cap K = \emptyset$ there is nothing to prove. If $X \cap K \neq \emptyset$ and $r = \infty$ then X must be already convex. Hence, $X_{-\varepsilon}$ is convex as well, and its reach is infinite everywhere. We will assume that $r < \infty$ in the rest of the proof.

Let (v, g, U) be a Lipschitz chart of X (see Definition 9.14) and let $\varepsilon > 0$ and $x, y \in X_{-\varepsilon}$ be such that $B(x, 2\varepsilon) \subset U$. We will show that

$$d_{\operatorname{Tan}(X_{-\varepsilon},x)}(y - x) \leq (2\eta(\varepsilon + r))^{-1}|y - x|^2, \tag{9.18}$$

where $\eta := (1 + (\operatorname{Lip} g)^2)^{-1/2}$. If $d_{\partial X}(x) > \varepsilon$ then $\operatorname{Tan}(X_{-\varepsilon}, x) = \mathbb{R}^d$ and (9.18) clearly holds. Assume thus that $d_{\partial X}(x) = \varepsilon$ and denote

$$\Sigma_X(x) := \{z \in \partial X : |z - x| = \varepsilon\}.$$

Observe that if a vector u belongs to the interior of the polar cone to $\Sigma_X(x) - x$ then the distance to ∂X increases in direction u at x, hence

$$(\Sigma_X(x) - x)^o \subset \operatorname{Tan}(X_{-\varepsilon}, x).$$

We will show that the vector

$$u := y - x - (2\eta(\varepsilon + r))^{-1}|y - x|^2 v \in \operatorname{Tan}(X_{-\varepsilon}, x),$$

which will imply (9.18). Let $z \in \Sigma_X(x)$. If $r > 0$ then $z - x \in \operatorname{Nor}(X, z)$ and $\mathring{B}(w, r) \cap X = \emptyset$, where $w := z + r\varepsilon^{-1}(z - x)$. Thus $\mathring{B}(w, \varepsilon + r) \cap X_{-\varepsilon} = \emptyset$, which clearly holds also if $r = 0$, and we get $|w - y| \geq \varepsilon + r$, thus

$$\begin{aligned}(\varepsilon + r)^2 \leq |w - y|^2 &= |w - x|^2 + |x - y|^2 + 2(w - x) \cdot (x - y) \\ &= (\varepsilon + r)^2 + |y - x|^2 - 2(w - x) \cdot (y - x),\end{aligned}$$

which implies

$$(w - x) \cdot (y - x) \leq \tfrac{1}{2}|y - x|^2.$$

From the Lipschitz property of g we infer that (since $w - x$ is a normal direction to the graph of g)

$$(w - x) \cdot v \geq \eta(\varepsilon + r).$$

Taking these two inequalities into account, we get

$$\begin{aligned} u \cdot (w - x) &= (y - x) \cdot (w - x) - (2\eta(\varepsilon + r))^{-1} |y - x|^2 \, v \cdot (w - x) \\ &\le \frac{1}{2}|y - x|^2 - (2\eta(\varepsilon + r))^{-1}|y - x|^2 \eta(\varepsilon + r) = 0, \end{aligned}$$

hence $u \in (\Sigma_X(x) - x)^o$ and $u \in \mathrm{Tan}(X_{-\varepsilon}, x)$ follows, proving (9.18).

Since K is compact there exists an $\varepsilon_0 > 0$ and a finite family

$$\{(v_i, g_i, \mathring{B}(x_i, 3\varepsilon_0)) : i = 1 \dots, k\}$$

of Lipschitz charts of X such that $K_1 \subset \bigcup_{i=1}^{k} B(x_i, \varepsilon_0)$, where $K_1 := K \oplus B(0, 1)$. Denoting

$$\eta := \inf_{1 \le i \le k} (1 + (\mathrm{Lip}\, g_i)^2)^{-1/2},$$

we have using (9.18)

$$d_{\mathrm{Tan}(X_{-\varepsilon}, x)}(y - x) \le (2\eta(\varepsilon + r))^{-1} |y - x|^2 \text{ whenever } x \in K_1 \cap X_{-\varepsilon},\, y \in X_{-\varepsilon}.$$

This already implies that $\mathrm{reach}\,(X_{-\varepsilon}, x) \ge \eta(\varepsilon + r)$ if $x \in K \cap X_{-\varepsilon}$. Indeed (proceeding as in the proof of Proposition 4.14), if $\mathrm{reach}\,(X_{-\varepsilon}, x) < \eta(\varepsilon + r)$ for some $x \in K \cap X_{-\varepsilon}$ there would exist a point $z \in \mathring{B}(x, \eta(\varepsilon + r))$ having two different nearest neighbours y_1, y_2 in $X_{-\varepsilon}$. Clearly $s := |y_1 - z| = |y_2 - z| < \eta(\varepsilon + r)$, and, as $\mathring{B}(z, s)$ does not intersect $X_{-\varepsilon}$, we have

$$d_{\mathrm{Tan}(X_{-\varepsilon}, y_1)}(y_2 - y_1) \ge (2s)^{-1} |y_2 - y_1|^2 > (2\eta(\varepsilon + r))^{-1} |y_2 - y_1|^2,$$

violating the above property. This completes the proof. □

In the following, the notation $X \sim Y$ will denote that the sets X, Y have the same homotopy type (in particular, $\chi(X) = \chi(Y)$ if it exists).

Proposition 9.20 *If $X \subset \mathbb{R}^d$ is a compact Lipschitz domain then for all sufficiently small $\varepsilon > 0$,*

$$X \sim X_{-\varepsilon}.$$

Proof We construct first a transversal unit vector field on the boundary ∂X, i.e., a Lipschitz mapping

$$v : \partial X \to S^{d-1}$$

such that $v(x)$ is admissible for X at x, $x \in \partial X$. The construction can be done as follows. Let $\{(v_i, g_i, \mathring{B}(x_i, 5\delta)) : i = 1, \dots, k\}$ be a finite family of Lipchitz charts of X (cf. Definition 9.14) such that

$$\partial X \subset \bigcup_{i=1}^{k} \mathring{B}(x_i, \delta),$$

and let $(h_i : i = 1, \dots, k)$ be a smooth partition of unity subordinated to the cover $\bigcup_{i=1}^{k} \mathring{B}(x_i, \delta)$. Set

$$\tilde{v}(x) := \sum_{i=i}^{k} h_i(x) v_i, \quad x \in \partial X.$$

For $x, y \in \partial X$ we have

$$|\tilde{v}(x) - \tilde{v}(y)| \leq \sum_{i=1}^{k} |h_i(x) - h_i(y)| \leq \left(\sum_{i=1}^{k} \operatorname{Lip} h_i \right) |x - y|,$$

thus $\tilde{v}$ is Lipschitz.

Set $L := \max_{1 \leq i \leq k} \operatorname{Lip} g_i$, $\gamma := \arctan L$ and consider a point $x \in \partial X$. Then, by Lemma 9.17 (v), $-V_{L,v_i} \subset C(X, x)$ whenever $x \in B(x_i, \delta)$, and by Lemma 9.17 (vi), $C(X, x)$ is contained in a halfspace $\{z : z \cdot w \geq 0\}$ with some $w \in S^{d-1}$. We claim that $v_i \cdot w \geq \cos \gamma$ whenever $x \in B(x_i, \delta)$. Indeed, otherwise the orthogonal projection $p_{w^\perp} v_i$ of v_i into $w^\perp$ would still belong to the interior of V_{L,v_i}, which would contradict the above inclusion in the halfspace. Thus we obtain $\tilde{v}(x) \cdot w = \sum_{i=1}^{k} v_i \cdot w \geq \cos \gamma$, hence $|\tilde{v}(x)| \geq \cos \gamma$. We define now

$$v(x) := \frac{\tilde{v}(x)}{|\tilde{v}(x)|}, \quad x \in \partial X,$$

and notice that v is Lipschitz.

Take a point $x \in \partial X$ and notice that $\mathring{B}(x, 2\delta) \subset \mathring{B}(x_i, 5\delta)$ whenever $x \in \mathring{B}(x_i, \delta)$. Hence, X is an (L, v_i)-Lipschitz subgraph in $\mathring{B}(x, 2\delta)$ for all such i and, using Lemma 9.16 (ii), we obtain that

$$X \text{ is an } (L, v(x))\text{-Lipschitz subgraph in } \mathring{B}(x, 2\delta), \quad x \in \partial X. \tag{9.19}$$

In particular,

$$(x - V_{L,v(x)}) \cap \mathring{B}(x, 2\delta) \subset X \text{ and } \operatorname{int}(x + V_{L,v(x)}) \cap \mathring{B}(x, 2\delta) \subset \mathbb{R}^d \setminus X \tag{9.20}$$

whenever $x \in \partial X$ by Lemma 9.16 (i), hence

$$(\cos \gamma) s \leq d_{\partial X}(x + s v(x)) \leq s, \quad 0 < |s| < \delta. \tag{9.21}$$

Consider the continuous mapping

$$\phi : (x, s) \mapsto x + s\upsilon(x), \quad (x, s) \in \partial X \times (-\delta, \delta),$$

and denote

$$c := \frac{1}{\sqrt{2}} \min\left\{\cos\gamma, \frac{1}{2}\right\} \text{ and } \tau := \frac{1}{2} \min\left\{\frac{c}{\operatorname{Lip}\upsilon}, \delta\right\}.$$

Let $x, x' \in \partial X$, $s, s' \in (-\tau, \tau)$ be given, and assume that $|x - x'| < \delta$. Equation (9.20) implies that

$$|\phi(x, s - s') - x'| \geq \max\{\cos\gamma|x - x'|, \cos\gamma|s - s'|\} \geq c|(x, s) - (x', s')|.$$

If $|x - x'| \geq \delta$ then

$$|\phi(x, s - s') - x'| \geq \frac{1}{2}|x - x'| \geq \frac{1}{2\sqrt{2}}|(x, s) - (x', s')| \geq c|(x, s) - (x', s')|$$

again. Since

$$|\upsilon(x) - \upsilon(x')| \leq (\operatorname{Lip}\upsilon)|x - x'| \leq (\operatorname{Lip}\upsilon)|(x, s) - (x', s')|,$$

we get

$$\begin{aligned}
|\phi(x, s) - \phi(x', s')| &= |(\phi(x, s - s') - x') + s'(\upsilon(x) - \upsilon(x'))| \\
&\geq |\phi(x, s - s') - x'| - \tau|\upsilon(x) - \upsilon(x')| \\
&\geq (c - \tau\operatorname{Lip}\upsilon)|(x, s) - (x', s')| \\
&\geq \frac{c}{2}|(x, s) - (x', s')|.
\end{aligned}$$

Thus ϕ is injective on $\partial X \times (-\tau, \tau)$ and the inverse ϕ^{-1} is Lipschitz on the open set $\phi(\partial X \times (-\tau, \tau))$. Let p, s denote the components of ϕ^{-1}, i.e.,

$$z \mapsto (p(z), s(z)) \in \partial X \times (-\tau, \tau), \quad z \in \phi(\partial X \times (-\tau, \tau)),$$

is a continuous mapping.

Fix some $0 < \varepsilon < \frac{1}{2}(\cos\gamma)\tau$ and denote

$$h^{\varepsilon}(x) := \inf\{h > 0 : d_{\partial X}(x - h\upsilon(x)) = \varepsilon\}, \quad x \in \partial X.$$

The above observation (9.21) implies that

$$0 < h^{\varepsilon} < \varepsilon / \cos\gamma < \tau/2.$$

If h^ε is continuous then the mapping

$$F:(z,s)\mapsto\begin{cases}(1-s)z+s(h^\varepsilon(p(z))-s(z))v(p(z)), & z\in\overline{X\setminus X_{-\varepsilon}},\\ z, & z\in X_{-\varepsilon},\end{cases}\quad s\in[0,1],$$

is a deformation retraction of X onto $X_{-\varepsilon}$, proving $X\sim X_{-\varepsilon}$.

It remains thus to show that h^ε is continuous on ∂X. We will show that h^ε is even locally q-Lipschitz with Lipschitz constant

$$q:=2\sin\gamma+1+\tau\mathrm{Lip}\,v.$$

Let $x,x'\in\partial X$ be given and assume that

$$|x'-x|<\tau/(2q)<\tau/2.$$

Denote $t:=h^\varepsilon(x)$, $t':=t+q|x'-x|$, $y:=x-tv(x)$ and $y':=x'-t'v(x')$. Note that $t<\tau/2$ and $t'<\tau/2+q|x'-x|<\tau$. We have

$$\begin{aligned}(y-y')\cdot v(x)&=((x-tv(x))-(x'-t'v(x')))\cdot v(x)\\&=(t'-t)v(x)\cdot v(x)+(x-x')\cdot v(x)+t'(v(x')-v(x))\cdot v(x)\\&\geq q|x-x'|-|x-x'|-t'|v(x')-v(x)|\\&\geq(q-1-\tau\mathrm{Lip}\,v)|x-x'|\\&=2\sin\gamma|x-x'|,\end{aligned}$$

and since

$$|y-y'|\leq|x-x'|+|t-t'||v(x)-v(x')|\leq|x-x'|(1+\tau\mathrm{Lip}\,v)\leq 2|x-x'|,$$

we get $(y-y')\cdot v(x)\geq\sin\gamma|y-y'|$, hence $y-y'\in V_{L,v(x)}$ and, consequently,

$$B(y',\varepsilon)\subset B(y,\varepsilon)-V_{L,v(x)}.$$

Since $|y'-x|+\varepsilon<t'+|x'-x|+\varepsilon<2\delta$, we have also $B(y',\varepsilon)\subset\mathring{B}(x,2\delta)$, hence

$$B(y',\varepsilon)\subset(B(y,\varepsilon)-V_{L,v(x)})\cap\mathring{B}(x,2\delta).$$

As $d_{\partial X}(y)=\varepsilon$ and $B(y,\varepsilon)\subset\mathring{B}(x,2\delta)$ we have $B(y,\varepsilon)\subset X$. Thus we deduce (using again Lemma 9.16 (i))

$$(B(y,\varepsilon)-V_{L,v(x)})\cap\mathring{B}(x,2\delta)\subset X,$$

which, together with the previous inclusion, yields that $B(y',\varepsilon) \subset X$. Thus $d_{\partial X}(y') \geq \varepsilon$, which implies that $h^\varepsilon(x') \leq t' \leq t + q|x'-x| = h^\varepsilon(x) + q|x'-x|$. Interchanging the role of x and x', we thus obtain that h^ε is locally q-Lipschitz and the proof is finished. □

We consider now a compact Lipschitz domain X. Since the sets $X_{-\varepsilon}$ have positive reach if $\varepsilon > 0$ is small enough, they admit normal cycles $N_{X_{-\varepsilon}}$. It is thus natural to try to define the normal cycle of X by approximation with $N_{X_{-\varepsilon}}$. This is, however, not possible in general; one can easily imagine examples in the plane of Lipschitz domains with unbounded Gauss curvature. Therefore, additional restrictions must be imposed.

Recall that $H_{v,t} = \{y : y \cdot v \leq t\}$, $\Lambda : (x,v) \mapsto (v, x \cdot v)$, cf. (9.16), and that $\operatorname{nor}^{\mathrm{Cl}} X$ is the Clarke normal bundle of X. Note that $\Lambda(\operatorname{nor}^{\mathrm{Cl}} X)$ can be interpreted as the set of all hyperplanes tangent to X.

Proposition 9.21 *Let X be a compact Lipschitz domain and assume that*

$$\mathcal{H}^d(\Lambda(\operatorname{nor}^{\mathrm{Cl}} X)) = 0. \tag{9.22}$$

Then, the following assertions hold for a.a. $(v,t) \in S^{d-1} \times \mathbb{R}$:

(i) *$X \cap H_{v,t}$ is a Lipschitz domain,*
(ii) *$X \cap H_{v,t} \sim X_{-\varepsilon} \cap H_{v,t}$ for all sufficiently small $\varepsilon > 0$.*

Proof First, note that X, $H_{v,t}$ do not touch whenever $(-v,t) \notin \Lambda(\operatorname{nor}^{\mathrm{Cl}} X)$, which is the case for a.a. $(v,t) \in S^{d-1} \times \mathbb{R}$ by (9.22). Assertion (i) follows thus by Lemma 9.18.

If X, $H_{v,t}$ do not touch then $Y := X \cap H_{v,t}$ is a compact Lipschitz domain. We construct a homotopy similarly as in the proof of Proposition 9.20. Let $v : \partial Y \to S^{d-1}$ be the transversal vector field and let the continuous mappings p, s be defined on some neighbourhood of ∂Y, as in the previous proof. Set

$$\tilde{h}^\varepsilon : x \mapsto \min\{h \geq 0 : d_{\partial X}(x - hv(x)) \geq \varepsilon\}, \quad x \in \partial Y$$

(i.e., $\tilde{h}^\varepsilon$ measures the distance from ∂Y to $\partial X_{-\varepsilon}$ along v). Again, $\tilde{h}$ is continuous and the mapping

$$\tilde{F} : (z,s) \mapsto \begin{cases} (1-s)z + s(\tilde{h}^\varepsilon(p(z)) - s(z))v(p(z)), & z \in \overline{Y \setminus X_{-\varepsilon}}, \\ z, & z \in Y \cap X_{-\varepsilon}, \end{cases} \quad s \in [0,1],$$

is a deformation retraction of Y onto $Y \cap X_{-\varepsilon}$, proving $X \cap H_{v,t} \sim X_{-\varepsilon} \cap H_{v,t}$. □

We are now ready to define normal cycles for compact Lipschitz manifolds.

Theorem 9.22 *Let a compact Lipschitz domain $X \subset \mathbb{R}^d$ satisfy* (9.22) *and assume that*

$$\liminf_{\varepsilon \to 0} \mathbf{M}(N_{X_{-\varepsilon}}) < \infty. \tag{9.23}$$

Then X admits a normal cycle N_X. Furthermore,

$$N_X = (\mathrm{F}) \lim_{i\to\infty} N_{X_{-\varepsilon_i}}$$

whenever $\varepsilon_i \to 0$, $(\mathbf{M}(N_{X_{-\varepsilon_i}}))$ is bounded and $(\mathrm{F}) \lim_{i\to\infty} N_{X_{-\varepsilon_i}}$ exists.

Remark 9.23

1. Buczolich [Buc98] presented an example of a C^1 function of two variables whose set of tangent planes has positive three-dimensional measure. Hence, condition (9.22) is not fulfilled automatically.
2. Note that since reach $X_{-\varepsilon} > 0$ we have

$$\mathbf{M}(N_{X_{-\varepsilon}}) = \mathcal{H}^{d-1}(\mathrm{nor}\, X_{-\varepsilon})$$

by the definition of the normal cycle, see Sect. 4.6.

Proof By (9.23), there exists a sequence $\varepsilon_i \to 0_+$ such that $(\mathbf{M}(N_{X_{-\varepsilon_i}}))$ is bounded. The Compactness theorem for currents (Theorem 1.57) and Lemma 1.60 imply that there exists a subsequence $\varepsilon_{i(j)}$ such that $N_{-\varepsilon_{i(j)}}$ converges in the flat norm to some current T. We will show that T is the normal cycle of X. Since the normal cycle is unique, the assertion will follow.

It remains thus to show that if $(\mathrm{F}) \lim_{i\to\infty} N_{X_{-\varepsilon_i}} = T$ for some $\varepsilon_i \to 0$ with $(\mathbf{M}(N_{X_{-\varepsilon_i}}))$ bounded then T is the normal cycle of X. Since $N_{X_{-\varepsilon_i}}$ is the normal cycle of $X_{-\varepsilon_i}$, we have

$$p_\# \langle N_{X_{-\varepsilon_i}}, \pi_1, -v\rangle (\mathbf{1}_{(-\infty,t]}) = \chi(X_{-\varepsilon_i} \cap H_{v,t}) \tag{9.24}$$

for a.a. $(v, t) \in S^{d-1} \times \mathbb{R}$. Proposition 1.66 implies that

$$(\mathrm{F}) \lim_{i\to\infty} \langle N_{X_{-\varepsilon_i}}, \pi_1, -v\rangle = \langle T, \pi_1, -v\rangle,$$

for a.a. $v \in S^{d-1}$ and it follows easily that also

$$(\mathrm{F}) \lim_{i\to\infty} p_\# \langle N_{-X_{\varepsilon_i}}, \pi_1, -v\rangle = p_\# \langle T, \pi_1, -v\rangle \tag{9.25}$$

for a.a. $v \in S^{d-1}$. Using subsequently Fatou's lemma, (1.34) and (9.23), we obtain

$$\begin{aligned}
\int_{S^{d-1}} & \liminf_{i\to\infty} \mathbf{M}\langle N_{X_{-\varepsilon_i}}, \pi_1, -v\rangle\, \mathcal{H}^{d-1}(\mathrm{d}v) \\
&\le \liminf_{i\to\infty} \int_{S^{d-1}} \mathbf{M}\langle N_{X_{-\varepsilon_i}}, \pi_1, -v\rangle\, \mathcal{H}^{d-1}(\mathrm{d}v) \\
&= \liminf_{i\to\infty} \mathbf{M}(N_{X_{-\varepsilon_i}} \llcorner (\pi_1)^\# \Omega_{S^{d-1}})
\end{aligned}$$

$$\leq \liminf_{i\to\infty} \mathbf{M}(N_{X_{-\varepsilon_i}}) < \infty,$$

where $\Omega_{S^{d-1}}$ denotes the natural $(d-1)$- form on the sphere. Consequently, for a.a. $v \in S^{d-1}$, a subsequence of $(\mathbf{M}\langle N_{X_{-\varepsilon_i}}, \pi_1, -v\rangle)_i$ is bounded, and, passing to a subsequence, we can assume that this is true for the whole sequence. Together with (9.25), this yields the weak convergence of signed measures $p_\#\langle N_{X_{-\varepsilon_i}}, \pi_1, -v\rangle$ to $p_\#\langle T, \pi_1, -v\rangle$, $i \to \infty$, on the real line and it is well-known that in such a case, the associated distribution functions converge almost everywhere:

$$\lim_{i\to\infty} p_\#\langle N_{X_{-\varepsilon_i}}, \pi_1, -v\rangle(\mathbf{1}_{(-\infty,t]}) = \langle T, \pi_1, -v\rangle(\mathbf{1}_{(-\infty,t]})$$

for a.a. $(v, t) \in S^{d-1} \times \mathbb{R}$. Hence, the left-hand side of (9.24) converges to the left-hand side of (9.15) in Definition 9.5. But the right-hand side converges as well by Proposition 9.21. This verifies condition (9.15) and the proof is finished. □

Our next aim is to extend the definition of the normal cycle to closed (unbounded) Lipschitz domains. If $X \subset \mathbb{R}^d$ is a closed Lipschitz domain and $\varepsilon > 0$ is small enough then $X_{-\varepsilon}$ has "locally positive reach" by Lemma 9.19. Hence, the unit normal bundle nor $X_{-\varepsilon}$ and normal cycle $N_{X_{-\varepsilon}}$ are defined as for sets with positive reach (in fact, both are defined locally). Instead of assumption (9.23), we will consider its local variant. For convenience, we introduce a notation for the family of closed Lipschitz domains for which normal cycles will be introduced.

Definition 9.24 Let $\mathcal{LD}_d$ denote the family of all closed Lipschitz domains $X \subset \mathbb{R}^d$ for which

$$\mathcal{H}^d(\mathrm{nor}^{\mathrm{Cl}}\, X) = 0 \tag{9.26}$$

and

$$\liminf_{\varepsilon\to 0} \mathbf{M}(N_{X_{-\varepsilon}} \llcorner \pi_0^{-1}(K)) < \infty, \quad K \subset \mathbb{R}^d \text{ compact} \tag{9.27}$$

(recall that $\pi_0 : (x, n) \mapsto x$).

Remark 9.25 Clearly (9.26) implies (9.22). Also, note that (9.27) is equivalent to

$$N_{X_{-\varepsilon_i}} \text{ have locally uniformly bounded mass norm for some } \varepsilon_i \to 0,$$

and that

$$\mathbf{M}(N_{X_{-\varepsilon}} \llcorner \pi_0^{-1}(K)) = \mathcal{H}^{d-1}(\mathrm{nor}\, X_{-\varepsilon} \cap \pi_0^{-1}(K)).$$

Remark 9.26 An important subclass of closed Lipschitz domains are closed *DC domains*, i.e., sets that can be locally represented as subgraphs of DC functions (differences of convex functions). Normal cycles of closed DC domains were constructed in [PR13]. Any closed DC domain satisfies (9.26) ($\operatorname{nor}^{\mathrm{Cl}} X$ has even locally finite $(d-1)$-dimensional Minkowski content) and an analogon to (9.27) holds, namely

$$\int_0^{\varepsilon_0} \mathbf{M}(N_{X_{-\varepsilon}} \llcorner \pi_0^{-1}(K))\, \mathrm{d}\varepsilon < \infty, \quad K \subset \mathbb{R}^d \text{ compact}, \ \varepsilon_0 > 0.$$

It is not known whether any closed DC domain belongs to $\mathcal{LD}_d$.

Let us recall some notions and facts from Sect. 6.1. If X, Y are sets with positive reach then the current $N_{X,Y}$ is defined as

$$N_{X,Y} = F_\#(((N_X \times N_Y) \llcorner R^0) \times [0,1])$$

(see (6.3) for the definition of F and R^0), provided that the mass norm of the push-forward is locally bounded. Note that F is Lipschitz on

$$R^\delta := \{(x,m,y,n) \in (\mathbb{R}^d)^4 : m \cdot n \geq (-1+\delta)|m||n|\}, \quad \delta > 0.$$

Let $\gamma^\delta \in C^\infty(R^0)$ be equal 0 on $R^0 \setminus R^{\delta/2}$ and 1 on $S^{d-1} \cap R^\delta$; the currents

$$N_{X,Y}^\delta := N_{X,Y} \llcorner \gamma^\delta, \quad \delta > 0 \tag{9.28}$$

have locally bounded mass norm (since F is Lipschitz on $\operatorname{spt} \gamma^\delta$) and (F) $\lim N_{X,Y}^\delta = N_{X,Y}$ whenever $N_{X,Y}$ has locally bounded mass norm. We also observe that

$$\begin{aligned}\mathbf{M}(N_{X,Y}^\delta \llcorner (K_1 \times K_2 \times S^{d-1}))& \\ &\leq (\operatorname{Lip}(F|R^{\frac{\delta}{2}}))^{2d-1}\, \mathbf{M}(N_X \llcorner \pi_0^{-1} K_1)\, \mathbf{M}(N_Y \llcorner \pi_0^{-1} K_2)\end{aligned}$$

whenever $K_1, K_2 \subset \mathbb{R}^d$ are compact.

The following technical lemma quantifies in certain way the important non-touching property.

Lemma 9.27 *If two closed Lipschitz manifolds X, Y do not touch and $K \subset \mathbb{R}^d$ is compact then there exist $\delta > 0$ such that $m \cdot n \geq -1 + \delta$ whenever $x \in K$, $(x,m) \in \operatorname{nor}^{\mathrm{Cl}} X$ and $(x,n) \in \operatorname{nor}^{\mathrm{Cl}} Y$ or $(x,m) \in \operatorname{nor} X_{-r}$, $(x,n) \in \operatorname{nor} Y_{-r}$ and $0 < r < \delta$.*

Proof Since $\operatorname{nor}^{\mathrm{Cl}} X$ and $\operatorname{nor}^{\mathrm{Cl}} Y$ are closed sets (see Lemma 9.17 (iii)), there exists $\delta > 0$ such that $\angle(m, n) < \pi - \delta$ whenever $x \in K$, $(x, m) \in \operatorname{nor}^{\mathrm{Cl}} X$ and $(x, n) \in \operatorname{nor}^{\mathrm{Cl}} Y$. We further claim that

$$\limsup_{r \to 0} \operatorname{nor} X_{-r} \subset \operatorname{nor}^{\mathrm{Cl}} X, \tag{9.29}$$

i.e., if $r_i \to 0$, $(x_i, n_i) \in \operatorname{nor} X_{-r_i}$ and $(x_i, n_i) \to (x, n)$, $i \to \infty$, then $(x, n) \in \operatorname{nor}^{\mathrm{Cl}} X$. This already implies the assertion. In order to show (9.29), denote

$$\Sigma_X(x) := \{y \in \partial X : |y - x| = r\}, \quad x \in \partial X_{-r},$$

and note that n_i belongs to the spherical convex hull of $r_i^{-1}(\Sigma_X(x_i) - x_i)$ (cf. the proof of Lemma 9.19). Using the Caratheodory's theorem and turning to a subsequence, we can achieve the existence of $y_i^{(0)}, \ldots, y_i^{(d)} \in \Sigma_X(x_i)$ such that $n_i^{(k)} := r_i^{-1}(y_i^{(k)} - x_i) \to n^{(k)} \in S^{d-1}$, $i \to \infty$, $0 \le k \le d$, and n_i is a spherical convex combination of $n_i^{(0)}, \ldots, n_i^{(d)}$, $i \in \mathbb{N}$. Lemma 9.17 (vii) implies that $n_i^{(k)} \in N(X, y_i)$, $k = 0, \ldots, d$, $i \in \mathbb{N}$, hence also $n^{(k)} \in N(X, x)$ by Lemma 9.17 (iii). Since n lies in the spherical convex hull of $n^{(0)}, \ldots, n^{(d)}$, also $n \in N(X, x)$. This proves (9.29). □

The following proposition states some generic properties of an intersection of two Lipschitz domains. Note that the assumptions on X, Y are not symmetric.

Proposition 9.28 *Let* $X, Y \in \mathcal{LD}_d$, $\operatorname{reach} Y > 0$. *Then, for almost all Euclidean motions* $g \in \mathcal{G}_d$,

(i) *X and gY do not touch,*
(ii) *$X \cap gY$ is a closed Lipschitz domain,*
(iii) $\mathcal{H}^d(\operatorname{nor}^{\mathrm{Cl}}(X \cap gY)) = 0$,
(iv) $X \cap gY \in \mathcal{LD}_d$.

Proof In order to show (i), we proceed similarly as in the proof of Proposition 6.13 (iv). Since Y has positive reach, $\operatorname{nor}^{\mathrm{Cl}} Y = \operatorname{nor} Y$ is locally $(d-1)$-rectifiable and, using (9.27), we obtain $\mathcal{H}^{2d-1}(\operatorname{nor}^{\mathrm{Cl}} X \times \operatorname{nor}^{\mathrm{Cl}} Y) = 0$. Lemma 1.73 implies that $\mathcal{H}^q(A) = 0$ for

$$A = \{(x, m, y, n, \rho) \in \operatorname{nor}^{\mathrm{Cl}} X \times \operatorname{nor}^{\mathrm{Cl}} Y \times \mathrm{SO}(d) : m + \rho n = 0\}$$

and

$$q = 2d - 1 + \frac{(d-1)(d-2)}{2} = d + \frac{d(d-1)}{2}.$$

Hence, also $\mathcal{H}^q(h(A)) = 0$ for $h : (x, m, y, n, \rho) \mapsto (x - \rho y, \rho)$, and since

$$h(A) = \{(z, \rho) \in \mathbb{R}^d \times \mathrm{SO}(d) : X \text{ and } z + \rho Y \text{ touch}\},$$

the conclusion follows.

(ii) follows from (i) using Lemma 9.18.

(iii) Take a point $x \in \partial X \cap \partial Y$. Using Lemma 9.17 (ii) one can easily show that $\operatorname{int} C(X,x) \cap \operatorname{int} C(Y,x) \subset \operatorname{int} C(X \cap Y, x)$. Further, if X and Y do not touch then it was shown in the proof of Lemma 9.18 that $\operatorname{int} C(X,x) \cap \operatorname{int} C(Y,x) \neq \emptyset$. Since $C(X,x)$ and $C(Y,x)$ are closed conves cones, we deduce that $C(X,x) \cap C(Y,x) \subset C(X \cap Y, x)$. Thus, using the properties of polar cones (2.7), we get

$$\begin{aligned} N(X,x) + N(Y,x) &= C(X,x)^o + C(Y,x)^o = (C(X,x) \cap C(Y,x))^o \\ &\supset C(X \cap Y, x)^o = N(X \cap Y, x). \end{aligned}$$

Thus, similarly as in Sect. 6.1, if X and $Y+z$ do not touch for some $z \in \mathbb{R}^d$ then

$$\begin{aligned} \operatorname{nor}^{\mathrm{Cl}}(X \cap (Y+z)) \subset{}& \operatorname{nor}^{\mathrm{Cl}} X \cup \operatorname{nor}^{\mathrm{Cl}}(Y+z) \\ &\cup (\hat{\pi} \circ F)\Big(((\operatorname{nor}^{\mathrm{Cl}} X \times \operatorname{nor}^{\mathrm{Cl}}(Y+z)) \cap R^0) \times [0,1]\Big) \cap G^{-1}\{z\}, \end{aligned}$$

where the mapping $F : R^0 \times [0,1] \to \mathbb{R}^{2d} \times S^{d-1}$ is defined in (6.3), $G : (x,y,u) \mapsto x-y$ and $\hat{\pi} : (x,y,u) \mapsto (x,u)$. Both $\operatorname{nor}^{\mathrm{Cl}} X$ and $\operatorname{nor}^{\mathrm{Cl}}(Y+z)$ and have d-dimensional measure zero by assumptions, and also $\mathcal{H}^{2d-1}(\operatorname{nor}^{\mathrm{Cl}} X \times \operatorname{nor}^{\mathrm{Cl}}(Y+z)) = 0$. Since F is locally Lipschitz,

$$\mathcal{H}^{2d}\Big((\hat{\pi} \circ F)\Big(((\operatorname{nor}^{\mathrm{Cl}} X \times \operatorname{nor}^{\mathrm{Cl}}(Y+z)) \cap R^0) \times [0,1]\Big)\Big) = 0.$$

An application of the Coarea formula then implies that

$$\mathcal{H}^d\Big\{z \in \mathbb{R}^d : X, Y+z \text{ do not touch and } \mathcal{H}^d(\operatorname{nor}^{\mathrm{Cl}}(X \cap (Y+z))) > 0\Big\} = 0.$$

Together with assertion (i), this already implies (iii).

(iv) In order to show (iv), it remains to verify (9.27) for $X \cap gY$. Let $\varepsilon_i \to 0_+$ be a sequence for which the normal cycles $N_{X_{-\varepsilon_i}}$ have locally uniformly bounded mass norms, and denote for brevity $X^i := X_{-\varepsilon_i}$, $Y^i := Y_{-\varepsilon_i}$. Any Euclidean motion $g \in \mathcal{G}_d$ can be uniquely decomposed into the shift $z \in \mathbb{R}^d$ and orthogonal part $\rho \in \mathrm{SO}(d)$, so that $g(\cdot) = z + \rho(\cdot)$. Using (6.7), Propositions 6.5 and 6.13 (iv), we can decompose the normal cycle of $X^i \cap gY^i$ for any $K \subset \mathbb{R}^d$ compact, almost all $\rho \in \mathrm{SO}(d)$, almost all $z \in \mathbb{R}^d$ as

$$\begin{aligned} N_{X^i \cap gY^i} \llcorner \pi_0^{-1}(K) = N_{X^i} &\llcorner \pi_0^{-1}(K \cap \operatorname{int} gY^i) \\ &+ N_{gY^i} \llcorner \pi_0^{-1}(K \cap \operatorname{int} X^i) \\ &+ (-1)^d \hat{\pi}_\# \langle N_{X^i, \rho Y^i}, G, z \rangle \llcorner \pi_0^{-1}(K). \end{aligned}$$

The mass norm of the first summand is uniformly bounded by assumption. We shall show that N_{Y^i} (and, hence, also N_{gY^i}) have locally uniformly bounded mass norms. Lemma 9.19 implies that $\liminf_{i\to\infty} \operatorname{reach} Y^i > 0$. If $B \subset \mathbb{R}^d$ is a ball not touching Y (which is the case for almost all balls), we get from Lemma 9.27 and Proposition 4.17 that

$$\liminf_{i\to\infty} \operatorname{reach}(Y^i \cap B) > 0.$$

Since $Y^i \cap B \to Y \cap B$ in the Hausdorff metric, Theorem 7.3 implies that the mass norms of $N_{Y^i \cap B}$ are uniformly bounded. Since $N_{Y^i} \llcorner \mathring{B} = N_{Y^i \cap B} \llcorner \mathring{B}$, the currents N_{Y^i} have locally uniformly bounded mass norms.

Using (1.34) we obtain that for almost all $\rho \in \mathrm{SO}(d)$ and any Borel subset $Z \subset \mathbb{R}^d$,

$$\int_Z \mathbf{M}\left(\hat{\pi}_\# \langle N_{X^i,\rho Y^i}, G, z\rangle \llcorner \pi_0^{-1} K\right) \mathrm{d}z \le \|N_{X^i,\rho Y^i}\|(K \times (K - Z) \times S^{d-1}).$$

This, using (6.14), we get

$$\limsup_{i\to\infty} \int_H \mathbf{M}(N_{X^i \cap g Y^i} \llcorner \pi_0^{-1}(K))\, \mathrm{d}g < \infty$$

for any bounded measurable set $H \subset \mathcal{G}_d$. The Fatou's lemma implies that

$$\begin{aligned} &\int_H \liminf_{i\to\infty} \mathbf{M}(N_{X^i \cap g Y^i} \llcorner \pi_0^{-1}(K))\, \mathrm{d}g \\ &\le \liminf_{i\to\infty} \int_H \mathbf{M}(N_{X^i \cap g Y^i} \llcorner \pi_0^{-1}(K))\, \mathrm{d}g < \infty, \end{aligned}$$

hence,

$$\liminf_{i\to\infty} \mathbf{M}(N_{X^i \cap g Y^i} \llcorner \pi_0^{-1}(K)) < \infty$$

for almost all $g \in H$. This completes the proof since H was any bounded measurable subset of $\mathcal{G}_d$. □

Recall that for a *compact* $X \in \mathcal{LD}_d$, the normal cycle N_X was already defined (cf. Theorem 9.22). We will show that these normal cycles are locally determined. This will enable us to extend the definition to closed (unbounded) sets from $\mathcal{LD}_d$.

Lemma 9.29 *Let $X, Y \in \mathcal{LD}_d$ be compact, and let $U \subset \mathbb{R}^d$ be an open set such that $X \cap U = Y \cap U$. Then*

$$N_X \llcorner \pi_0^{-1}(U) = N_Y \llcorner \pi_0^{-1}(U).$$

Proof Let $\varepsilon_i, \delta_i \to 0_+$ be two sequences such that $(N_{X_{-\varepsilon_i}})$, $(N_{Y_{-\delta_i}})$ have uniformly bounded mass norms and

$$\text{(F)} \lim_{i\to\infty} N_{X_{-\varepsilon_i}} = N_X, \quad \text{(F)} \lim_{i\to\infty} N_{Y_{-\delta_i}} = N_Y.$$

Given $r > 0$ there exists a compact C^2-domain Z such that $U_{-r} \subset Z \subset U$. Using Lemma 9.28, we can even achieve (applying a "small" motion) that $W := X \cap Z = Y \cap Z$ is a compact Lipschitz domain satisfying (9.23) and (9.26), and, turning to suitable subsequences of (ε_i) and (δ_i) (for which we do not change notation),

$$N_W = \text{(F)} \lim_{i\to\infty} N_{W_{-\varepsilon_i}} = \text{(F)} \lim_{i\to\infty} N_{W_{-\delta_i}}.$$

Note that

$$W_{-\varepsilon} \cap U_{-r} = X_{-\varepsilon} \cap U_{-r} = Y_{-\varepsilon} \cap U_{-r}, \quad \varepsilon < r,$$

thus

$$N_W \llcorner \pi_0^{-1}(U_{-r}) = N_X \llcorner \pi_0^{-1}(U_{-r}) = N_Y \llcorner \pi_0^{-1}(U_{-r}),$$

and since this is true for all $r > 0$, we get also

$$N_W \llcorner \pi_0^{-1}(U) = N_X \llcorner \pi_0^{-1}(U) = N_Y \llcorner \pi_0^{-1}(U).$$

□

Theorem 9.30 *If $X \in \mathcal{LD}_d$ then there exists a unique Legendrian cycle N_X such that $N_X \llcorner \pi_0^{-1}(U) = N_Y \llcorner \pi_0^{-1}(U)$ whenever $Y \in \mathcal{LD}_d$ is compact and $U \subset \mathbb{R}^d$ an open set with $X \cap U = Y \cap U$.*

Proof Note that for any bounded open set $U \subset \mathbb{R}^d$ there is a compact $Y \in \mathcal{LD}_d$ satisfying $N_X \llcorner \pi_0^{-1}(U) = N_Y \llcorner \pi_0^{-1}(U)$. Indeed, we can find such an Y by intersecting X with a finite number of halfspaces, using Proposition 9.28. We can thus define

$$N_X \llcorner \pi_0^{-1}(U) := N_Y \llcorner \pi_0^{-1}(U).$$

The uniqueness follows from Lemma 9.29. □

Remark 9.31 It follows from the definition and from (9.29) that

$$\operatorname{spt} N_X \subset \limsup_{i\to\infty} \operatorname{spt} N_{X_{\varepsilon_i}} \subset \operatorname{nor}^{\mathrm{Cl}} X.$$

Remark 9.32 A weak point of Theorem 9.22 is that it is difficult to determine which Lipchitz domains fulfill the assumption (9.27). In fact, aside of domains constructed from convex bodies or sets with positive reach and their differences, we do not know other examples for which Theorem 9.22 can be applied. The method of approximation using the uniqueness theorem and continuity of Euler characteristic is, however, canonical and used in other variants in different situation, e.g. for d.c. domains (with a different approximation), see Bibliographical note No. 2.

Finally, we show a Principal kinematic formula (cf. Theorem 6.1) for intersections of sets from $\mathcal{LD}_d$ with sets of positive reach. We start with two auxiliary lemmas.

Lemma 9.33 *Let X, Y be as in Proposition 9.28. Then for almost all $g \in \mathcal{G}_d$,*

$$N_{X\cap gY} = N_X \llcorner \pi_0^{-1}(\operatorname{int} gY) + N_{gY} \llcorner \pi_0^{-1}(\operatorname{int} X) + (-1)^d \hat{\pi}_\#\langle N_{X,\rho Y}, G, z\rangle,$$

where $z \in \mathbb{R}^d$ and $\rho \in SO(d)$ are the shift and orthogonal part of g.

Proof Let X, Y and (ε_i) be as in Proposition 9.28 and denote $X^i := X_{-\varepsilon_i}$, $Y^i := Y_{-\varepsilon_i}$, $i \in \mathbb{N}$. We know from (6.7) and Proposition 6.5, Proposition 6.13 (iv) and Lemma 6.18 that for all i and almost all $g \in \mathcal{G}_d$,

$$N_{X^i\cap gY^i} = N_{X^i} \llcorner \pi_0^{-1}(\operatorname{int} gY^i) + N_{gY^i} \llcorner \pi_0^{-1}(\operatorname{int} X^i) + (-1)^d \hat{\pi}_\#\langle N_{X^i,\rho Y^i}, G, z\rangle.$$

Let ψ be a smooth function with compact support K, assume that X and gY do not touch and let $\delta > 0$ be the number from Lemma 9.27. Then

$$\begin{aligned} N_{X^i\cap gY^i} \llcorner \psi = {} & (N_{X^i} \llcorner \psi) \llcorner \pi_0^{-1}\operatorname{int} gY^i + (N_{gY^i} \llcorner \psi) \llcorner \pi_0^{-1}\operatorname{int} X^i \\ & + (-1)^d \hat{\pi}_\#\langle N^\delta_{X^i,\rho Y^i}, G, z\rangle \llcorner \psi \end{aligned}$$

fur sufficiently large i. Proposition 9.28 implies that, turning to a subsequence, the left side converges to $N_{X\cap gY} \llcorner \psi$. We will show that (again for a subsequence)

$$(N_{X^i} \llcorner \psi) \llcorner \pi_0^{-1}\operatorname{int} gY^i \to (N_X \llcorner \psi) \llcorner \pi_0^{-1}\operatorname{int} gY \text{ weakly.} \tag{9.30}$$

Denoting $T_i := N_{X^i} \llcorner \psi$, $T := N_X \llcorner \psi$ and $W_i := \operatorname{int} Y \setminus \operatorname{int} Y^i$, we have

$$T_i \llcorner \pi_0^{-1}\operatorname{int} gY^i - T \llcorner \pi_0^{-1}\operatorname{int} gY = -T_i \llcorner \pi_0^{-1} gW_i + (T_i - T) \llcorner \pi_0^{-1}\operatorname{int} gY.$$

We will show that the first term tends to zero in mass norm, and the second one weakly. Concerning the first one, note that

$$\mathbf{M}(T_i \llcorner \pi_0^{-1} gW_i) \le \mu_i(gW_i)$$

with $\mu_i := \|T_i\| \circ \pi_0^{-1}$. If $C > 0$ and $\mathcal{G}_d(C)$ denotes the set of all Euclidean motions from $\mathcal{G}_d$ with shift $z \in B(0, C)$ and orthogonal part $\rho \in \mathrm{SO}(d)$, we have using Fubini's theorem

$$\begin{aligned}\int_{\mathcal{G}_d(C)} \mu_i(gW_i)\,\mathrm{d}g &= \int\int_{\mathrm{SO}(d)}\int_{B(0,C)} \mathbf{1}_{z+\rho W_i}(x)\,\mathrm{d}z\,\vartheta_d(\mathrm{d}\rho)\,\mu_i(\mathrm{d}x)\\ &= \int\int_{\mathrm{SO}(d)} \mathcal{L}^d(B(0,C)\cap(x-\rho W_i))\,\vartheta_d(\mathrm{d}\rho)\,\mu_i(\mathrm{d}x)\\ &\le \mathcal{L}^d(W_i\cap B(0,C'))\,\mu_i(\mathbb{R}^d)\end{aligned}$$

with $C' = C + \sup_{x\in K}|x|$. Since $W^i \searrow \emptyset$ and $\mu_i(\mathbb{R}^d)$ are uniformly bounded, we get

$$\lim_{i\to\infty}\int_{\mathcal{G}_d(C)} \mu_i(gW_i)\,\mathrm{d}g = 0,$$

and Fatou's lemma gives

$$\int_{\mathcal{G}_d(C)} \liminf_{i\to\infty} \mu_i(gW_i)\,\mathrm{d}g = 0.$$

Hence, for any $C > 0$ and almost all $g \in \mathcal{G}_d(C)$ (and, consequently, also for almost all $g \in \mathcal{G}_d$), $\mu_{i_k}(gW_{i_k}) \to 0$ for some subsequence (i_k). Similarly we can show that $\mu(\partial gY) = 0$ for almost all g, thus $\pi_0^{-1}(\operatorname{int} Y)$ is a continuity set for $\|T\|$ and, since $T_i \to T$ weakly, also $(T_i - T) \llcorner \pi_0^{-1}\operatorname{int} gY \to 0$ weakly. This verifies (9.30).

By a similar procedure we can show that for almost all g,

$$(N_{gY^i} \llcorner \psi) \llcorner \pi_0^{-1}\operatorname{int} X^i \to (N_{gY} \llcorner \psi) \llcorner \pi_0^{-1}\operatorname{int} X \text{ weakly.}$$

Finally, we have (F) $\lim_{i\to\infty} N^\delta_{X^i,\rho Y^i} = N^\delta_{X,\rho Y}$ for any $\rho \in \mathrm{SO}(d)$ and $\delta > 0$. Hence, using Proposition 1.66, also

$$\text{(F)}\lim_{i\to\infty}\langle N^\delta_{X^i,\rho Y^i}, G, z\rangle = \langle N^\delta_{X,\rho Y}, G, z\rangle$$

for almost all z. Thus we have

$$\begin{aligned}N_{X\cap gY} \llcorner \psi = (N_X \llcorner \psi) \llcorner \pi_0^{-1}\operatorname{int} gY + (N_{gY} \llcorner \psi) \llcorner \pi_0^{-1}\operatorname{int} X\\ +(-1)^d\hat{\pi}_\#\langle N^\delta_{X,\rho Y}, G, z\rangle \llcorner \psi.\end{aligned}$$

Using again Lemma 9.27 (as at the beginning of the proof), we infer

$$\begin{aligned}N_{X\cap gY} \llcorner \psi &= (N_X \llcorner \psi) \llcorner \pi_0^{-1}\operatorname{int} gY + (N_{gY} \llcorner \psi) \llcorner \pi_0^{-1}\operatorname{int} X \\ &\quad +(-1)^d \hat{\pi}_\#\langle N_{X,\rho Y}, G, z\rangle \llcorner \psi.\end{aligned}$$

for almost all g, which is already a local version of the formula to be shown. □

Lemma 9.34 *Let X, Y be as in Proposition* 9.28 *and let $N_{X_{-\varepsilon_i}}, N_{Y_{-\varepsilon_i}}$ have locally uniformly bounded mass norms for some $\varepsilon_i \to 0$. Then for any $K \subset \mathbb{R}^{3d}$ compact*

$$\lim_{\delta\to 0}\sup_i \int_{SO(d)} \mathbf{M}\left((N_{X_{-\varepsilon_i},\rho Y_{-\varepsilon_i}} - N^\delta_{X_{-\varepsilon_i},\rho Y_{-\varepsilon_i}}) \llcorner K\right) \vartheta_d(\mathrm{d}\rho) = 0$$

and

$$\sup_{\delta>0}\int_{SO(d)} \mathbf{M}\left(N^\delta_{X,\rho Y} \llcorner K\right) \vartheta_d(\mathrm{d}\rho) < \infty.$$

Consequently, $N_{X,\rho Y} =$ (F) $\lim_{\delta\to 0} N^\delta_{X,\rho Y}$ is a well-defined integral current for almost all $\rho \in SO(d)$.

Proof It is sufficient to prove the statement for X, Y compact. Denote $X^i := X_{-\varepsilon_i}$, $Y^i := Y_{-\varepsilon_i}$, $i \in \mathbb{N}$, for brevity. Using Lemma 6.18, we have for some finite constant c

$$\begin{aligned}&\mathbf{M}\left(N_{X^i,\rho Y^i} - N^\delta_{X^i,\rho Y^i}\right) \\ &\le c\int_{\operatorname{nor} X^i\times \operatorname{nor} \rho Y^i} \mathbf{1}_{m\cdot n\ge -1+\delta}(\sin\angle(m,n))^{3-d}\,\mathcal{H}^{2d-2}(\mathrm{d}(x,m,y,n)) \\ &= c\int_{\operatorname{nor} X^i\times \operatorname{nor} Y^i} \mathbf{1}_{m\cdot n\ge -1+\delta}(\sin\angle(m,\rho^{-1}n))^{3-d}\,\mathcal{H}^{2d-2}(\mathrm{d}(x,m,y,n)).\end{aligned}$$

A direct computation reveals that

$$\begin{aligned}&\lim_{\delta\to 0}\int_{SO(d)} \mathbf{1}_{m\cdot n\ge -1+\delta}(\sin\angle(m,\rho^{-1}n))^{3-d}\,\vartheta_d(\mathrm{d}\rho) \\ &= \mathcal{O}_{d-1}^{-1}\lim_{\delta\to 0}\int_{S^{d-1}} \mathbf{1}_{m\cdot n\ge -1+\delta}(\sin\angle(m,v))^{3-d}\mathcal{H}^{d-1}(\mathrm{d}v) = 0,\end{aligned}$$

and since $\mathcal{H}^{2d-2}(\operatorname{nor} X^i \times \operatorname{nor} Y^i) = \mathbf{M}(N_{X^i})\mathbf{M}(N_{Y^i})$ is uniformly bounded, the first assertion follows.

For the second statement, note that for almost all $\rho \in SO(d)$, $N^\delta_{X^i,\rho Y^i} \to N_{X^i,\rho Y^i}$ weakly, thus

$$\mathbf{M}(N^\delta_{X,\rho Y}) \le \liminf_{i\to\infty} \mathbf{M}(N^\delta_{X^i,\rho Y^i}),$$

and Fatou's lemma yields

$$\int_{SO(d)} \mathbf{M}(N^{\delta}_{X,\rho Y})\,\vartheta_d(\mathrm{d}\rho) \le \liminf_{i\to\infty} \int_{SO(d)} \mathbf{M}(N^{\delta}_{X^i,\rho Y^i})\,\vartheta_d(\mathrm{d}\rho),$$

which is, due to (6.14), bounded by

$$\liminf_{i\to\infty} \mathrm{const}\,\mathbf{M}(N_{X^i})\mathbf{M}(N_{Y^i}) < \infty.$$

□

We are now able to formulate and prove the Principal kinematic formula for intersections of closed Lipschitz domains from $\mathcal{LD}_d$ with sets of positive reach.

Theorem 9.35 *Let $X, Y \in \mathcal{LD}_d$ and assume that* reach $Y > 0$. *Then $X \cap gY \in \mathcal{LD}_d$ for almost all Euclidean motions $g \in \mathcal{G}_d$, and for any bounded Borel sets $A, B \subseteq \mathbb{R}^d$ and any integer $0 \le k \le d$,*

$$\int_{\mathcal{G}_d} C_k\big(X \cap gY, A \cap gB\big)\mathrm{d}g = \sum_{\substack{1\le r,s\le d\\ r+s=d+k}} \gamma(d,r,s)C_r(X,A)C_s(Y,B).$$

Proof The first statement has already been proved in Proposition 9.28. We will prove a version of the kinematic formula for smooth functions:

$$\int_{\mathcal{G}_d}\int f(z)h(g^{-1}z)\,C_k\big(X\cap gY,\mathrm{d}z\big)\vartheta_d(\mathrm{d}g) \tag{9.31}$$

$$= \sum_{\substack{1\le p,q\le d\\ p+q=d+k}} \gamma(d,p,q)\int f(x)\,C_p(X,\mathrm{d}x)\int h(y)\,C_q(Y,\mathrm{d}y)$$

whenever f, h are smooth function on $\mathbb{R}^d$ with compact supports; this is clearly equivalent to the Principal kinematic formula. Using localization, we can assume that both X and Y are compact.

The left hand side of (9.31) can be written as

$$I := \int_{\mathcal{G}_d} (N_{X\cap gY} \llcorner \psi_g)(\varphi_k)\,\mathrm{d}g$$

with $\psi_g = (f\circ\pi_0)(h\circ g^{-1}\circ\pi_0)$. Using Lemma 9.33 we get

$$I = \int_{\mathcal{G}_d}\Big(N_X \llcorner (\mathbf{1}_{\pi_0^{-1}\mathrm{int}\, gY}\cdot\psi_g)\Big)(\varphi_k)\,\mathrm{d}g$$

$$+\int_{\mathcal{G}_d}\Big(N_{gY} \llcorner (\mathbf{1}_{\pi_0^{-1}\mathrm{int}\, X}\cdot\psi_g)\Big)(\varphi_k)\,\mathrm{d}g$$

$$+\int_{\mathcal{G}_d} (-1)^d \left((\hat{\pi}_\# \langle N_{X,\rho Y}, G, z \rangle \llcorner \psi_g) \right) (\varphi_k) \, \mathrm{d}g$$
$$=: I_1 + I_2 + I_3.$$

Again, z and ρ are the shift and orthogonal part of $g \in \mathcal{G}_d$. The current $N_{X,\rho Y}$ is well defined for almost all $\rho \in \mathrm{SO}(d)$ by Lemma 9.34. Using the same procedure as in the proof of Theorem 6.10, we get

$$I_1 = (N_X \llcorner (f \circ \pi_0)) (\varphi_k) \int h(y) \, \mathrm{d}y = \int f \, \mathrm{d}C_k(X, \cdot) \int h \, \mathrm{d}C_d(Y, \cdot),$$

$$I_2 = (N_Y \llcorner (h \circ \pi_0)) (\varphi_k) \int f(x) \, \mathrm{d}x = \int f \, \mathrm{d}C_d(X, \cdot) \int h \, \mathrm{d}C_k(Y, \cdot).$$

Applying (1.33), we obtain

$$\begin{aligned}
(-1)^d I_3 &= \int_{\mathcal{G}_d} \langle N_{X,\rho Y}, G, z \rangle \left(\hat{\pi}^\# (\psi_g \cdot \varphi_k) \right) \mathrm{d}g \\
&= \int_{\mathrm{SO}(d)} \left(N_{X,\rho Y} \llcorner G^\# \Omega_d \right) \left(\hat{\pi}^\# (\psi_g \cdot \varphi_k) \right) \vartheta_d(\mathrm{d}\rho) \\
&= \int_{\mathrm{SO}(d)} N_{X,\rho Y} \left((\psi_g \circ \hat{\pi}) G^\# \Omega_d \wedge \hat{\pi}^\# \varphi_k \right) \vartheta_d(\mathrm{d}\rho) \\
&=: \int_{\mathrm{SO}(d)} N_{X,\rho Y}(\phi) \, \vartheta_d(\mathrm{d}\rho).
\end{aligned}$$

Let $\varepsilon_i \to 0$ be a sequence for which $(\mathbf{M}(N_{X_{-\varepsilon_i}}))$ is bounded and

$$\text{(F)} \lim_{i \to \infty} N_{X_{-\varepsilon_i}} = N_X.$$

Since reach $Y > 0$, also $(\mathbf{M}(N_{Y_{-\varepsilon_i}}))$ is bounded and (F) $\lim_{i \to \infty} N_{Y_{-\varepsilon_i}} = N_X$. We will write $X^i := X_{-\varepsilon_i}$ and $Y^i := Y_{-\varepsilon_i}$ for brevity. If we replace X, Y with X^i, Y^i, respectively, in the above formula for I_3, we obtain

$$I_3^i = \sum_{\substack{r+s=d+k \\ 1 \le r,s \le d-1}} \gamma(d, r, s) \int f(x) \, C_p(X^i, \mathrm{d}x) \int h(y) \, C_q(Y^i, \mathrm{d}y),$$

see Chapter 6, Corollary 6.12 and Lemma 6.19, and this converges to

$$\sum_{\substack{r+s=d+k \\ 1 \le r,s \le d-1}} \gamma(d, r, s) \int f(x) \, C_p(X, \mathrm{d}x) \int h(y) \, C_q(Y, \mathrm{d}y)$$

as $i \to \infty$. Thus, the proof will be finished by showing that

$$\lim_{i\to\infty} \int_{SO(d)} (N_{X,\rho Y} - N_{X^i,\rho Y^i})(\phi)\, \vartheta_d(d\rho) = 0. \tag{9.32}$$

We can estimate

$$\begin{aligned}\left|\int_{SO(d)} (N_{X,\rho Y} - N_{X^i,\rho Y^i})(\phi)\, \vartheta_d(d\rho)\right| &\le \int_{SO(d)} |(N_{X,\rho Y} - N^{\delta}_{X,\rho Y})(\phi)|\, \vartheta_d(d\rho)\\ &+ \int_{SO(d)} |(N^{\delta}_{X,\rho Y} - N^{\delta}_{X^i,\rho Y^i})(\phi)|\, \vartheta_d(d\rho)\\ &+ \int_{SO(d)} |(N^{\delta}_{X^i,\rho Y^i} - N_{X^i,\rho Y^i})(\phi)|\, \vartheta_d(d\rho).\end{aligned}$$

The first and third integrals tend to zero with $\delta \to 0$ by Lemma 9.34 uniformly in $i \in \mathbb{N}$ (we apply the Lebesgue dominated theorem for the first integral). The second integral tends to zero as $i \to \infty$ for any fixed $\delta > 0$. This proves (9.32). □

9.4 Bibliographical Notes

1. The uniqueness theorem for normal cycles was first proved by Fu in 1989 [Fu89b] and further developed in [Fu94] to introduce normal cycles for subanalytic sets in $\mathbb{R}^d$. Fu introduced the following technique to construct normal cycles. A function on $\mathbb{R}^d$ is called *Monge-Ampère* if it admits a so-called "gradient cycle" (current corresponding to the integration over the graph of the gradient in the smooth case). The gradient cycle is then used to define normal cycles for weakly regular sublevel sets of a Monge-Ampère function. It can be shown that weakly regular sublevel sets of semiconvex functions are exactly the sets with positive reach [Ban82, Kle81].
2. A real function is called DC if it can be represented as a difference of two convex functions. Normal cycles for compact *DC domains* (sets locally representable as subgraphs of d.c. functions) were introduced in [PR13]. In fact, a larger family of sets was considered, namely weakly regular sublevel sets of DC functions on $\mathbb{R}^d$, called (locally) *WDC sets*. Since DC functions are Monge-Ampère, the construction of Fu described in 1 can be applied. The case of compact DC domains in dimension $d = 3$ was already solved by Fu in [Fu00], following the approach of A.D. Alexandrov. Some properties of (locally) WDC sets, including a full characterization in $\mathbb{R}^2$, were given in [PRZar].
3. Geometry of sets in Euclidean spaces from a certain o-minimal system (including, in particular, subanalytic sets) was considered by several authors. Bröcker, Kuppe [BK00] and Bernig, Bröcker [BB02] introduced curvature measures and proved the Principal kinematic formula under certain technical assumptions, the

basic technique was the stratified Morse theory. Bernig [Ber07] defined normal cycles for sets from an o-minimal system using a technique of "definable cycles". Kashiwara and Shapiro [KS94] used a different method of sheaves.

4. A natural question is whether the sets admitting normal cycles fulfill integral geometric formulas as Croftom formula (see (2.5)) or Principal kinematic formula (see (2.4)). For (locally) WDC sets (see 2), the Crofton formula was shown in [PR13, Theorem 1.3], and the Principal kinematic formula in [FPR17]. In fact, the kinematic formula in [FPR17] works in a much more general setting, namely in Riemannian isotropic spaces (including, i.e., spherical case). For a survey on normal cycles and kinematic formulas, see [Fu17].

Chapter 10
Fractal Versions of Curvatures

10.1 Introduction

In this chapter we will give a brief introduction to certain fractal versions of Lipschitz-Killing curvature measures. We present only some of the proofs giving insight into certain techniques which are applied in this and different fields. For the other proofs we refer to the literature.

Throughout this section we use the notation $|A| := \operatorname{diam} A$ for $A \subset \mathbb{R}^d$.

The geometric sets considered in the previous chapters admit only singularities which guarantee an extension of the notions and tools from geometric integration theory, multilinear algebra and topology. More precisely, they all possess some second order rectifiability properties expressed in terms of unit normal bundles. For the more irregular fractal structures such concepts do not work anymore. However, the well-known Hausdorff measures and other covering or packing measures of arbitrary dimensions may be interpreted as fractal extensions of the corresponding classical first order analysis and geometry. They have been studied in the literature for a long time. Special related notions are the following.

Definition 10.1 The *Minkowski dimension* of a bounded set $F \subset \mathbb{R}^d$ is given by

$$\dim_M F = D := d - \lim_{\varepsilon \to 0} \frac{\log \mathcal{L}^d(F_\varepsilon)}{\log \varepsilon},$$

provided the limit exists. Then the values

$$\overline{\mathcal{M}}^D(F) := \limsup_{\varepsilon \to 0} \frac{\mathcal{L}^d(F_\varepsilon)}{\omega_{d-D}\varepsilon^{d-D}},$$

$$\underline{\mathcal{M}}^D(F) := \liminf_{\varepsilon \to 0} \frac{\mathcal{L}^d(F_\varepsilon)}{\omega_{d-D}\varepsilon^{d-D}}$$

J. Rataj, M. Zähle, *Curvature Measures of Singular Sets*, Springer Monographs in Mathematics, https://doi.org/10.1007/978-3-030-18183-3_10

are called *upper (resp. lower) Minkowski content* of such a D-dimensional F. If $\underline{\mathcal{M}}^D(F) = \overline{\mathcal{M}}^D(F) =: \mathcal{M}^D(F)$ then $\mathcal{M}^D(F)$ is called *Minkowski content* of F. F is said to be *Minkowski measurable* if it has positive and finite Minkowski content.

It is well known that for bounded smooth sets and rectifiable versions the Minkowski content agrees with the total Hausdorff measure of the corresponding integral dimension. This does not remain valid, in general, for more irregular sets. However, Minkowski dimension and content play an important role, e.g., in spectral analysis on fractal sets or domains with fractal boundary. If one slightly extends the above notions via average convergence, fractal versions for a large class of so-called self-similar and self-conformal sets can be obtained.

Definition 10.2

$$\widetilde{\mathcal{M}}^D(F) := \lim_{\delta\to 0} \frac{1}{|\ln\delta|} \int_\delta^1 \frac{\mathcal{L}^d(F_\varepsilon)}{\omega_{d-D}\varepsilon^{d-D}} \frac{1}{\varepsilon} d\varepsilon$$

is called *D-dimensional average Minkowski content* of the set F, if the limit exists.

Note that for Minkowski measurable sets the Minkowski content $\mathcal{M}^D(F)$ coincides with its average version. Furthermore, other types of averaging measures than the logarithmic one could be used. However, the latter is appropriate for the fractals considered in this chapter. This also concerns the gauge function ε^{d-D} for the volume of the parallel sets.

For compact sets $X \subset \mathbb{R}^d$ with positive reach of dimension m the Minkowski content may be considered as a marginal case of the associated family of total curvatures $C_k(X)$ from the previous chapters, namely

$$\mathcal{M}^m(X) = C_m(X) = \mathcal{H}^m(X).$$

The first equality follows here immediately from the Steiner formula (see Corollary 4.33). Moreover, in this case we have

$$\mathbf{C}_m(X) = \lim_{\varepsilon\to 0} \binom{d-m}{d-k}^{-1} \frac{\omega_{d-k}}{\omega_{d-m}} \varepsilon^{m-k} \mathbf{C}_k(X_\varepsilon) \tag{10.1}$$

for all $m \le k \le d$ in view of Proposition 4.34.

An essential background are the scaling properties of the associated measures. Having this in mind a natural extension of all total Lipschitz-Killing curvatures to certain fractals can be considered.

Definition 10.3 We call a set $F \subset \mathbb{R}^d$ *tube regular* if the following conditions are fulfilled.

1. **(TN)** for Lebesgue almost all $r > 0$ the parallel sets F_r admit a normal cycle in the sense of Definition 9.5

2. **(TM)** for all $k \leq d-2$ and bounded Borel sets B the curvatures $C_k(F_r, B)$ are measurable functions in r, where we set $C_k(F_r, \cdot) := 0$ for the exceptional ε.

For general $F \subset \mathbb{R}^d$ and $r > 0$ we use throughout this chapter the notation

$$C_d(F_r, \cdot) := \mathcal{L}^d(F_r \cap (\cdot)), \text{ and } C_{d-1}(F_r, \cdot) := \frac{1}{2}\mathcal{H}^{d-1}(\partial F_r, \cap(\cdot)),$$

For $k = d$ this is compatible with all former variants, and for $k = d-1$ see Remark 10.5 below. For these marginal cases we need the following results which complement the notion of tube regularity.

Proposition 10.4

(i) *For $F \subset \mathbb{R}^d$, $k \in \{d-1, d\}$ and all bounded Borel sets B the mappings $r \to C_k(F_r, B)$, $r > 0$, are measurable.*
(ii) *If F is bounded, then for all $r > 0$, ∂F_r is upper Ahlfors $(d-1)$-regular.*

Proof For $k = d$ the first assertion follows from continuity of Lebesgue measure with respect to the Hausdorff distance. In the case $k = d-1$ we can use the measurability statement for the section integrals in the Coarea theorem 1.15 applied to the Lipschitz mapping $f : F_R \to [0, R]$ with $f(x) := d(x, F)$ for arbitrary $R > 0$.

In order to show (ii) assume first that $r > 0$ is a regular value of the distance function $x \mapsto \operatorname{dist}(x, F)$. Then ∂F_r is a $(d-1)$-dimensional Lipschitz manifold by the inverse function theorem for Lipschitz mappings (see Clarke [Cla76]). This appears, in particular, if $r > \sqrt{d/(2d+1)}|F|$ (cf. Fu [Fu85, Theorem 4.1]). In such a case, ∂F_r is Ahlfors $(d-1)$-regular by Proposition 1.12.

Let now $r > 0$ be arbitrary and partition $F = F^1 \cup \cdots \cup F^n$ into finitely many pieces with $|F^i| < r$, $i = 1, \ldots, n$. Then, each $\partial(F^i)_r$ is upper Ahlfors $(d-1)$-regular by the above observation and, since clearly $\partial F_r \subset \partial(F^1)_r \cup \cdots \cup \partial(F^n)_r$, ∂F_r is $(d-1)$-rectifiable and upper Ahlfors $(d-1)$-regular. □

Remark 10.5

(i) A sufficient condition for the tube regularity of F is that a.a. $r > 0$ are *regular values of the distance function* $d(\cdot, F)$, which is fulfilled for all $r > \sqrt{d/(2d+2)}|F|$ according to Fu [Fu85, Theorem 4.1]. (To see this implication use that for such r the sets $\overline{(F_{r'})^c}$ have positive reach for all r' in some neighbourhood of r, see [Fu85, Theorem 4.1]. Then for any continuous function f with compact support the integrals $\int f(z)C_k(F_r, dz)$ are continuous at r, see [Zäh11, Lemma 2.3.4] which refers to [RZ05, Proposition 6]. Since the indicator function of a bounded Borel set B can be approximated by such f the measurability property (TM) is a consequence.)
(ii) In space dimensions $d \leq 3$ any compact F is tube regular, since in this case a.a. $r > 0$ are regular values of the distance function $d(\cdot, F)$ in view of [Fu85, Theorem 4.1].

(iii) If r is a regular value of the distanced function, then the measure $C_{d-1}(F_r,\cdot)$ in the sense of Definition 9.5 agrees with half the surface area measure on the boundary, i.e., this is compatible with the above notation. We conjecture that this remains valid in the general case when F_r admits a normal cycle.

Suppose now that F is bounded, and if $k \leq d-2$ that it is also tube regular. Let $C_k^{\text{var}}(F_\varepsilon,\cdot)$ be the *variation measures* of the corresponding curvature measures. Concerning the scaling exponents we use the notions from Pokorný and Winter [PW14]. For $k \in \{0,\ldots,d\}$ introduce

$$s_k := \inf\left\{s \geq 0 : \operatorname*{ess\,lim}_{\varepsilon\to 0} \varepsilon^{s-k}\mathbf{C}_k^{\text{var}}(F_\varepsilon) = 0\right\}, \tag{10.2}$$

$$a_k := \inf\left\{s \geq 0 : \lim_{\delta\to 0}\frac{1}{|\ln\delta|}\int_\delta^{\varepsilon_0} \varepsilon^{s-k}\mathbf{C}_k^{\text{var}}(F_\varepsilon)\,\frac{1}{\varepsilon}\mathrm{d}\varepsilon = 0\right\}. \tag{10.3}$$

Here and in the sequel the essential limit is taken with respect to Lebesgue measure.

Definition 10.6 For $0 \leq s \leq d$,

$$\mathbf{C}_k^s(F) := \operatorname*{ess\,lim}_{\varepsilon\to 0} \varepsilon^{s-k}\mathbf{C}_k(F_\varepsilon)$$

and

$$\overline{\mathbf{C}}_k^s(F) := \lim_{\delta\to 0}\frac{1}{|\ln\delta|}\int_\delta^{\varepsilon_0} \varepsilon^{s-k}\mathbf{C}_k(F_\varepsilon)\,\frac{1}{\varepsilon}\mathrm{d}\varepsilon$$

are called the *s-dimensional (resp. average) total fractal (Lipschitz-Killing) curvature of order* k of the set F, provided the corresponding limit exists.

Remark 10.7 In order to get geometrically meaningful notions one should consider for fixed k the scaling exponents $s = s_k$ and $s = a_k$, respectively. Below we are mainly interested in the cases where s_k or a_k coincide for all k with the Minkowski dimension D of the set F. The Minkowski content and its average version are then contained as marginal cases

$$\mathcal{M}^D(F) = \frac{1}{\omega_{d-D}}\mathbf{C}_d^D(F) \text{ and } \widetilde{\mathcal{M}}^D(F) = \frac{1}{\omega_{d-D}}\overline{\mathbf{C}}_d^D(F),$$

where no curvature properties are involved. (However, we will not treat here conditions for $s_k = a_k = D$, see also the discussion in [PW14]). We conjecture that similar properties as in (10.1) are fulfilled for the values $\overline{\mathbf{C}}_k^D(F)$ with $k \geq \lfloor D \rfloor$ (cf. Theorem 10.15 below). For completeness we call them for all k fractal Lipschitz-Killing curvatures.

Similarly one can introduce the localized versions, i.e., *fractal curvature measures* on the Borel σ-algebra of $\mathbb{R}^d$ with the above total values:

Definition 10.8

$$C_k^s(F,\cdot) := \operatorname*{ess\,lim}_{\varepsilon\to 0} \varepsilon^{s-k} C_k(F_\varepsilon,\cdot)$$

is called *s-dimensional fractal curvature measure of order k*, and

$$\overline{C}_k^s(F,\cdot) := \lim_{\delta\to 0} \frac{1}{|\ln\delta|} \int_\delta^{\varepsilon_0} \varepsilon^{s-k} C_k(F_\varepsilon,\cdot)\,\frac{1}{\varepsilon}\mathrm{d}\varepsilon$$

its *average version* if the corresponding limits exist in the sense of weak convergence of signed measures.

Again, we are interested in the cases $s = s_k = D$ or $s = a_k = D$.

Remark 10.9 In the last definition F need not be bounded if weak convergence is replaced by vague convergence.

In all definitions introduced above the curvature measures $C_k(F_\varepsilon,\cdot)$ of the parallel sets could be replaced by their extensions, the curvature-direction measures $\widetilde{C}_k(F_\varepsilon,\cdot)$ on $\mathbb{R}^d \times S^{d-1}$ (cf. Definition 9.5) in order to obtain the exponents $\widetilde{s}_k, \widetilde{a}_k$ and the fractal curvature-direction measures $\widetilde{C}_k^s(F,\cdot)$ with total values $\widetilde{\mathbf{C}}_k^s(F) = \mathbf{C}_k^s(F)$ in the same manner as before, provided the limits exist.

For $X \subset \mathbb{R}^d$ with positive reach and more general sets from the previous sections the curvature-direction measures $\widetilde{C}_k(X,\cdot)$, $k = 0,\ldots,d-1$, possess signed densities with respect to Hausdorff measure $\mathcal{H}^{d-1}$ on the associated unit normal bundle $\operatorname{nor} X$. For $\mathcal{PR}$-sets these are the generalized local Lipschitz-Killing curvatures $s_{d-1-k}(X; x, n)$ at $(x,n) \in \operatorname{nor} X$, cf. Definitions 4.25 and 4.28. In the special case of m-dimensional smooth submanifolds this corresponds to the classical $(m-k)$-th order mean curvatures at $x \in X$ via projections from $\operatorname{nor} X$ to X, see Remark 4.32. The latter are then the signed densities of the k-th curvature measures with respect to Hausdorff measure $\mathcal{H}^m$ on X. Therefore both versions of these pointwise curvatures can be obtained from Lebesgue's density theorem at almost all points. It turned out that for large classes of self-similar sets and some extensions one can also obtain relationships between local and global fractal curvatures in form of densities with respect to Hausdorff measure. Moreover, the latter can be determined by means of local versions of the above approximations with parallel sets, however with different scaling behaviour. To this aim we consider a localization procedure by means of appropriate bump functions as follows. For brevity we write

$$C_k(F_\varepsilon, f) := \int f(z)\, C_k(F_\varepsilon, \mathrm{d}z),$$

if f is an integrable function. Then we introduce the following notions. Let $F \subset \mathbb{R}^d$ and μ_F be a Radon measure on F which is lower Ahlfors D-regular (recall Definition 1.5). In particular, if F is bounded then $\mu_F(F) < \infty$.

Definition 10.10 For any $\varepsilon_0 > 0$ and $0 < a < 1$ and any nonnegative bounded measurable function g with support in $[-1, 1]$ which is bounded from below by a positive constant in a neighbourhood of 0 the functions

$$A_F(x, \varepsilon)(z) := \varepsilon^D \frac{g\left(\frac{|x-z|}{a\varepsilon}\right)}{\int_F g\left(\frac{|y-z|}{a\varepsilon}\right) \mu_F(\mathrm{d}y)}, \quad z \in F_\varepsilon,$$

with parameters $x \in F$, $0 < \varepsilon < \varepsilon_0$, form a *localizing family of bump functions* associated with F and μ_F.

Note that lower Ahlfors regularity of μ_F and the properties of g imply that the localizing functions A_F are uniformly bounded. The following measurability property for such families of bump functions A_F will be used in the sequel.

Proposition 10.11 *Let $F \subset \mathbb{R}^d$ and $k \in \{0, \ldots, d\}$. If $d \geq 4$ and $k \leq d-2$ suppose additionally that the set F is tube regular in the sense of Definition* 10.3. *Then the curvature integral $C_k\big(F_\varepsilon, A_F(x, \varepsilon)\big)$ (with value* 0 *at the exceptional ε) is a measurable function in $(x, \varepsilon) \in F \times (0, \varepsilon_0)$.*

Proof It suffices to consider $x \in F \cap K$ for arbitrary compact K. First note that the function $A_F(x, \varepsilon)(z)$ is product measurable in (x, ε, z). Therefore it may be approximated by a sequence of uniformly bounded functions which are representable by linear combinations of measurable functions of the form $f_1(x) f_2(\varepsilon) f_3(z)$ with supports in $K \times [0, \varepsilon_0] \times K_{a\varepsilon_0}$. Thus, their integrals with respect to $C_k(F_\varepsilon, \cdot)$ converge to $C_k(F_\varepsilon, A_F(x, \varepsilon))$.

Furthermore, by tube regularity (TM) and Remark 10.5 for any bounded measurable function f with compact support the integral $C_k(F_\varepsilon, f)$ is a measurable function in $\varepsilon \in (0, \varepsilon_0)$.

Combining both these arguments we obtain the assertion. □

This enables us to introduce the notion of *fractal curvature densities*.

Definition 10.12 Let $F \subset \mathbb{R}^d$ and $k \in \{0, \ldots, d\}$. If $d \geq 4$ and $k \leq d-2$ suppose additionally that F is tube regular in the sense of Definition 10.3. Then for localizing functions A_F as in Definition 10.10,

$$\mathcal{D}_k(F, x) := \lim_{\delta \to 0} \frac{1}{|\ln \delta|} \int_\delta^{\varepsilon_0} \varepsilon^{-k} C_k(F_\varepsilon, A_F(x, \varepsilon)) \frac{1}{\varepsilon} \mathrm{d}\varepsilon$$

is called k-th *curvature density* of F w.r.t. μ_F at point x, provided the integrals and their limit exist.

There is a close relationship between such fractal densities and associated fractal curvature measures if $D_k(F, x)$ exists at μ_F-a.a. $x \in F$. In this case the densities do not depend on the choice of the constants ε_0, a and the bump function g in the localizing procedure. This will be shown in the next theorem.

Recall that on a finite measure space for a family of functions that converges in measure, uniform integrability is equivalent to L_1-convergence (see, e.g., Doob [Doo94, Theorem 6.18]).

Theorem 10.13 *Let $k \in \{0, 1, \ldots, d-1\}$, F be a compact subset of $\mathbb{R}^d$ such that there exists a lower Ahlfors D-regular finite measure μ_F with $\operatorname{spt} \mu_F = F$, and let $A_F(x, \varepsilon)$ be a family of localizing functions as in Definition 10.10. For $k \le d-2$ let F be tube regular in the sense of Definition 10.3 (for $d \le 3$ this is always fulfilled). Suppose that the corresponding curvature density $\mathcal{D}_k(F, x)$ exists at μ_F-a.a. $x \in F$ and that the functions*

$$x \mapsto \frac{1}{|\ln \delta|} \int_\delta^{\varepsilon_0} \frac{C_k^{\mathrm{var}}(F_\varepsilon, A_F(x, \varepsilon))}{\varepsilon^k} \frac{1}{\varepsilon} \mathrm{d}\varepsilon \,, \quad x \in F \,, \tag{10.4}$$

with parameter δ are uniformly integrable w.r.t. μ_F. Then the k-th average fractal curvature measure in the sense of Definition 10.8 exists and is equal to

$$\overline{C}_k^D(F, \cdot) = \int_F \mathbf{1}_{(\cdot)}(x) \mathcal{D}_k(F, x) \, \mu_F(\mathrm{d}x) \,.$$

Consequently, in this case the density $\mathcal{D}_k(F, x)$ does not depend on the choice of the localizing functions $A_F(x, \varepsilon)$ with property (10.4).
For $k = d-1, d$ the uniform integrability is always fulfilled.

Proof According to Proposition 10.11 the function $C_k(F_\varepsilon, A_F(x, \varepsilon))$ is measurable in $(x, \varepsilon) \in F \times (0, \varepsilon_0)$. Then in view of the integrability condition (10.4) Fubini's theorem can be applied to the integrals below. Using that $\int_F A_F(x, \varepsilon)(z) \, \mu_F(\mathrm{d}x) = \varepsilon^D$ we obtain for an arbitrary continuous function f and $\varepsilon < \varepsilon_0$,

$$\begin{aligned} \varepsilon^{D-k} \int f(z) \, C_k(F_\varepsilon, dz) &= \varepsilon^{-k} \int f(z) \int_F A_F(x, \varepsilon)(z) \, \mu_F(dx) \, C_k(F_\varepsilon, \mathrm{d}z) \\ &= \int_F \varepsilon^{-k} \int_{F_\varepsilon} f(z) \, A_F(x, \varepsilon)(z) \, C_k(F_\varepsilon, dz) \, \mu_F(\mathrm{d}x) \,. \end{aligned}$$

We are interested in the average limit of the left hand side with respect to the logarithmic measure $\frac{1}{|\ln \delta|} \int_\delta^{\varepsilon_0} \mathbf{1}_{(\cdot)}(\varepsilon) \, \frac{1}{\varepsilon} \mathrm{d}\varepsilon$. Since $\operatorname{spt} A_F(x, \varepsilon) \subset B(x, a\varepsilon)$ and f is uniformly continuous on compact sets, the last expression can be replaced asymptotically by

$$\int_F f(x) \, \varepsilon^{-k} \int_{F_\varepsilon} A_F(x, \varepsilon)(z) \, C_k(F_\varepsilon, dz) \, \mu_F(\mathrm{d}x) \,,$$

i.e., the average limit of the difference vanishes (see the arguments below). Therefore the primary average limit exists and equals

$$\begin{aligned}\int f(z)\,\overline{C}_k^D(F,\mathrm{d}z) &= \lim_{\delta\to 0}\frac{1}{|\ln\delta|}\int_\delta^{\varepsilon_0}\int_F f(x)\,\varepsilon^{-k}C_k(F_\varepsilon, A_F(x,\varepsilon))\,\mu_F(\mathrm{d}x)\,\frac{1}{\varepsilon}\mathrm{d}\varepsilon\\ &= \int_F f(x)\lim_{\delta\to 0}\frac{1}{|\ln\delta|}\int_\delta^{\varepsilon_0}\varepsilon^{-k}C_k(F_\varepsilon, A_F(x,\varepsilon))\,\frac{1}{\varepsilon}\mathrm{d}\varepsilon\,\mu_F(\mathrm{d}x)\\ &= \int_F f(x)\,\mathcal{D}_k(F,x)\,\mu_F(\mathrm{d}x)\,,\end{aligned}$$

since a.e. convergence and uniform integrability imply L_1-convergence.
Using that f is an arbitrary continuous function we get

$$C_k^D(F,\cdot) = \mathcal{D}_k(F,\cdot)\,\mu$$

and hence, $\mathcal{D}_k(F,\cdot)$ does not depend on the choice of A_F.
In order to complete the above arguments, for given $\varepsilon' > 0$ choose $0 < \delta' = \delta'(\varepsilon) < \varepsilon_0$ small enough so that $|f(z) - f(x)| \le \varepsilon'$ for all $z \in B(x, a\delta')$ uniformly in $x \in F$. Then for $\delta < \delta'$ the above difference, say $R(\delta)$, can be estimated as follows.

$$\begin{aligned}R(\delta) \le &\int_F \frac{1}{|\ln\delta|}\int_\delta^{\delta'}\varepsilon^{-k}\int |f(z) - f(x)|A_F(x,\varepsilon)(z)C_k^{\mathrm{var}}(F_\varepsilon,\mathrm{d}z)\,\frac{1}{\varepsilon}\mathrm{d}\varepsilon\,\mu_F(\mathrm{d}x)\\ +&\frac{|\ln\delta'|}{|\ln\delta|}\int_F\frac{1}{|\ln\delta'|}\int_{\delta'}^{\varepsilon_0}\varepsilon^{-k}\int |f(z) - f(x)|A_F(x,\varepsilon)(z)C_k^{\mathrm{var}}(F_\varepsilon,\mathrm{d}z)\,\frac{1}{\varepsilon}\mathrm{d}\varepsilon\,\mu_F(\mathrm{d}x)\\ &\le c\big(\varepsilon' + \frac{|\ln\delta'|}{|\ln\delta|}\big)\end{aligned}$$

for some constant c, since the integrals

$$\int_F\frac{1}{|\ln\delta|}\int_\delta^{\varepsilon_0}\frac{C_k^{\mathrm{var}}(F_\varepsilon, A_F(x,\varepsilon))}{\varepsilon^k}\,\frac{1}{\varepsilon}\mathrm{d}\varepsilon\mu_f(\mathrm{d}x)$$

are uniformly bounded. Taking first $\delta \to 0$ and then $\varepsilon' \to 0$, one obtains that $\lim_{\delta\to 0} R(\delta) = 0$, i.e., the above assertion.

Finally, for $k = d$ the uniform integrability condition (i) is obviously fulfilled, since $C_d(F_\varepsilon,\cdot)$ agrees with Lebesgue measure on F_ε, A_F is uniformly bounded and $\operatorname{spt} A_F(x,\varepsilon) \subset B(x, a\varepsilon)$. Similarly, $C_{d-1}(F_\varepsilon,\cdot)$ is a constant multiple of the $(d-1)$-dimensional Hausdorff measure on the boundary of F_ε. In view of Proposition 10.4 it is upper Ahlfors $(d-1)$-regular, i.e., $\mathcal{H}^{d-1}(\partial F_\varepsilon \cap B(x, a\varepsilon)) < \mathrm{const}\,\varepsilon^{d-1}$ which implies (i). □

Below we will see that the conditions of the last theorem are satisfied for a large class of self-similar fractal sets with $\mu_F = \mathcal{H}^D(F\cap(\cdot))$. Moreover, we will summarize different methods for determining their fractal curvature measures introduced above. The marginal cases $k = d - 1, d$ will be considered in the more general context of Minkowski content.

Remark 10.14 A special bump function is associated with small balls. Namely,

$$A_F(x, \varepsilon)(z) := \varepsilon^D \left(\mu_F(B(x, a\varepsilon))\right)^{-1} \mathbf{1}_{B(x,a\varepsilon)}(z)$$

is the case where $g(u) = \mathbf{1}_{[-1,1]}(u)$.
The continuous tent function $g(u) := \max(1 - |u|, 0)$ first suggested by T. Bohl is also applied to related problems in the self-conformal case [Boh12]. Furthermore, for self-similar sets satisfying the Open Set Condition (see Sect. 10.2 below) in [RZ12] for $\varepsilon_0 := \mathcal{H}^D(F)^{1/D}$ instead of $A_F(x, \varepsilon)$ the indicator function of the following set (denoted by the same symbol) was used, which yields the same limits for $\mu_F := \mathcal{H}^D(F \cap (\cdot))$:

$$A_F(x, \varepsilon) := \{z \in F_\varepsilon : |x - z| \leq \rho_F(z, \varepsilon)\}$$

where

$$\rho_F(z, \varepsilon) := \min\left\{\rho : \mathcal{H}^D(F \cap B(z, \rho)) = \varepsilon^D\right\}.$$

10.2 Self-Similar Sets

We now briefly introduce some basic notions and results from the literature on self-similar sets needed in the sequel. A first general approach to such sets, which included the special cases and famous examples from the former literature like Cantor set, Koch curve, Sierpinski gasket and carpet, Menger sponge etc., was given in Hutchinson [Hut81]. (For subsets on the real line it was already developed in Moran [Mor46].) The facts mentioned below without a reference can be found e.g. in [Hut81] together with Falconer [Fal90], [Fal97] and Kigami [Kig01].

For $N \geq 2$ let $S_1, \dots, S_N$ be an *iterated function system* (IFS) consisting of *contracting similitudes* in $\mathbb{R}^d$ with *contraction ratios* $r_1, \dots, r_N$. It is called *lattice*, if there exists a constant a such that $\ln r_i \in a\mathbb{Z}$ for all i. (This is, for example, the case when all contraction ratios are the same.) In the general case the associated *self-similar set* F may be constructed by means of the *code space* $W := \{1, \dots, N\}^{\mathbb{N}}$, i.e., the set of all infinite words over the alphabet $\{1, \dots, N\}$. We write $W_0 := \emptyset$ and $W_n := \{1, \dots, N\}^n$ for the set of all words of length $|w| = n$, $W_* := \bigcup_{n=0}^{\infty} W_n$ for the set of all finite words and $w|n := w_1 w_2 \dots w_n$, if $w = w_1 w_2 \dots w_n w_{n+1} \dots$, for the *restriction* of a (finite or infinite) word to the first n letters. If $w = w_1 \dots w_n \in W_n$, we also use the abbreviations $S_w := S_{w_1} \circ \dots \circ S_{w_n}$ and $r_w := r_{w_1} \dots r_{w_n}$ for the contraction ratio of this mapping. In these terms the set F is determined by

$$F = \bigcap_{n=1}^{\infty} \bigcup_{w \in W_n} S_w K$$

for an arbitrary non-empty compact K such that $S_1 K \cup \ldots \cup S_N K \subset K$. (For brevity the corresponding brackets are omitted.) F is characterized as the unique non-empty compact set with the *self-similarity property*

$$F = S_1 F \cup \ldots \cup S_N F . \tag{10.5}$$

It can also be obtained by a contraction principle for the mapping

$$K \mapsto \mathbf{S}K := \bigcup_{i=1}^{N} S_i K$$

on the space of nonempty compact subsets of $\mathbb{R}^d$ equipped with the Hausdorff metric.

Alternatively, the self-similar set F is the image of the code space W under the *projection* π given by

$$\pi(w) := \lim_{n \to \infty} S_{w|n} x_0$$

for an arbitrary starting point x_0. We now assume the *strong open set condition* (SOSC) for the IFS, i.e., there exists a non-empty bounded open set O such that for all $i \neq j$,

$$S_i O \subset O , \quad S_i O \cap S_j O = \emptyset ,$$

which is called *open set condition* (OSC), and

$$F \cap O \neq \emptyset$$

for the associated self-similar set F.

(According to a result of Schief (see [Sch94]), (OSC) for some O already implies (SOSC) for a possibly different open set. Basing on this paper a characterization of (SOSC) in algebraic terms of the $S_1, \ldots, S_N$ was given in Bandt and Graf [BG92].)

An open set O from (OSC) or (SOSC) for F is called *feasible* or *strong feasible*, respectively.

(OSC) implies that the *Hausdorff dimension* D of F is determined by

$$\sum_{j=1}^{N} r_j^D = 1 \tag{10.6}$$

and that it coincides with the Minkowski dimension and other types of fractal dimensions of F. Moreover,

$$0 < \mathcal{H}^D(F) < \infty \, .$$

(In fact, the last property is equivalent to (SOSC), see [BG92].) The self-similar set F is even Ahlfors D-regular, i.e., there exist positive constants c_F and C_F such that

$$c_F\, r^D \leq \mathcal{H}^D(F \cap B(x,r)) \leq C_F\, r^D \, , \quad x \in F, \; r \leq \operatorname{diam} F \, .$$

Furthermore, under (SOSC) the above projection mapping $w \mapsto x = \pi(w)$ from the code space W onto F is biunique at $\mathcal{H}^D$-a.a. of points $x \in F$. Therefore we *identify* almost every point $x \in F$ with its *coding sequence* and write

$$x_1 x_2 \ldots := \pi^{-1}(x)$$

for this infinite word.

If ν denotes the infinite product measure on W determined by the probability measure on the alphabet $\{1, \ldots, N\}$ which assigns to i the probability $r_i^D, i = 1, \ldots N$, then the normalized D-dimensional Hausdorff measure on F equals

$$\mathcal{H}^D(F)^{-1} \mathcal{H}^D(F \cap (\cdot)) = \nu \circ \pi^{-1} \, . \tag{10.7}$$

It agrees with the unique normalized *self-similar measure* μ on F such that

$$\mu = \sum_{i=1}^{N} r_i^D \mu \circ S_i^{-1} \, . \tag{10.8}$$

In particular,

$$\mu(\partial O) = 0$$

for any strong feasible open set O. Iterated applications of (10.5) and (10.8) lead to

$$F = \bigcup_{w \in W_n} S_w F \quad \text{and} \quad \mu = \sum_{w \in W_n} r_w^D \mu \circ S_w^{-1} \, ,$$

respectively, for all $n \in \mathbb{N}$. This extends to arbitrary *Markov stoppings*, i.e., mappings $\tau : W \mapsto \mathbb{N}$ such that the value $\tau(w) = n$ is determined by the first n letters $w_1, \ldots, w_n$ of the infinite word w. For such τ denote

$$W(\tau) := \left\{w_1 \ldots w_{\tau(w)} : w \in W\right\} .$$

Then one infers

$$F = \bigcup_{w \in W(\tau)} S_w F , \tag{10.9}$$

$$\mu = \sum_{w \in W(\tau)} r_w^D \mu \circ S_w^{-1} , \tag{10.10}$$

in particular,

$$\sum_{w \in W(\tau)} r_w^D = 1 . \tag{10.11}$$

An example for a Markov stopping, which will be used in Sect. 10.3.3, is the following. Recall that $r_1, \ldots, r_N$ denote the contraction ratios of the similarities $S_1, \ldots, S_N$. Then for given $0 < r < \operatorname{diam} F$ the value $\tau(w) := \tau_r(w)$ is determined by

$$r_{w_1} \ldots r_{w_{\tau_r(w)}} |F| < r \leq r_{w_1} \ldots r_{w_{\tau_r(w)-1}} |F| . \tag{10.12}$$

Finally, for any strong feasible open set O and all $p \in \mathbb{N}$ one obtains

$$\int |\ln d(y, O^c)|^p \mu(dy) < \infty . \tag{10.13}$$

(See Graf [Gra95, proof of Proposition 3.4] for $p = 1$. The proof for general p is similar.)

These notions and results for dimensions and measures can also be obtained as special linear cases from the so-called *thermodynamic formalism* in the sense of Bowen and Ruelle. (We refer to Falconer [Fal97] for an introduction and to Mauldin and Urbanski [MU03] as well as the references therein for more general versions concerning self-conformal sets.) In particular, if we consider the mapping $T : F \to F$ with the inverse branches $S_1, \ldots, S_N$ defined for μ-a.a. x by

$$T(x) := S_i^{-1}(x) \,, \text{ if } x \in S_i F \,, \; i = 1, \ldots, N \,,$$

taking into regard that $\mu(S_i F \cap S_j F) = 0$ for $i \neq j$, then (F, T, μ) is an *ergodic dynamical system*. This follows here directly from the relationship $\mu = \nu \circ \pi^{-1}$, since the product measure ν on the code space is shift invariant and ergodic.

In the sequel we also consider tilings associated with the self-similar set F and their generators. Recall that $\overline{A}$ denotes the closure of a set A in $\mathbb{R}^d$. Suppose that for a feasible open set O the open set $G := O \setminus \mathbf{S}\overline{O}$ is not empty. (For example, G is the inner open triangle in the first construction step of the Sierpinski gasket.) Then the family of iterates of G under the corresponding IFS,

$$\mathcal{T} = \mathcal{T}(O) := \{S_w G : w \in W_*\}$$

is called *tiling* of O with *generator* G. The tiles, i.e., the elements of $\mathcal{T}$ are pairwise disjoint and

$$\overline{O} = \overline{\bigcup_{R \in \mathcal{T}} R}\,.$$

Such a tiling is called *compatible* if $\partial G \subset F$, which is equivalent to $\partial O \subset F$ (see Pearse and Winter [PW12, Theorem 6.2]. (For these and further relationships and examples see [PW12] and Winter [Win15].)

Classical examples of self-similar sets are e.g. the Sierpinski triangle (Fig. 10.1), the Sierpinski carpet (Fig. 10.2) and its modified version (Fig. 10.3), or the Koch curve (Fig. 10.4) together with a strong feasible open set (Fig. 10.5).

Fig. 10.1 First iterations of the *Sierpinski triangle* generated by $N = 3$ similitudes with contraction factors $r = 1/2$. The interior of the primary black triangle is a strong feasible open set O and the inner white triangle in the first iteration is the generator G of a compatible tiling. The Hausdorff dimension of the limit set F equals $D = \log 3/\log 2$

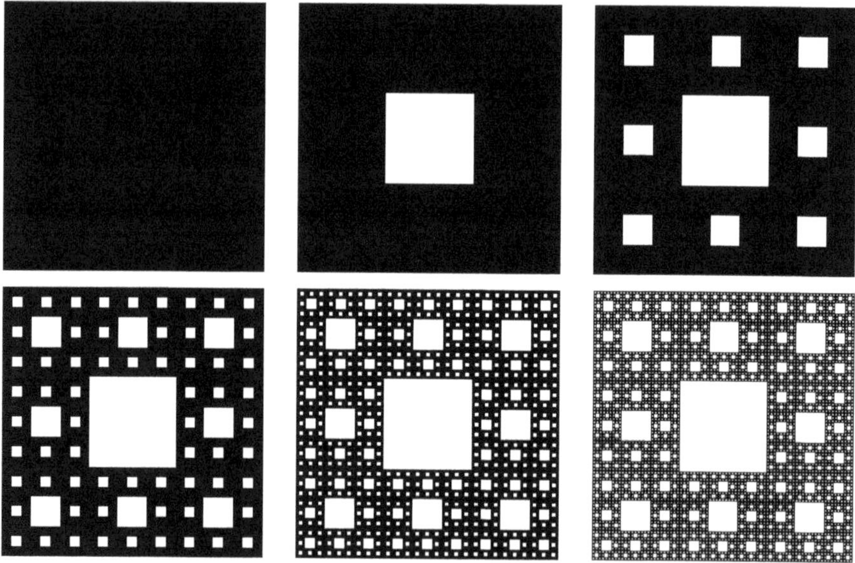

Fig. 10.2 First iterations of the *Sierpinski carpet* generated by 8 similitudes with contraction factors $1/3$. The interior of the primary black square is a feasible open set O and the inner white square G in the first iteration generates a compatible tiling. The Hausdorff dimension of the limit set F equals $\log 8/\log 3$

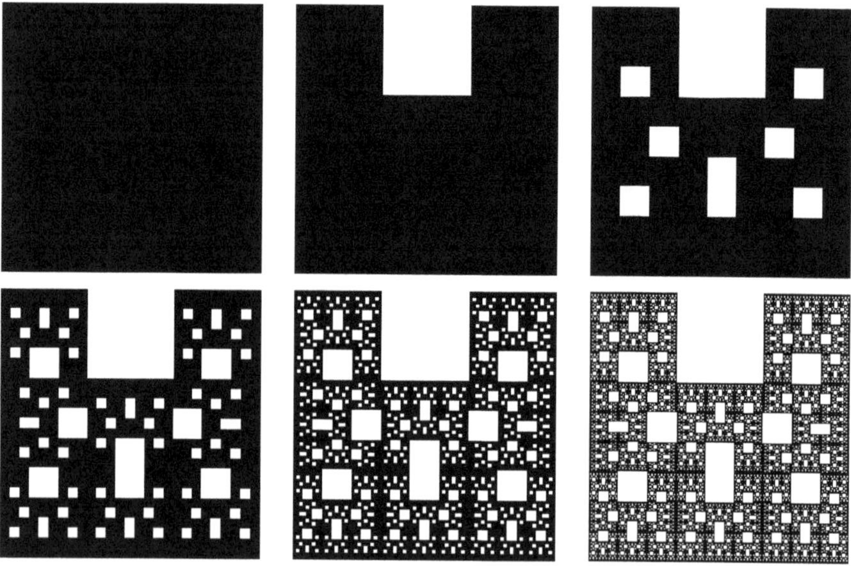

Fig. 10.3 First iterations of the *modified Sierpinski carpet*. Here the interior of the primary square is also a feasible open set. In order to get a compatible tiling one should consider the interior of the polygon in the first iteration as feasible open set O and the inner white set in the second iteration as associated generator. The Hausdorff dimension is again $\log 8/\log 3$

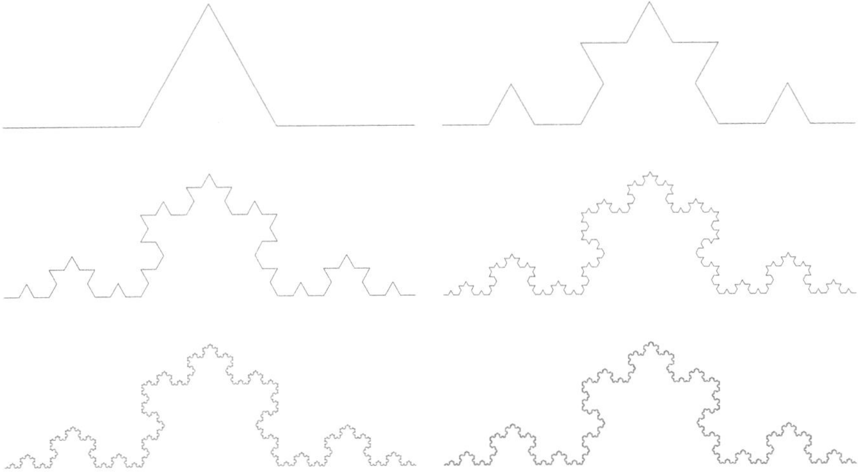

Fig. 10.4 First iterations of the *Koch curve*. Here the primary interval with the same boundary points as the curves is omitted. The iterations by 4 similitudes with contraction factors $1/3$ are presented. The following consideration allows to include this in the above general construction

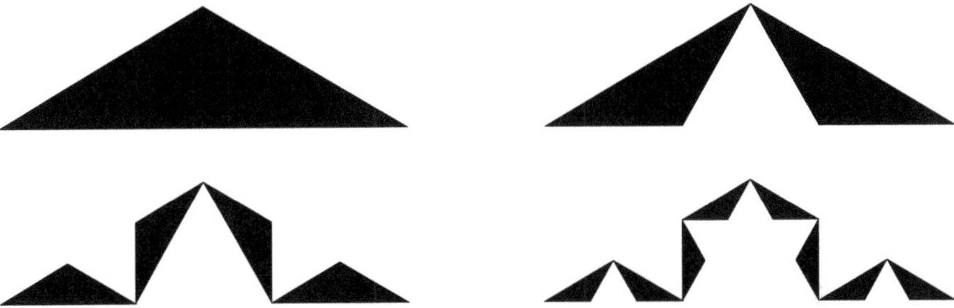

Fig. 10.5 The interior of the primary black triangle is a strong feasible open set for the IFS of 2 similitudes with contraction factors $1/\sqrt{3}$, for which the first three iterates are presented. In the above construction only every second iterate of this type is considered, i.e., one has an IFS of 4 similitudes with contraction factors $1/3$ generating the same limit set. Note that here the associated tilings are not compatible. The Hausdorff dimension of the Koch curve equals $\log 4/\log 3$

10.3 (Average) Minkowski Content and Localized Versions

10.3.1 Characterization in Terms of Surface Area

Recall that the measures $C_d(F_\varepsilon, \cdot)$ and $C_{d-1}(F_\varepsilon, \cdot)$ correspond to volume and surface area of the parallel sets. They are included as marginal cases, though they do not reflect curvature properties. However, the Lipschitz-Killing curvature (measures) C_k, $k = 0, \ldots, d$, form a complete system of Euclidean invariants with certain properties, cf. Chap. 8, which underlines their geometric meaning. Since the marginal cases $k = d - 1, d$ concern only the first order infinitesimal behaviour, their fractal versions exist under much weaker conditions than for the other fractal curvatures which are obtained by analytically second order properties of the approximating sets.

We start with the observation that in the approaches mentioned above the (average) limits of the appropriately rescaled version for arbitrary bounded F with $\mathcal{L}^d(F) = 0$ in the cases $k = d - 1$ and $k = d$ coincide provided one of them exists. This means that the (average) Minkowski content can be obtained by means of the volume as well as of the surface area of the parallel sets:

Theorem 10.15 *Let F be a bounded subset of $\mathbb{R}^d$, $D \in [0, d)$, or $D = d$ and $\mathcal{L}^d(F) = 0$, $M \in (0, \infty)$ and $\varepsilon_0 > 0$. Then we have the following.*

(i)

$$\lim_{\varepsilon \to 0} \frac{\mathcal{L}^d(F_\varepsilon)}{\omega_{d-D}\varepsilon^{d-D}} = M$$

if and only if

$$\lim_{\varepsilon \to 0} \frac{\mathcal{H}^{d-1}(\partial F_\varepsilon)}{(d-D)\omega_{d-D}\varepsilon^{d-1-D}} = M\,.$$

(ii) *If*

$$\limsup_{\delta \to 0} \frac{1}{|\ln \delta|} \int_\delta^{\varepsilon_0} \frac{\mathcal{L}^d(F_\varepsilon)}{\varepsilon^{d-D}} \frac{1}{\varepsilon} \mathrm{d}\varepsilon < \infty$$

then

$$\liminf_{\delta \to 0} \frac{1}{|\ln \delta|} \int_\delta^{\varepsilon_0} \frac{\mathcal{L}^d(F_\varepsilon)}{\varepsilon^{d-D}} \frac{1}{\varepsilon} \mathrm{d}\varepsilon = \liminf_{\delta \to 0} \frac{1}{|\ln \delta|} \int_\delta^{\varepsilon_0} \frac{\mathcal{H}^{d-1}(\partial F_\varepsilon)}{(d-D)\varepsilon^{d-1-D}} \frac{1}{\varepsilon} \mathrm{d}\varepsilon$$

and

$$\limsup_{\delta \to 0} \frac{1}{|\ln \delta|} \int_\delta^{\varepsilon_0} \frac{\mathcal{L}^d(F_\varepsilon)}{\varepsilon^{d-D}} \frac{1}{\varepsilon} \mathrm{d}\varepsilon = \limsup_{\delta \to 0} \frac{1}{|\ln \delta|} \int_\delta^{\varepsilon_0} \frac{\mathcal{H}^{d-1}(\partial F_\varepsilon)}{(d-D)\varepsilon^{d-1-D}} \frac{1}{\varepsilon} \mathrm{d}\varepsilon\,.$$

Part (i) is shown in [RW10] for self-similar sets and in [RW13, Theorem 2.4] for the general case. The average limits in part (ii) are treated in [RW13, Lemma 4.6].

10.3.2 Densities and Measures via Ergodic Theory

For the case of self-similar sets with (OSC) both the *average Minkowski content* and its density with respect to Hausdorff measure always exist. Recall Definition 10.12

for the pointwise localizing function A_F with related constants a, ε_0 and the notations from Sect. 10.2. In particular, $y_1 := (\pi^{-1}(y))_1$ is the first letter in the coding sequence of a.a. $y \in F$.

Theorem 10.16 *Let $F \subset \mathbb{R}^d$ be a self-similar set satisfying (OSC) with Hausdorff dimension D and let O be an associated strong feasible open set.*

(i) *At $\mathcal{H}^D$-a.a. $x \in F$ the average Minkowski densities $((d-D)\omega_{d-D})^{-1}\mathcal{D}_{d-1}(F,x)$ and $(\omega_{d-D})^{-1}\mathcal{D}_d(F,x)$ exist and are both equal to a constant positive value $\mathcal{D}_M(F)$. Moreover, at these x,*

$$\mathcal{D}_d(F,x) = \frac{1}{\eta}\int\int_{b^{-1}d(y,\partial S_{y_1}O)}^{b^{-1}d(y,\partial O)} \frac{1}{\varepsilon^d}\int_{F_\varepsilon} A_F(y,\varepsilon)(z)\mathcal{L}^d(\mathrm{d}z)\frac{1}{\varepsilon}\,\mathrm{d}\varepsilon\,\mu(\mathrm{d}y)$$

$$\mathcal{D}_{d-1}(F,x) = \frac{1}{\eta}\int\int_{b^{-1}d(y,(\partial S_{y_1}O)}^{b^{-1}d(y,\partial O)} \frac{1}{\varepsilon^{d-1}}\int_{\partial F_\varepsilon} A_F(y,\varepsilon)(z)\mathcal{H}^{d-1}(\mathrm{d}z)\frac{1}{\varepsilon}\,\mathrm{d}\varepsilon\,\mu(\mathrm{d}y)$$

where $\eta := \sum_{i=1}^N r_i^D|\ln r_i|$ and $b := \max(2a, \varepsilon_0^{-1}|O|)$.

(ii) *The average Minkowski content of F exists and satisfies*

$$\widetilde{\mathcal{M}}^D(F) = \mathcal{D}_M(F)\mathcal{H}^D(F)\,.$$

Proof In Sect. 10.4 we will show in the more general context of fractal curvatures the existence of the average densities $\mathcal{D}_{d-1}(F,x)$ and $\mathcal{D}_d(F,x)$ at a.a. $x \in F$. By means of an embedded dynamical system the problem can be reduced to the application of Birkhoff's ergodic theorem, which provides the desired integral representations for the constant values. Choosing then in the proof of Theorem 10.13 $\mu_F := \mathcal{H}^D(F\cap(\cdot))$, $k = d-1, d$, and $f \equiv 1$ we infer that the average Minkowski content defined by means of volume or surface area of the parallel sets exists and equals the product of the corresponding constant average density and $\mathcal{H}^D(F)$. Therefore Theorem 10.15 for the total values implies that the average Minkowski densities in (i) coincide. □

Remark 10.17 It is not difficult to see that $\mathcal{H}^D(F)$ does not exceed the lower Minkowski content of F, hence $\mathcal{D}_M(F) \geq 1$. For the classical examples of the Cantor set and the Sierpinski gasket one can show the strict inequality.

More generally, Theorem 10.13 for the measure $\mu_F := \mathcal{H}^D(F\cap(\cdot))$ and $k = d$ (or $k = d-1$), together with Theorem 10.16 yield the following measure version, i.e., the *localized Minkowski content.*

Corollary 10.18 *Under the conditions of Theorem* 10.16 *the weak limit*

$$\widetilde{\mathcal{M}}^D(F,\cdot) := \lim_{\delta\to 0}\frac{1}{|\ln\delta|}\int_\delta^{\varepsilon_0}\frac{\mathcal{L}^d(F_\varepsilon\cap(\cdot))}{\omega_{d-D}\varepsilon^{d-D}}\frac{1}{\varepsilon}\mathrm{d}\varepsilon$$

exists and satisfies

$$\widetilde{\mathcal{M}}^D(F,\cdot) = \mathcal{D}_M(F)\mathcal{H}^D(F \cap (\cdot)).$$

Like for the average densities the statement remains valid up to the constant $(d - D)^{-1}$ *if the Lebesgue measure on the parallel sets is replaced by the surface area measure* $\mathcal{H}^{d-1}$ *on their boundaries.*

10.3.3 Total Values, Minkowski Measures and the Renewal Theorem

The average Minkowski content $\widetilde{\mathcal{M}}^D(F)$ was first obtained by D. Gatzouras [Gat00] using methods from classical renewal theory. This approach leads to an integral representation which is more useful for numerical computations (see Theorem 10.21 below). Moreover, under an additional condition the case of ordinary limits can be treated. For $d = 1$ the latter was already considered in Lapidus [Lap93] and Falconer [Fal95]. We now briefly present these results.

Throughout this section we use the notions and results from Sect. 10.2. Let $S_1, \ldots, S_N$ be an IFS satisfying (OSC) with contraction ratios $r, \ldots, r_N$ generating the self-similar set F with Hausdorff dimension D as before. Recall that μ denotes the unique self-similar probability measure on F, which coincides with the normalized Hausdorff measure, i.e., in the above notations we have $\mu_F = \mathcal{H}^D(F)\,\mu$. Recall that the IFS is said to be *non-lattice* if the set $\{\ln r_1, \ldots, \ln r_N\}$ has this property, i.e., these numbers do not concentrate on an additive subgroup of $\mathbb{R}$.

Classical renewal theory and some extensions provide useful tools for determining Minkowski content and fractal curvatures for a large class of self-similar sets and some self-conformal versions. To this aim we first need an adapted version of the *Renewal theorem* in Feller [Fel71, p. 363] substituting there the variable t by $-\ln\varepsilon$ and the distribution function F by that of a discrete random variable with values in $\{|\ln r_1|, \ldots, |\ln r_N|\}$ and corresponding probabilities $(r_1^D, \ldots, r_N^D)$. (For more details see Gatzouras [Gat00] or Winter [Win08].)

Theorem 10.19 *Suppose that for* $r_1, \ldots, r_N$, D *as above the function* $Z(\varepsilon)$ *with* $\varepsilon \in (0, 1]$ *satisfies the renewal equation*

$$Z(\varepsilon) = \sum_{i=1}^{N} r_i^D \mathbf{1}_{(0,r_i]}(\varepsilon) Z(\varepsilon/r_i) + z(\varepsilon),$$

and the error term $t \mapsto z(e^{-t})$ *is directly Riemann integrable on* $(0, \infty)$. *Then the function Z is bounded and the following holds:*

(i) *The average limit* $\lim_{\varepsilon\to 0}\frac{1}{|\ln\delta|}\int_\delta^1 Z(\varepsilon)\frac{1}{\varepsilon}\,\mathrm{d}\varepsilon$ *exists and equals*

$$\frac{1}{\eta}\int_0^1 z(\varepsilon)\frac{1}{\varepsilon}\,\mathrm{d}\varepsilon\,,$$

where $\eta := \sum_{i=1}^N r_i^D|\ln r_i|$.

(ii) *If the IFS is non-lattice, then the ordinary limit* $\lim_{\varepsilon\to 0} Z(\varepsilon)$ *exists and agrees with the average limit.*

Remark 10.20 According to a result of Asmussen [Asm87, Proposition 4.1, p.118] any a.e. continuous function which is bounded by a directly Riemann integrable function is also directly Riemann integrable. In particular, if the above $z(\varepsilon)$ is continuous at Lebesgue-a.a. $\varepsilon \in (0,1)$ and $|z(\varepsilon)| \le c\,\varepsilon^\gamma$ for some constants c and $\gamma > 0$ these conditions are fulfilled.

This can be applied to both the *Minkowski content and its average version.*

Theorem 10.21 *Let the IFS and the associated self-similar set F be as above. Then in the non-lattice case the Minkowski content*

$$\mathcal{M}^D(F) = \lim_{\varepsilon\to 0}\frac{\mathcal{L}^d(F_\varepsilon)}{\omega_{d-D}\varepsilon^{d-D}} = \lim_{\varepsilon\to 0}\frac{\mathcal{H}^{d-1}(\partial F_\varepsilon)}{(d-D)\omega_{d-D}\varepsilon^{d-1-D}}$$

exists and is equal to

$$\begin{aligned}\mathcal{M}^D(F) &= \frac{1}{\sum_{i=1}^N r_i^D|\ln r_i|}\int_0^1 \frac{R_d(\varepsilon)}{\omega_{d-D}\varepsilon^{d-D}}\,\frac{1}{\varepsilon}\mathrm{d}\varepsilon\\ &= \frac{1}{\sum_{i=1}^N r_i^D|\ln r_i|}\int_0^1 \frac{R_{d-1}(\varepsilon)}{(d-D)\omega_{d-D}\varepsilon^{d-1-D}}\,\frac{1}{\varepsilon}\mathrm{d}\varepsilon\,,\end{aligned}$$

where

$$R_d(\varepsilon) := \mathcal{L}^d(F_\varepsilon) - \sum_{i=1}^N \mathbf{1}_{(0,r_i]}(\varepsilon)\mathcal{L}^d((S_iF)_\varepsilon)\,,$$

$$R_{d-1}(\varepsilon) := \mathcal{H}^{d-1}(\partial F_\varepsilon) - \sum_{i=1}^N \mathbf{1}_{(0,r_i]}(\varepsilon)\mathcal{H}^{d-1}(\partial(S_iF)_\varepsilon)\,.$$

In the lattice case the average Minkowski content $\widetilde{\mathcal{M}}^D(F)$*, which always exists, also agrees with the above integral representations.*

Proof We first consider the function

$$Z(\varepsilon) := \mathbf{1}_{(0,1]}(\varepsilon)\frac{\mathcal{L}^d(F_\varepsilon)}{\omega_{d-D}\varepsilon^{d-D}}\,.$$

It satisfies the renewal equation in the last theorem with

$$z(\varepsilon) := \frac{1}{\omega_{d-D}\varepsilon^{d-D}}\left(\mathbf{1}_{(0,1]}(\varepsilon)\mathcal{L}^d(F_\varepsilon) - \sum_{i=1}^{N} r_i^D \mathbf{1}_{(0,r_i]}(\varepsilon)\mathcal{L}^d(F_{\varepsilon/r_i})\right)$$
$$= \frac{1}{\omega_{d-D}\varepsilon^{d-D}}\left(\mathbf{1}_{(0,1]}(\varepsilon)\mathcal{L}^d(F_\varepsilon) - \sum_{i=1}^{N} \mathbf{1}_{(0,r_i]}(\varepsilon)\mathcal{L}^d((S_i F)_\varepsilon)\right).$$

The last equality follows from the scaling property of Lebesgue measure.

First observe that this function z is piecewise continuous. Furthermore, below after the proof of the next theorem, which uses similar estimates, we will show that

$$|z(\varepsilon)| \le c\varepsilon^{\gamma} \text{ for some constants } c \text{ and } \gamma > 0. \tag{10.14}$$

Therefore the above Renewal theorem together with Remark 10.20 implies the assertion for the volume part.

The proof for the limit of the surface area version is similar, and equality follows from Theorem 10.15. □

For the volume this was proved in Falconer [Fal95] for $d = 1$ and in Gatzouras [Gat00] for general d using a result of Lalley [Lal88]. The surface area was considered in Winter [Win08], Rataj and Winter [RW10] in a more general context of the above curvature measures. Furthermore, in [Win08] and [WZ13] ordinary convergence of the corresponding measures for the non-lattice case was treated in this sense. We will give now a slightly shorter proof for the special case of the *Minkowski measure*. (The average version is already shown in Theorem 10.18.) Recall that μ is the normalized Hausdorff measure on F.

Theorem 10.22 *Under the above conditions in the non-lattice case the limit*

$$M^D(F,\cdot) = \lim_{\varepsilon\to 0} \frac{1}{\omega_{d-D}\varepsilon^{d-D}}\mathcal{L}^d(F_\varepsilon \cap (\cdot))$$
$$= \lim_{\varepsilon\to 0} \frac{1}{(d-D)\omega_{d-D}\varepsilon^{d-1-D}}\mathcal{H}^{d-1}(\partial F_\varepsilon \cap (\cdot))$$

exists in the sense of weak convergence of measures and satisfies

$$M^D(F,\cdot) = \mathcal{M}^D(F)\,\mu(\cdot).$$

Proof Denote

$$\mu_\varepsilon := \frac{1}{\omega_{d-D}\varepsilon^{d-D}}\mathcal{L}^d(F_\varepsilon \cap (\cdot)).$$

Since the self-similar set F is compact, $\{\mu_\varepsilon, \varepsilon < |F|\}$ is a family of tight Borel measures. Theorem 10.21 implies that these measures are uniformly bounded, i.e.,

$$\sup_{\varepsilon<\varepsilon_0} \mu_\varepsilon(F_\varepsilon) < \infty . \tag{10.15}$$

This can also be seen directly: For such ε consider the Markov stopping $\tau_\varepsilon(w)$ of $w \in W$ from (10.12) determined by

$$r_{w_1} \dots r_{w_{\tau_\varepsilon(w)}} |F| < \varepsilon \leq r_{w_1} \dots r_{w_{\tau_\varepsilon(w)-1}} |F| .$$

From the corresponding self-similarity properties (10.9) and (10.10) and scaling and monotonicity of the Lebesgue measure we infer

$$\begin{aligned} \mathcal{L}^d(F_\varepsilon) &= \mathcal{L}^d\Big(\bigcup_{w\in W(\tau_\varepsilon)} (S_w F)_\varepsilon\Big) \leq \sum_{w\in W(\tau_\varepsilon)} \mathcal{L}^d((S_w F)_\varepsilon) = \sum_{w\in W(\tau_\varepsilon)} r_w^d \mathcal{L}^d(F_{\varepsilon/r_w}) \\ &\leq \varepsilon^d \mathcal{L}^d(F_c)\, \#(W(\tau_\varepsilon)) , \end{aligned}$$

where $c := (\min(r_1, \dots, r_N))^{-1}|F|$. Hence, for (10.15) it suffices to show that the number of elements $\#(W(\tau_\varepsilon))$ of $W(\tau_\varepsilon)$ is bounded from above by ε^{-D} up to a constant, which follows from (10.11): This equation and the definition of τ_ε yield

$$1 = \sum_{w\in W(\tau_\varepsilon)} r_w^D \geq c^{-D} \varepsilon^D \#(W(\tau_\varepsilon)) .$$

Consequently, the measures μ_ε are uniformly bounded. Then by Prokhorov's theorem they are sequentially compact in the sense of weak convergence of measures. Hence, for the volume part of the theorem it remains to show that the limit measure ν of every weakly converging subsequence $\nu_n := \mu_{\varepsilon_n}$ agrees with the measure $\mathcal{M}^D(F)\mu$.

Theorem 10.21 implies that

$$\lim_{n\to\infty} \nu_n(\mathbb{R}^d) = \mathcal{M}^D(F) = \mathcal{M}^D(F)\mu(\mathbb{R}^d) .$$

Therefore by a characterization of weak convergence of normalized measures it is sufficient to show that

$$\liminf_{n\to\infty} \nu_n(G) \geq \mathcal{M}^D(F)\mu(G) \ \text{ for any open set } G . \tag{10.16}$$

In Winter [Win08, Lemma 6.1.1] one can find an intersection stable generator for the Borel σ-algebra of the form

$$\mathcal{A}_F = \{S_w O : w \in W_*\} \cup \mathcal{C}_F ,$$

where O is an arbitrary feasible open set for F and $\mu(C) = 0$ for all $C \in \mathcal{C}_F$. Then G admits a representation

$$G = \bigcup_{i=1}^{\infty} A_i \cup \bigcup_{j=1}^{\infty} C_j$$

with

$$A_i = S_{w(i)} O \ \text{ for some finite word } w(i) \in W_* \ \text{ and } \ C_j \in \mathcal{C}_F .$$

Note that by (OSC) the different A_i can be chosen pairwise disjoint. Then we obtain

$$\liminf_{n\to\infty} \nu_n(G) \ge \liminf_{n\to\infty} \nu_n\big(\bigcup_{i=1}^{\infty} A_i\big) = \liminf_{n\to\infty} \sum_{i=1}^{\infty} \nu_n(A_i) \ge \sum_{i=1}^{\infty} \liminf_{n\to\infty} \nu_n(A_i) .$$

Below we will see that

$$\lim_{\varepsilon\to 0} \mu_\varepsilon(S_w O) = \mathcal{M}^D(F)\mu(S_w O) \ \text{ for any } \ w \in W_* . \tag{10.17}$$

Hence, the last sum is equal to

$$\mathcal{M}^D(F) \sum_{i=1}^{\infty} \mu(A_i) = \mathcal{M}^D(F)\mu\big(\bigcup_{i=1}^{\infty} A_i\big) = \mathcal{M}^D(F)\mu(G) ,$$

since $\mu(C_j) = 0$, which proves (10.16).

In order to show (10.17) we decompose

$$\mu_\varepsilon(S_w O) = \mu_\varepsilon(S_w O \setminus (\partial S_w O)_\varepsilon) + \mu_\varepsilon(S_w O \cap (\partial S_w O)_\varepsilon) =: \Sigma_1(\varepsilon) + \Sigma_2(\varepsilon) .$$

By (OSC) we have $F_\varepsilon \cap (S_w O \setminus (\partial S_w O)_\varepsilon) = (S_w F)_\varepsilon \cap (S_w O \setminus (\partial S_w O)_\varepsilon)$. Therefore locality and the scaling property of Lebesgue measure under the mapping S_w yield

$$\begin{aligned}\Sigma_1(\varepsilon) &= r_w^D \mu_{\varepsilon/r_w}(F_{\varepsilon/r_w} \cap (O \setminus (\partial O)_{\varepsilon/r_w})) \\ &= r_w^D \mu_{\varepsilon/r_w}(F_{\varepsilon/r_w}) - r_w^D \mu_{\varepsilon/r_w}(F_{\varepsilon/r_w} \cap (\partial O)_{\varepsilon/r_w}) =: \Sigma_3(\varepsilon) - r_w^D \Sigma_4(\varepsilon) .\end{aligned}$$

Furthermore, $\Sigma_3(\varepsilon) = r_w^D \mu_{\varepsilon/r_w}(\mathbb{R}^d)$ which converges to

$$r_w^D \mathcal{M}^D(F) = \mathcal{M}^D(F)\mu(S_w F) = \mathcal{M}^D(F)\mu(S_w O) ,$$

as $\varepsilon \to 0$, in view of Theorem 10.21 and the self-similarity (10.8) of μ. Hence, for (10.17) it remains to prove that for $\Sigma_2'(\varepsilon) = \mu_\varepsilon((\partial S_w O)_\varepsilon)$ and $\Sigma_4(\varepsilon) = \mu_\varepsilon((\partial O)_\varepsilon)$ we get

$$\lim_{\varepsilon\to 0} \Sigma_2'(\varepsilon) = \lim_{\varepsilon\to 0} \Sigma_4(\varepsilon) = 0$$

if O is a strong feasible open set. Since in this case $F \cap O \neq \emptyset$, there exists a finite word u such that $S_u F \subset O$. $S_u F$ is compact and O is open. Therefore we infer for some ε_0 and all $\varepsilon < \varepsilon_0$,

$$(S_u F)_\varepsilon \subset O \setminus (\partial O)_\varepsilon \,, \tag{10.18}$$

which implies

$$F_\varepsilon \cap (\partial O)_\varepsilon = \bigcup_{v \in W(\tau_\varepsilon)} (S_v F)_\varepsilon \cap (\partial O)_\varepsilon = \bigcup_{v \in W(\tau_\varepsilon), u \not\prec v} (S_v F)_\varepsilon \cap (\partial O)_\varepsilon \,,$$

where $u \not\prec v$ means that $v \neq uv'$ for each finite word v'. In the sequel let c denote a varying constant. From the definition of the Markov stopping τ_ε we get for $v \in W(\tau_\varepsilon)$ that $\operatorname{diam}(S_v F)_\varepsilon < c\,\varepsilon$ and hence, by locality and scaling of Lebesgue measure,

$$\mu_\varepsilon((S_v F)_\varepsilon) \leq c\,\varepsilon^D \,. \tag{10.19}$$

Hence,

$$\Sigma_4(\varepsilon) = \mu_\varepsilon(F_\varepsilon \cap (\partial O)_\varepsilon) \leq \sum_{v \in W(\tau_\varepsilon), u \not\prec v} \mu_\varepsilon\big((S_v F)_\varepsilon\big) \leq c\,\varepsilon^D \# \{v \in W(\tau_\varepsilon), u \not\prec v\} \,.$$

If we show that the last cardinality does not exceed $c\,\varepsilon^{-D+\gamma}$ for some $\gamma > 0$, it follows that

$$\lim_{\varepsilon \to 0} \Sigma_4(\varepsilon) = 0 \,.$$

This estimate can be derived similarly as at the beginning of the proof. Let D' be the unique number such that

$$\sum_{v \in W(\tau_\varepsilon), u \not\prec v} r_v^{D'} = 1 \,.$$

Since $\sum_{v \in W(\tau_\varepsilon)} r_v^D = 1$ (see (10.11)), we get $D' < D$ and

$$1 = \sum_{v \in W(\tau_\varepsilon), u \not\prec v} r_v^{D'} \geq c\varepsilon^{D'} \#(\{v \in W(\tau_\varepsilon), u \not\prec v\}) \,,$$

i.e., the desired estimate for the above cardinality setting $\gamma := D - D'$.

The proof for the summand $\Sigma_2'(\varepsilon)$ instead of $\Sigma_4(\varepsilon)$ is analogous if considering the word wu instead of u and the set $S_w O$ instead of O.

This completes the proof for the volume part in the assertion of the theorem. The arguments for the boundary version are similar, when replacing $\mathcal{L}^d$ on F_ε by $\mathcal{H}^{d-1}$ on ∂F_ε and regarding the corresponding scaling behaviour. □

Now we will complete the proof of Theorem 10.21 for the Minkowski content.

Proof of (10.14) At the end of the last proof we have seen for sufficiently small ε the estimate

$$\mu_\varepsilon((\partial O)_\varepsilon) \le c\varepsilon^\gamma \tag{10.20}$$

and the same (with different constants c and $\gamma > 0$) for $\mu_\varepsilon((\partial S_w O)_\varepsilon)$, where

$$\mu_\varepsilon = \frac{\mathcal{L}^d(F_\varepsilon \cap (\cdot))}{\omega_{d-D}\varepsilon^{d-D}}.$$

Similarly we can get the auxiliary inequality

$$\mu_\varepsilon((\partial \mathbf{S}O)_\varepsilon) \le c\varepsilon^\gamma \tag{10.21}$$

for such constants by the following arguments. (Denote $\mathbf{S}A = \cup_{i=1}^N S_i A$, for general $A \subset \mathbb{R}^d$.) From (10.18) we infer

$$(\mathbf{S}S_u F)_\varepsilon \subset \mathbf{S}O \setminus (\partial \mathbf{S}O)_\varepsilon$$

for all sufficiently small ε. Like in the estimate of the summand $S_4(\varepsilon)$ in the previous proof we conclude

$$\mu_\varepsilon((\partial \mathbf{S}O)_\varepsilon) \le \sum \mu_\varepsilon((S_v F)_\varepsilon),$$

where the sum runs over all words $v \in W(\tau_\varepsilon)$ which are not representable as $v = iuv'$ for some $i \in \{1, \ldots, N\}$ and $v' \in W_*$. The remaining arguments for (10.21) are similar as in the above case with a different $\gamma > 0$.

We now consider $z(\varepsilon)$ from the assertion (10.14). In terms of μ_ε we get

$$z(\varepsilon) = \mathbf{1}_{(0,1]}\mu_\varepsilon(F_\varepsilon) - \sum_{i=1}^N \mathbf{1}_{(0,r_i]}(\varepsilon)\mu_\varepsilon((S_i F)_\varepsilon).$$

It suffices to choose $\varepsilon \le r_{\min} := \min(r_1, \ldots, r_N)$, since $z(\varepsilon)$ is bounded on $[b, 1]$ for any $0 < b \le 1$. Denote

$$U(\varepsilon) := \bigcup_{i \ne j} (S_i F)_\varepsilon \cap (S_j F)_\varepsilon \quad \text{and} \quad B_i(\varepsilon) := (S_i F)_\varepsilon \setminus U(\varepsilon).$$

Then $F_\varepsilon = \bigcup_{i=1}^N B_i(\varepsilon) \cup U(\varepsilon)$ is a disjoint union and hence,

$$\mu_\varepsilon(F_\varepsilon) = \sum_{i=1}^N \mu_\varepsilon(B_i(\varepsilon)) + \mu_\varepsilon(U(\varepsilon)) .$$

Similarly,

$$\mu((S_i F)_\varepsilon) = \mu_\varepsilon(B_i(\varepsilon)) + \mu_\varepsilon((S_i F)_\varepsilon \cap U(\varepsilon)).$$

Substituting these expressions in the above equation for $z(\varepsilon)$ we infer

$$|z(\varepsilon)| \le \mu_\varepsilon(U(\varepsilon)) + \sum_{i=1}^N \mu_\varepsilon((S_i F)_\varepsilon \cap U(\varepsilon)) \le (N+1)\mu_\varepsilon(U(\varepsilon)) .$$

From (OSC) we will see that $U(\varepsilon) \subset (\partial \mathbf{S}O)_\varepsilon$. Consequently, for sufficiently small ε we can use (10.21) and obtain the desired upper estimate $c\varepsilon^\gamma$.

In order to show the above set inclusion note that for any $y \in U(\varepsilon)$ there exist $i \neq j$ and $x_i \in S_i F \subset S_i \overline{O}$, $x_j \in S_j F \subset S_j \overline{O}$ with $|x_i - y| \le \varepsilon$, $|x_j - y| \le \varepsilon$. Because of (OSC) the segment $[x_i, x_j]$ between these points intersects $\partial S_i O \cup \partial S_j O$. Since any point $x \in [x_i, x_j]$ fulfills $|x - y| \le \varepsilon$, and since $\partial S_i O \cup \partial S_j O \subset \partial \mathbf{S}O$, we get $y \in (\partial \mathbf{S}O)_\varepsilon$. □

In Freiberg and Kombrink [FK12] the last two theorems are extended to images of self-similar sets with (OSC) under conformal $C^{1+\alpha}$-mappings.

Relationships between the (average) Minkowski contents of a self-similar set F as above and its associated tiling $\mathcal{T}(O)$ for certain feasible open sets O are studied in Winter [Win15] with similar methods. The following particular result for compatible tilings $\mathcal{T}$, which has been known for special cases before, provides a nice formula for computing the (average) Minkowski content of F in terms of the generator G. We write $G_{-\varepsilon} := \{x \in G : d(x, \partial G) \le \varepsilon\}$ for the *inner parallel sets* of G.

Theorem 10.23 (cf. [Win15, Theorem 3.10]) *If the self-similar set F with Minkowski dimension $D \in (d-1, d)$ admits a strong feasible open set O such that $\partial G \subset F$ for the corresponding generator $G = O \setminus \overline{\mathbf{S}O}$, then the average Minkowski content, and in the non-lattice case the Minkowski content, of F exists and is equal to*

$$\frac{1}{\eta} \int_0^\infty \frac{\mathcal{L}^d(G_{-\varepsilon})}{\omega_{d-D}\varepsilon^{d-D}} \frac{1}{\varepsilon} \mathrm{d}\varepsilon$$

where $\eta = \sum_{i=1}^N r_i^D |\ln r_i|$.

The proof of existence via the Renewal theorem for the corresponding tiling is here more direct than in the general situation of Theorem 10.21. The result can be applied to many classical cases like the Cantor set, the Sierpinski gasket or the

Sierpinski carpet with evident choices of G. (See also the references in [Win15] to former literature.) In Kombrink [Kom11] under some additional conditions on the generator such a formula is shown even for certain self-conformal sets, where symbolic dynamics and a related Renewal theorem of Lalley [Lal12] is used.

S. Winter also derived a generator type expression for the (average) Minkowski content of a general self-similar set F with (OSC) and $D < d$ for any strong feasible O satisfying the so-called *projection condition* (PC) (see [Win15, Theorem 3.17]):

$$\widetilde{\mathcal{M}}^D(F) = \frac{1}{\eta} \int_0^\infty \frac{\mathcal{L}^d(F_\varepsilon \cap \Gamma)}{\omega_{d-D}\varepsilon^{d-D}} \frac{1}{\varepsilon} \mathrm{d}\varepsilon \tag{10.22}$$

($= \mathcal{M}^D(F)$ in the non-lattice case), where $\Gamma := O \setminus \mathbf{S}O$. The conditions $D < d$ and (PC) are also necessary for the validity of this formula. E. Pearse observed that the *central open set* studied in Bandt, Hung and Rao [BHR06] provides an example for such an O (cf. [Win15, Proposition 3.17]). However, its concrete construction may be complicated.

10.3.4 Relationships to Fractal Zeta Functions

A different approach to the (average) Minkowski content has been developed by M. Lapidus and his collaborators within a very general theory of fractal zeta functions and complex dimensions, see Lapidus, Radunović, Žubrinić [LRŽ16, Chapter 2] and related references to the former literature therein. Applied to self-similar sets with (OSC) and dimension D under some additional conditions on the asymptotic behaviour of $\varepsilon^{-(d-D)}\mathcal{L}^d(F_\varepsilon)$ as $\varepsilon \to 0$ one obtains

$$\mathcal{M}^D(F) = \frac{\operatorname{res}(\tilde{\zeta}_F, D)}{\omega_{d-D}} \tag{10.23}$$

in the non-lattice case and the same formula for $\widetilde{\mathcal{M}}^D(F)$) in the lattice case. Here $\operatorname{res}(\tilde{\zeta}_F, D)$ denotes the residue of the *tube zeta function*

$$\tilde{\zeta}_F(s) := \int_0^\delta \varepsilon^{s-d-1}\mathcal{L}^d(F_\varepsilon)\, d\varepsilon\,,\ s \in \mathbb{C}\,, \operatorname{Re} s > \dim_H F\,,$$

of F at the pole $s = D$, where $\delta > 0$ is a fixed number (cf. [LRŽ16, Theorems 2.3.18, 2.4.3].) Similar results are derived in terms of the *distance zeta function*

$$\zeta_F(s) := \int_{F_\delta} d(x, F)^{s-D} dx$$

(cf. [LRŽ16, Theorems 2.2.3, 2.3.37]). Moreover, under further conditions in [LRŽ16, Theorem 4.1.14] the relative Minkowski content $\mathcal{M}^D(F, G)$ for open $G \subset \mathbb{R}^d$ is expressed by the residue of the associated *relative distance zeta function* at pole D. Note that again more general sets F are considered. The authors conjecture that the corresponding conditions are fulfilled for all self-similar sets with (OSC) in the non-lattice (resp. lattice) case.

10.4 Extension to Fractal Curvature Measures

10.4.1 Application of Renewal Theory

We will present now extensions of the above results for the (average) Minkowski content of self-similar sets F as above and its localized versions to fractal curvatures and associated measures C_k^D as well as their densities in the sense of Definitions 10.6, 10.8 and 10.12, respectively. Here we additionally assume the regularity of the parallel sets for almost all distances less than $R|F|$ for an $R > \sqrt{2}$ and some integrability properties of their curvature measures. In contrast to the previous section we start with the approach to the total fractal curvatures by means of renewal theory. Some ideas of the proof of Theorem 10.21 are used for the case of curvatures taking into regard Proposition 10.24 below. For completeness we include the former in the formulations as marginal cases $k = d$ or $k = d - 1$. However, since the curvature measures $C_k(F_\varepsilon, \cdot)$ of the parallel sets are determined by the corresponding unit normal bundles, one has to be more careful here concerning intersections. For the geometric relationships only the following properties of the curvature measures $C_k(F_\varepsilon, \cdot)$ at regular ε are involved. They are *motion invariant*, *scaling with exponent k* and *locally determined*:

$$C_k(g(F_\varepsilon)), g(\cdot)) = C_k(F_\varepsilon, \cdot) \;\; \text{for any Euclidean motion } g\,,$$

$$C_k(r(F_\varepsilon), r(\cdot)) = r^k C_k(F_\varepsilon, \cdot)\;, r > 0\,,$$

$$C_k(F_\varepsilon, G) = C_k((S_w F)_\varepsilon), G) \;\; \text{if } G \text{ is open and } F_\varepsilon \cap G = (S_w F)_\varepsilon \cap G\,,$$

$$w \in W_*\,, \;\; \text{provided that } \varepsilon \text{ is also regular for } S_w F\,.$$

(The last property is already specified to the smaller copies of F under the similarities S_w.)

An important tool is the following nice behaviour of the curvature measures of parallel sets of sufficiently large distances.

Proposition 10.24 ([Zäh11, Theorem 4.1]) *For any $R > \sqrt{2}$ and $k \in \{0, 1, \ldots, d\}$ there exists a constant $c_k(R)$ such that for all compact $K \subset \mathbb{R}^d$*

and $r \geq R\,|K|$,

$$\operatorname{reach} \overline{(K_r)^c} \geq |K|\sqrt{R^2-1} \quad and$$

$$\sup_{r \geq R\,|K|} \frac{C_k^{\mathrm{var}}(K_r, \mathbb{R}^d)}{r^k} \leq c_k(R)\,.$$

The proof of the first statement is based on Fu [Fu85]. For $k = d$, i.e. the case of Lebesgue measure, the second inequality applied to the smaller copies of $K := F$ with respect to the former Markov stopping corresponds to the estimate (10.19). It plays a similar role for general k: Recall that for F as above we have

$$W(\tau_{r/R}) = \{w \in W_* : r_{w_1} \dots r_{w_{|w|}} R\,|F| \leq r < r_{w_1} \dots r_{w_{|w|-1}} R\,|F|\}$$

and set for a feasible open set O,

$$W_{O,R}(r) := \left\{w \in W(\tau_{r/R}) : (S_w F)_r \cap (\mathbf{S}O)_r \neq \emptyset\right\}\,.$$

Proposition 10.24 implies for $w \in W(\tau_{r/R})$,

$$C_k^{\mathrm{var}}((S_w F)_r) = r_w^k C_k^{\mathrm{var}}(F_{r/r_w}) \leq r_w^k c_k(R)(r^k/r_w^k) = c\,r^k\,.$$

This is used in the proof of the following extension of Theorem 10.21 to *(average) fractal curvatures*.

Theorem 10.25 *Let* $k \in \{0, 1, \dots, d\}$ *and* $F \subset \mathbb{R}^d$ *be a self-similar set with Minkowski dimension* D *and contraction ratios* $r_1, \dots, r_N$ *of the corresponding similarities satisfying (OSC). If* $k \leq d-2$ *suppose additionally the following two conditions:*

(i) *If* $d \geq 4$, *then almost all* $r \in (0, \sqrt{2}|F|)$ *are regular for* F.
For $k \leq 3$ *this is always fulfilled.*
(ii) *There exist a strong feasible open set* O *and constants* c, $R > \sqrt{2}$ *such that for almost all* $r \in (0, R\,|F|)$ *and all* $w \in W_{O,R}(r)$,

$$C_k^{\mathrm{var}}\left(F_r, \partial(S_w F)_r \cap \partial \bigcup_{v \in W_{O,R}(r)\setminus\{w\}} (S_v F)_r\right) \leq c\,r^k\,.$$

Set

$$R_k(\varepsilon) := \mathbf{1}_{(0,R|F|]}(\varepsilon) C_k(F_\varepsilon) - \sum_{i=1}^N \mathbf{1}_{(0,r_i R|F|]} C_k((S_i F)_\varepsilon)\,.$$

Then the average fractal curvatures exist and are given by

$$\overline{\mathbf{C}}_k^D(F) = \lim_{\delta\to 0} \frac{1}{|\ln\delta|} \int_\delta^1 \varepsilon^{D-k} C_k(F_\varepsilon) \frac{1}{\varepsilon} \mathrm{d}\varepsilon = \frac{1}{\eta} \int_0^{R|F|} r^{D-k} R_k(r) \frac{1}{r} \mathrm{d}r\,,$$

where $\eta = \sum_{i=1}^N r_i^D |\ln r_i|$*, and in the non-lattice case the fractal curvatures are given by*

$$\mathbf{C}_k^D(F) = \operatorname*{ess\,lim}_{\varepsilon\to 0} C_k(F_\varepsilon) = \overline{\mathbf{C}}_k^D(F)\,.$$

Remark 10.26 In Winter [Win11, Corollary 4.9] it is shown that Condition (ii) is independent of the choice of O.

The corresponding *measure version* as a refinement is a complement of Theorem 10.22. Because of the self-similarity all fractal curvature measures F are constant multiples of the normalized Hausdorff measure μ on F:

Theorem 10.27 *Under the conditions of Theorem* 10.25 *we get*

$$\overline{C}_k^D(F,\cdot) = \lim_{\delta\to 0} \frac{1}{|\ln\delta|} \int_\delta^1 \varepsilon^{D-k} C_k(F_\varepsilon,\cdot) \frac{1}{\varepsilon} \mathrm{d}\varepsilon = \overline{\mathbf{C}}_k^D(F)\,\mu$$

and in the non-lattice case

$$C_k^D(F,\cdot) = \operatorname*{ess\,lim}_{\varepsilon\to 0} C_k(F_\varepsilon,\cdot) = \mathbf{C}_k^D(F)\,\mu$$

in the sense of weak convergence.

The whole approach was worked out in Winter [Win08] for the special case, when the parallel sets F_ε are polyconvex for one and thus all $\varepsilon > 0$. In this case it can be shown that all conditions of the theorems are fulfilled. Moreover, for some classical examples the total values are computed. Theorem 10.25 is the special deterministic version of [Zäh11, Theorem 2.3.8, Corollary 2.3.9], where such curvatures for random self-similar sets are considered. Theorem 10.27 is derived in [WZ13, Theorem 2.3]. As in the case of the Minkowski measure the proof essentially relies on Theorem 10.25. The techniques for applying the Renewal theorem in the proofs of Theorems 10.25 and 10.27 are similar to the Minkowski case, but much more extensive, since signed measures and geometric boundary properties are involved.

As in the case of the Minkowski content (cf. Theorem 10.23), for *compatible tilings* (the boundary of their generator G belongs to F) the assumptions and the formulas for the (average) fractal curvatures can be simplified:

Theorem 10.28 ([Win15, Theorem 4.5]) *Let F be a self-similar set satisfying (OSC) and let O be a strong feasible open set for F such that the associated tiling $\mathcal{T}(O)$ with generator $G = O\setminus\mathbf{S}\overline{O}$ is compatible. Let $k \in \{0,\ldots,d-1\}$. If $k \le d-2$*

assume additionally the following two conditions:

(i) *For $d \geq 4$ almost all $\varepsilon < \sqrt{2}|F|$ are regular values for F.*
(For $d \leq 3$ this condition is always fulfilled.)

(ii') *There are constants $c, \gamma > 0$ and $R > \sqrt{2}$ such that for almost all $0 < \varepsilon < R|F|$,*

$$C_k^{\mathrm{var}}(F_\varepsilon, ((\mathbf{S}O)^c)_\varepsilon) \leq c\varepsilon^{k-D+\gamma}\,.$$

Then the average fractal curvatures of F exist and are given by

$$\overline{\mathbf{C}}_k^D(F) = \frac{1}{\eta}\int_0^\rho \varepsilon^{D-k}\mathbf{C}_k(G_{-\varepsilon})\,\mathrm{d}\varepsilon\,,$$

where ρ is the inradius of G. In the non-lattice case the fractal curvatures $\mathbf{C}_k^D(F)$ exist and coincide with $\overline{\mathbf{C}}_k^D(F)$.

Remark 10.29 Condition (i) agrees with that from Theorem 10.25. Furthermore, from the arguments in the proof of this theorem it follows that the integrability condition (ii) from there implies Condition (ii').

A measure version of Theorem 10.28 can also be obtained. Under slightly stronger assumptions such a result was proved in Kombrink [Kom11, Theorem 2.37] with methods from Perron-Frobenius theory and renewal theory for shift dynamical system in the sense of Lalley [Lal12]. This approach is more complicated, however it was used in [Kom11] for extensions to certain *self-conformal fractals* with compatible tilings.

For self-similar tilings *without the compatibility condition* in Winter [Win15, Theorem 4.9] an analogous formula is proved by choosing any strong feasible open set O satisfying the projection condition (PC) and assuming additionally that $C_k^{\mathrm{var}}(F_\varepsilon, \partial O) = 0$ for almost all $\varepsilon < \rho$. If for $k \leq d-2$ the conditions (i) and (ii') from the last theorem are fulfilled, then the average fractal curvatures (and in the non-lattice case the fractal curvatures) for $k \leq d-1$ exist and can be expressed in terms of the generator $G = O \setminus \mathbf{S}\overline{O}$:

$$\overline{\mathbf{C}}_k^D(F) = \frac{1}{\eta}\int_0^\rho \varepsilon^{D-k}C_k(F_\varepsilon, G)\frac{1}{\varepsilon}\mathrm{d}\varepsilon\,. \tag{10.24}$$

This is the complement to formula (10.22) for the Minkowski content.

The representation formulas for the fractal curvatures from Theorems 10.25 and 10.28 as well as 10.24 admit to compute *numerical values* for some classical cases.

Example 10.30 In Winter [Win08] the following values have been obtained.

$$\overline{\mathbf{C}}_0^D(F) = -0,016,\ \overline{\mathbf{C}}_1^D(F) = 0,0725,\ \overline{\mathbf{C}}_2^D(F) = 1,352$$

for the *Sierpinski carpet* (Fig. 10.2) and

$$\overline{\mathbf{C}}_0^D(F) = -0,014,\ \overline{\mathbf{C}}_1^D(F) = 0,0720,\ \overline{\mathbf{C}}_2^D(F) = 1,344$$

for the *modified Sierpinski carpet* (Fig. 10.3). This shows that self-similar sets with equal Hausdorff dimension and different geometric features can be distinguished by means of these parameters. In particular, the corresponding fractal Euler-type number $\overline{\mathbf{C}}_0^D(F)$ reflect the topological property that the Sierpinski carpet has more holes than the modified version.

10.4.2 Methods from Ergodic Theory: Curvature Densities and Measures

In the previous sections we have seen that in the self-similar case the (average) Minkowski measure and fractal curvature measures can be obtained from the classical Renewal theorem. Next we will show how *Birkhoff's ergodic theorem* (see, e.g., Walters [Wal82]) for the embedded shift dynamical system can be used in order to prove that all these measures possess a.e. constant densities with respect to the normalized Hausdorff measure. Moreover, measure convergences can also be derived from this approach.

First recall the notion of a localizing family of bump functions (cf. Definition 10.12) specified to a self-similar set F with a strong feasible open set O and $\mu_F := \mathcal{H}^D(F \cap (\cdot))$:

$$A_F(x,\varepsilon)(z) := \varepsilon^D \frac{g\left(\frac{|x-z|}{a\varepsilon}\right)}{\int_F g\left(\frac{|y-z|}{a\varepsilon}\right) \mathcal{H}^D(\mathrm{d}y)}\,, \quad z \in \mathbb{R}^d\,,$$

$x \in F$, $\varepsilon \in (0\varepsilon_0)$ for some $\varepsilon_0 > 0$, $a > 1$ and any nonnegative bounded measurable function g with support in $[-1, 1]$ which is bounded from below by a positive constant in a neighbourhood of 0. Such a family fulfills the following three conditions: Denote $b := \max(2a, \varepsilon_0^{-1}|O|)$.

$$\operatorname{spt} A_F(x,\varepsilon) \subset F_\varepsilon \cap B(x, a\varepsilon)\,; \tag{10.25}$$

$$A_F(x,\varepsilon)(z) = A_F(S_j^{-1}x, r_j^{-1}\varepsilon)(S_j^{-1}z)\,, \tag{10.26}$$

if $x \in S_j F$, $\varepsilon < b^{-1} d(x, \partial S_j O)$ and $1 \le j \le N$ (*Covariance*);

$$A_F(x,\varepsilon)(z) \text{ is measurable in } (x,\varepsilon,z) \in F \times (0,\varepsilon_0) \times \mathbb{R}^d\,. \tag{10.27}$$

The covariance property follows from the strong open set condition and the self-similarity of the measure μ. (Note that in 10.26 the condition on ε implies $r_j^{-1}\varepsilon < \varepsilon_0$ because of the choice of b.)

Now we can formulate the *main result* of this section concerning *fractal curvature densities of self-similar sets* including the Minkowski densities as marginal case. This was proved in [RZ12] for the special choice of the localizing functions from the end of Sect. 10.1. The general case is completely analogous. In this approach we identify a.a. points of a self-similar set satisfying (SOSC) with their coding sequences using the corresponding notations from Sect. 10.2.

Theorem 10.31 *Let $k \in \{0, 1, \dots, d\}$ and suppose that the self-similar set F in $\mathbb{R}^d$ with contraction ratios $r_1, \dots, r_N$ and Hausdorff dimension D satisfies the strong open set condition w.r.t. O. If $d \geq 4$ and $k \leq d-2$, we additionally suppose that a.a. $\varepsilon < \varepsilon_0$ are regular for F in the sense of Definition* 10.3. *Let $\{A(x,\varepsilon) : x \in F,\ \varepsilon < \varepsilon_0\}$ be a localizing family of bump functions with constants $a > 1$, $\varepsilon_0 > 0$, and $b = \max\left(2a, \varepsilon_0^{-1}|O|\right)$ as above. Then for $\mathcal{H}^D$-a.a. $x \in F$ the curvature density*

$$\mathcal{D}_k(F,x) := \lim_{\delta\to 0} \frac{1}{|\ln\delta|} \int_\delta^{b^{-1}d(x,\partial O)} \varepsilon^{-k} C_k\big(F_\varepsilon, A_F(x,\varepsilon)\big)\, \varepsilon^{-1} \mathrm{d}\varepsilon \tag{10.28}$$

exists and equals the constant

$$\mathcal{D}_k(F) := \mathcal{H}^D(F)^{-1} \frac{1}{\eta} \int_F \int_{b^{-1}d(y,\partial S_{y_1}O)}^{b^{-1}d(y,\partial O)} \varepsilon^{-k} C_k\big(F_\varepsilon, A_F(y,\varepsilon)\big)\, \varepsilon^{-1} \mathrm{d}\varepsilon\, \mathcal{H}^D(\mathrm{d}y)\,, \tag{10.29}$$

where $\eta = \sum_{j=1}^N r_j^D |\ln r_j|$, provided the last double integral converges absolutely if $k \leq d-2$, and for $k \in \{d-1, d\}$ this is always true.

Proof As an essential auxiliary tool we use the ergodic *shift dynamical system* $[W, \nu, \theta]$ on the code space W for the shift operator $\theta : W \to W$ with $\theta(w_1 w_2 \dots) := (w_2 w_3 \dots)$. According to (10.7) it induces the ergodic dynamical system $[F, \mu, T]$, where the transformation $T : F \to F$ is defined for μ-a.a. x by

$$Tx : S_j^{-1}x \text{ if } x \in S_j F\,,\ j = 1, \dots, N\,,$$

taking into regard that $\mu(S_i F \cap S_j F) = 0$, $i \neq j$. (More general references on this subject may be found, e.g., in Falconer [Fal97], Mauldin and Urbanski [MU03].) In the above identification of a.a. points with their coding sequences we have for such x,

$$Tx = \theta(x_1 x_2 \dots)\,.$$

For brevity we write $x|i := x_1 \dots x_i$ for the corresponding concatenation. Next note that $\varepsilon < b^{-1}d(x, \partial S_{x|i}O)$ implies $\varepsilon < r_{x|i}^{-1}\varepsilon < \varepsilon_0$, since $d(x, \partial S_{x|i}O) =$

$r_{x|i}\, d(T^i x, \partial O)$. From this and $A_F(x,\varepsilon) \subset B(x, a\varepsilon)$ we obtain for Lebesgue-a.a. ε, μ-a.a. x, and $i \in \mathbb{N}$ satisfying the first condition the equalities

$$\begin{aligned} &C_k\big(F_\varepsilon, A_F(x,\varepsilon)\big) = C_k\big((S_{x|i}F)_\varepsilon, A_F(x,\varepsilon)\big) \\ &= C_k\big((S_{x|i}F)_\varepsilon, A_F(T^i x, r_{x|i}^{-1}\varepsilon) \circ S_{x|i}^{-1}\big) = r_{x|i}^k C_k\big(F_{r_{x|i}^{-1}\varepsilon}, A_F(T^i x, r_{x|i}^{-1}\varepsilon)\big)\,. \end{aligned}$$

Here we have used the locality of the curvature measure C_k, the covariance property (10.26) of the functions $A_F(x,\varepsilon)$, and the scaling behaviour of C_k under similarities.

In Proposition 10.11 it is shown that $C_k\big(F_\varepsilon, A_F(x,\varepsilon)\big)$ (with value 0 if ε is not regular) is a measurable function in $(x,\varepsilon) \in F \times (0,\varepsilon_0)$. Now we can verify the limit

$$\begin{aligned} &\lim_{\delta\to 0} \frac{1}{|\ln\delta|} \int_\delta^{b^{-1}d(x,\partial O)} \varepsilon^{-k}\, C_k\big(F_\varepsilon, A_F(x,\varepsilon)\big)\, \varepsilon^{-1} d\varepsilon \\ &= \lim_{\delta\to 0} \frac{n(x,\delta)}{|\ln\delta|} \frac{1}{n(x,\delta)} \Bigg(\sum_{i=0}^{n(x,\delta)-1} \int_{b^{-1}d(x,\partial S_{x|(i+1)}O)}^{b^{-1}d(x,\partial S_{x|i}O)} \varepsilon^{-k} C_k\big(F_\varepsilon, A_F(x,\varepsilon)\big)\, \varepsilon^{-1} d\varepsilon \\ &\qquad + \int_\delta^{b^{-1}d(x,\partial S_{x|n(x,\delta)}O)} \varepsilon^{-k} C_k\big(F_\varepsilon, A_F(x,\varepsilon)\big)\, \varepsilon^{-1} d\varepsilon \Bigg), \end{aligned}$$

where

$$n(x,\delta) := \max\{n \in \mathbb{N} : b^{-1} d(x, \partial S_{x|n}O) \geq \delta\}\,.$$

By the above relation the integrand in the ith integral may be replaced by

$$r_{x|i}^k\, \varepsilon^{-k}\, C_k\big(F_{r_{x|i}^{-1}\varepsilon}, A_F(T^i x, r_{x|i}^{-1}\varepsilon)\big)\, \varepsilon^{-1}\,.$$

For the integral bounds we use

$$\begin{aligned} d(x, \partial S_{x|i}O) &= r_{x|i}\, d(T^i x, \partial O), \\ d(x, \partial S_{x|(i+1)}O) &= r_{x|i}\, d(T^i x, \partial S_{(T^i x)_1}O)\,. \end{aligned}$$

Substituting then in the i-th integral $r_{x|i}^{-1}\varepsilon$ by ε, we obtain the expression

$$\int_{b^{-1}d(T^i x, \partial S_{(T^i x)_1}O)}^{b^{-1}d(T^i x,\partial O)} \varepsilon^{-k} C_k\big(F_\varepsilon, A_F(T^i x,\varepsilon)\big)\, \varepsilon^{-1} d\varepsilon\,.$$

Therefore it suffices to show that for μ-a.a. $x \in F$ the following integrals and limit relationships exist:

$$\lim_{n\to\infty}\frac{1}{n}\sum_{i=1}^{n}\int_{b^{-1}d(T^ix,\partial S_{(T^ix)_1}O)}^{b^{-1}d(T^ix,\partial O)} \varepsilon^{-k}C_k\big(F_\varepsilon, A_F(T^ix,\varepsilon)\big)\,\varepsilon^{-1}\mathrm{d}\varepsilon$$

$$= \int_F\int_{b^{-1}d(y,\partial S_{y_1}O)}^{b^{-1}d(y,\partial O)} \varepsilon^{-k}C_k\big(F_\varepsilon, A_F(y,\varepsilon)\big)\,\varepsilon^{-1}\mathrm{d}\varepsilon\,\mu(\mathrm{d}y)\,, \tag{10.30}$$

$$\lim_{n\to\infty}\frac{1}{n}\int_{b^{-1}d(T^nx,\partial S_{(T^nx)_1}O)}^{b^{-1}d(T^nx,\partial O)} \varepsilon^{-k}\big|C_k\big(F_\varepsilon), A_F(T^nx,\varepsilon)\big)\big|\,\varepsilon^{-1}\mathrm{d}\varepsilon = 0\,, \tag{10.31}$$

(note that under the above conditions $b^{-1}d(T^nx, \partial S_{(T^nx)_1}O) < \delta$) for all n, and

$$\lim_{\delta\to 0}\frac{|\ln\delta|}{n(x,\delta)} = \sum_{j=1}^{N} r_j^D\,|\ln r_j|\,. \tag{10.32}$$

Under the integrability assumption of our theorem, (10.30) follows from Birkhoff's ergodic theorem applied to the ergodic dynamical system $[F,\mu,T]$. Here the curvature measures may also be replaced by their absolute values. Taking into regard that $a_n = \sum_{i=1}^{n} a_i - \sum_{i=1}^{n-1} a_i$ for any real sequence, (10.31) is a consequence.

In order to use these arguments for (10.32) note that for $\delta(x,n) := b^{-1}d(x, \partial S_{x|n}O)$ we get

$$\lim_{\delta\to 0}\frac{|\ln\delta|}{n(x,\delta)} = \lim_{n\to\infty}\frac{|\ln\delta(x,n)|}{n}$$

provided the last limit exists. Since

$$\delta(x,n) = r_{x|n}\,b^{-1}d(T^nx,\partial O) = \prod_{i=1}^{n} r_{x_i}\,b^{-1}d(T^nx,\partial O)$$

and $x_i = (T^ix)_1\,,\ i\in\mathbb{N}$, Birkhoff's ergodic theorem implies for μ-a.a. $x \in F$,

$$\lim_{n\to\infty}\frac{1}{n}\,|\ln\prod_{i=1}^{n} r_{x_i}| = \lim_{n\to\infty}\frac{1}{n}\sum_{i=1}^{n}|\ln r_{(T^ix)_1}| = \int_F|\ln r_{y_1}|\,\mu(\mathrm{d}y) = \sum_{j=1}^{N}|\ln r_j|\,r_j^D\,,$$

as well as

$$\lim_{n\to\infty}\frac{1}{n}\,|\ln d(T^nx,\partial O)| = 0\,,$$

since $\int |\ln d(y, \partial O)|\, \mu(dy) < \infty$ (cf. (10.13)). This shows (10.32). It remains to consider the cases $k \in \{d-1, d\}$. Here we have $C_k(F_\varepsilon, B(x, a\varepsilon)) \le c_k \varepsilon^k$ (see the end of the proof of Theorem 10.13). Since the last integral for the logarithmic distance is finite, the above integrability condition is fulfilled and hence, the preceding arguments for convergence remain valid. □

Recall that Theorem 10.13 establishes a general relationship between curvature densities and measures. This can be used in order to conclude from Theorem 10.31 that under the stronger uniform integrability condition (10.4) the limit values $\mathcal{D}_k(F)$ are always the *densities of the average fractal curvature measures with respect to Hausdorff measure* $\mathcal{H}^D$ on the self-similar set F with (OSC). In particular, they do not depend on the choice of the localizing functions A_F satisfying (10.4). More precisely, we infer the following.

Corollary 10.32 *If the integrability condition in Theorem* 10.31 *is replaced by the uniform integrability of the mappings*

$$x \mapsto \frac{1}{|\ln \delta|} \int_\delta^{\varepsilon_0} \frac{C_k^{\mathrm{var}}(F_\varepsilon, A_F(x, \varepsilon))}{\varepsilon^k} \frac{1}{\varepsilon} \mathrm{d}\varepsilon\,,$$

$0 < \delta < \varepsilon_0$, *with respect to* $\mathcal{H}^D(F \cap (\cdot))$ *for some family of localizing functions* A_F *as before, then the* k*-th average fractal curvature measure in the sense of Definition* 10.8 *exists and is equal to*

$$\overline{C}_k^D(F, \cdot) = \mathcal{D}_k(F)\mathcal{H}^D(F \cap (\cdot))\,.$$

Remark 10.33

(i) The whole approach was extended in [BZ13] to average curvature-direction measures of self-similar sets. Moreover, in [BZ13, Proposition 3.12] it is shown that the uniform integrability condition (10.4) is also necessary for the corresponding convergence, if the curvature measures $C_k(F_\varepsilon, \cdot)$ are replaced by the two nonnegative measures of their Hahn decomposition, which also possess almost everywhere constant fractal densities.

(ii) Sufficient conditions for the uniform integrability are

$$\int_F \sup_{\delta < \varepsilon_0} \frac{1}{|\ln \delta|} \int_\delta^{\varepsilon_0} \frac{C_k^{\mathrm{var}}(F_\varepsilon, A_F(x, \varepsilon))}{\varepsilon^k} \frac{1}{\varepsilon} \mathrm{d}\varepsilon\, \mathcal{H}^D(\mathrm{d}x) < \infty\,, \qquad (10.33)$$

and consequently, since $\mathrm{spt}\, A_F(x, \varepsilon) \subset B(x, a\varepsilon)$,

$$\operatorname*{ess\,sup}_{\varepsilon < \varepsilon_0} \sup_{x \in F} \varepsilon^{-k} C_k^{\mathrm{var}}\big(F_\varepsilon, B(x, a\varepsilon)\big) < \infty\,. \qquad (10.34)$$

The latter is fulfilled for the special case of polyconvex neighbourhoods. Moreover, for $k \in \{d-1, d\}$ this estimate is always true. In general, from Proposition 10.24 one can derive that the integrability condition (ii) in Theorem 10.25 in terms of overlaps is equivalent to (10.34). Therefore Corollary 10.32 extends the first part of that Theorem concerning average limits, which was obtained there by renewal theory. Moreover, the equality

$$\varepsilon^{d-k}\mathbf{C}_k(F_\varepsilon) = \int_F \varepsilon^{-k} C_k(F_\varepsilon; A_F(x,\varepsilon))\mathcal{H}^D(\mathrm{d}x)$$

can also be used for a shorter proof of ordinary convergence of the total curvatures in the non-lattice case by means of the Renewal theorem (see [Zäh13, Theorem 3.1]). However, the integral representation of the limit obtained from this proof

$$\mathbf{C}_k^D(F) = \frac{1}{\eta}\int_F \int_{b^{-1}d(y,\partial S_{y_1}O)}^{b^{-1}d(y,\partial O)} \varepsilon^{-k} C_k\big(F_\varepsilon, A_F(y,\varepsilon)\big)\, \varepsilon^{-1}\mathrm{d}\varepsilon\, \mathcal{H}^D(\mathrm{d}y)$$

is less comfortable for numerical computations than those from Theorem 10.25 for overlaps or from Theorem 10.28 for the case of compatible tilings in terms of the generator.

(iii) Extensions to *average curvature-direction measures of self-conformal sets* can be found in Bohl [Boh12] under rather general conditions.

References

[AFP00] L. Ambrosio, N. Fusco, D. Pallara, *Functions of Bounded Variation and Free Discontinuity Problems* (Oxford University Press, Oxford, 2000)

[Asm87] S. Asmussen, *Applied Probabilities and Queues* (Wiley, Chichester, 1987)

[AW43] C.B. Allendoerfer, A. Weil, The Gauss-Bonnet theorem for Riemannian polyhedra. Trans. Am. Math. Soc. **53**, 101–129 (1943)

[Bad80] A.J. Baddeley, Absolute curvatures in integral geometry. Math. Proc. Camb. Philos. Soc. **88**, 45–58 (1980)

[Ban67] T. Banchoff, Critical points and curvature for embedded polyhedra. J. Differ. Geom. **1**, 245–256 (1967)

[Ban82] V. Bangert, Sets with positive reach. Arch. Math. (Basel) **38**, 54–57 (1982)

[BB02] A. Bernig, L. Bröcker, Lipschitz-killing invariants. Math. Nachr. **245**, 5–25 (2002)

[BCY18] J.-D. Boissonnat, F. Chazal, M. Yvinec, *Geometric and Topological Inference* (Cambridge University Press, Cambridge, 2018)

[Ber07] A. Bernig, The normal cycle of a compact definable set. Israel J. Math. **159**, 373–411 (2007)

[BG92] C. Bandt, S. Graf, A characterization of self-similar fractals with positive Hausdorff measure. Proc. Am. Math. Soc. **114**, 995–1001 (1992)

[BHR06] C. Bandt, N.V. Hung, H. Rao, On the open set condition for self-similar fractals. Proc. Am. Math. Soc. **134**, 1369–1374 (2006)

[BK00] L. Bröcker, M. Kuppe, Integral geometry of tame sets. Geom. Dedicata **82**, 285–323 (2000)

[Boh12] T. Bohl, Fractal curvatures and Minkowski content of self-conformal sets. arXiv:1211.3421v1 (2012)

[Buc98] Z. Buczolich, Functions of two variables with large tangent plane sets. J. Math. Anal. Appl. **220**, 562–570 (1998)

[Bud89] L. Budach, Lipschitz-Killing curvatures of angular partially ordered sets. Adv. Math. **78**, 140–167 (1989)

[BZ13] T. Bohl, M. Zähle, Curvature-direction measures of self-similar sets. Geom. Dedicata **167**, 215–231 (2013)

[Che66] S.S. Chern, On the kinematic formula in integral geometry. J. Math. Mech. **16**, 101–118 (1966)

[Che83] J. Cheeger, Spectral geometry of singular Riemannian spaces. J. Differ. Geom. **18**, 575–657 (1983)

[Che04] B. Chen, A simplified elementary proof of Hadwiger's volume theorem. Geom. Dedicata **105**, 107–120 (2004)

J. Rataj, M. Zähle, *Curvature Measures of Singular Sets*, Springer Monographs in Mathematics, https://doi.org/10.1007/978-3-030-18183-3

[Cla76] F.H. Clarke, On the inverse function theorem. Trans. Am. Math. Soc. **93**, 97–102 (1976)

[Cla83] F.H. Clarke, *Optimization and Nonsmooth Analysis* (Wiley, New York, 1983)

[CMS84] J. Cheeger, W. Müller, R. Schrader, On the curvature of piecewise flat spaces. Commun. Math. Phys. **92**, 405–454 (1984)

[CMS86] J. Cheeger, W. Müller, R. Schrader, Kinematic and tube formulas for piecewise linear spaces. Indiana Univ. Math. J. **35**, 737–754 (1986)

[CSM06] D. Cohen-Steiner, J.-M. Morvan, Second fundamental measure of geometric sets and local approximation of curvatures. J. Differ. Geom. **74**, 363–394 (2006)

[Dol72] A. Dold, *Lectures on Algebraic Geometry* (Springer, Berlin, 1972)

[Doo94] J.L. Doob, *Measure Theory* (Springer, New York, 1994)

[Fal90] K. Falconer, *Fractal Geometry* (Wiley, Chichester, 1990)

[Fal95] K. Falconer, On the Minkowski measurability of fractals. Proc. Am. Math. Soc. **123**, 1115–1124 (1995)

[Fal97] K. Falconer, *Techniques in Fractal Geometry* (Wiley, Chichester, 1997)

[Fed59] H. Federer, Curvature measures. Trans. Am. Math. Soc. **93**, 418–491 (1959)

[Fed69] H. Federer, *Geometric Measure Theory* (Springer, New York, 1969)

[Fel71] W. Feller, *An Introduction to Probability Theory and Its Applications*, vol. II, 2nd edn. (Wiley, New York, 1971)

[FK12] U. Freiberg, S. Kombrink, Minkowski content and local Minkowski content for a class of self-conformal sets. Geom. Dedicata **159**, 307–325 (2012)

[FPR17] J.H.G. Fu, D. Pokorný, J. Rataj, Kinematic formulas for sets defined by differences of convex functions. Adv. Math. **311**, 796–832 (2017)

[FS13] J.H.G. Fu, R.C. Scott, Piecewise linear approximation of smooth functions of two variables. Adv. Math. **248**, 229–241 (2013)

[Fu85] J.H.G. Fu, Tubular neighborhoods in Euclidean spaces. Duke Math. J. **52**, 1025–1046 (1985)

[Fu89a] J.H.G. Fu, Curvature measures and generalized Morse theory. J. Differ. Geom. **30**, 619–642 (1989)

[Fu89b] J.H.G. Fu, Monge-Ampère functions I. Indiana Univ. Math. J. **38**, 745–771 (1989)

[Fu93] J.H.G. Fu, Convergence of curvatures in secant approximations. J. Differ. Geom. **37**, 177–190 (1993)

[Fu94] J.H.G. Fu, Curvature measures for subanalytic sets. Am. J. Math. **116**, 819–880 (1994)

[Fu00] J.H.G. Fu, Stably embedded surfaces of bounded integral curvature. Adv. Math. **152**, 28–71 (2000)

[Fu17] J.H.G. Fu, Integral geometric regularity, in *Tensor Valuations and Their Applications in Stochastic Geometry and Imaging*, ed. by E.B. Vedel Jensen, M. Kiderlen (Springer, Cham, 2017), pp. 261–300

[Gat00] D. Gatzouras, Lacunarity of self-similar sets and stochastically self-similar sets. Trans. Am. Math. Soc. **352**, 1953–1983 (2000)

[Gla97] S. Glasauer, A generalization of intersection formulae of integral geometry. Geom. Dedicata **68**, 101–121 (1997)

[Gra95] S. Graf, On Bandts' tangential distribution for self-similar measures. Mh. Math. **120**, 223–246 (1995)

[Gro78] H. Groemer, On the extension of additive functionals on classes of convex sets. Pac. J. Math. **45**, 525–533 (1978)

[Had57] H. Hadwiger, Vorlesungen über Inhalt, Oberfläche und Isoperimetrie (Springer, Berlin, 1957)

[Har77] R.M. Hardt, Uniqueness of nonparametric area minimizing currents. Indiana Univ. Math. J. **26**, 65–71 (1977)

[Hat02] A. Hatcher, *Algebraic Topology* (Cambridge University Press, Cambridge, 2002)

[Hir76] M.W. Hirsch, *Differential Topology* (Springer, New York, 1976)

[HLW04] D. Hug, G. Last, W. Weil, A local Steiner-type formula for general closed sets and applications. Math. Z. **246**, 237–272 (2004)

[HP16] J. Harrison, H. Pugh, Plateau's problem, in *Open Problems in Mathematics*, ed. by J. Nash Jr., M. Rassias (Springer, Cham, 2016), pp. 273–302

[HR18] D. Hug, J. Rataj, Mixed curvature measures of translative integral geometry. Geom. Dedicata **195**, 101–120 (2018)

[HS08] D. Hug, R. Schneider, Integral geometry of tensor valuations. Adv. Appl. Math. **41**, 482–509 (2008)

[Hug99] D. Hug, *Measures, Curvatures and Currents in Convex Geometry*, Habilitation Thesis, Universität Freiburg, 1999

[Hut81] J. Hutchinson, Fractals and self-similarity. Indiana Univ. Math. J. **30**, 713–747 (1981)

[Jen98] E.B. Vedel Jensen, *Local Stereology* (World Scientific, Singapore, 1998)

[JK17] E.B. Vedel Jensen, M. Kiderlen, Rotation invariant valuations, in *Tensor Valuations and Their Applications in Stochastic Geometry and Imaging*, ed. by E.B. Vedel Jensen, M. Kiderlen (Springer, Cham, 2017), pp. 185–212

[JR08] E.B. Vedel Jensen, J. Rataj, A rotation integral formula for intrinsic volumes. Adv. Appl. Math. **41**, 530–560 (2008)

[Kig01] J. Kigami, *Analysis on Fractals* (Cambridge University Press, Cambridge, 2001)

[Kla95] D.A. Klain, A short proof of Hadwiger's characterization theorem. Mathematika **42**, 329–339 (1995)

[Kle81] N. Kleinjohann, Nächste Punkte in der Riemanschen Geometrie. Math. Z. **176**, 327–344 (1981)

[KLZ17] Z. Kabluchko, G. Last, D. Zaporozhets, Inclusion–exclusion principles for convex hulls and the Euler relation. Discret. Comput. Geom. **58**, 417–434 (2017)

[Kom11] S. Kombrink, *Fractal Curvature Measures and Minkowski Content for Limit Sets of Conformal Function Systems*. Ph.D. thesis, University of Bremen, 2011

[KP08] S.G. Krantz, H.R. Parks, *Geometric Integration Theory* (Birkhäuser, Boston, 2008)

[KR97] D.A. Klain, G.-C. Rota, *Introduction to Geometric Probability* (Cambridge University Press, Cambridge, 1997)

[KS94] M. Kashiwara, P. Schapira, *Sheaves on Manifolds* (Springer, Berlin, 1994)

[Lal88] S. Lalley, The packing and covering functions of some self-similar fractals. Indiana Univ. Math. J. **37**, 699–709 (1988)

[Lal12] S. Lalley, Renewal theorems in symbolic dynamics, with applications to geodesic flows, noneuclidean tessellations and their fractal limits. Acta Math. **230**, 2474–2512 (2012)

[Lap93] M.L. Lapidus, Vibrations of fractal drums, the Riemann hypothesis, waves in fractal media and the Weyl-Berry conjecture, in *Ordinary and Partial Differential Equations (Dundee, 1992)*, vol. IV (Longman Scientific & Technical, Harlow, 1993), pp. 126–209

[LRŽ16] M. Lapidus, G. Radunović, D. Žubrinić, *Fractal Zeta Functions and Fractal Drums. Higher-Dimensional Theory of Complex Dimensions* (Springer International Publishing, Switzerland, 2016)

[Lyt04] A. Lytchak, On the geometry of subsets of positive reach. Manuscripta Math. **115**, 199–205 (2004)

[Lyt05] A. Lytchak, Almost convex subsets. Geom. Dedicata **115**, 201–218 (2005)

[Mat75] G. Matheron, *Random Sets and Integral Geometry* (Wiley, New York, 1975)

[Mat95] P. Mattila, *Geometry of Sets and Measures in Euclidean Spaces* (Cambridge University Press, Cambridge, 1995)

[MC69] M. Morse, S.S. Cairns, *Critical Point Theory in Global Analysis and Differential Topology* (Academic, New York, 1969)

[Mor46] P.A.P. Moran, Additive functions of intervals and Hausdorff measure. Proc. Camb. Philos. Soc. **42**, 15–23 (1946)

[Mor88] F. Morgan, *Geometric Measure Theory, a Beginner's Guide* (Academic, Boston, 1988)

[Mor08] J.-M. Morvan, *Generalized Curvatures* (Springer, Berlin/Heidelberg, 2008)

[MP10] P. Mörters, Y. Peres, *Brownian Motion* (Cambridge University Press, Cambridge, 2010)

[MS83] P. McMullen, R. Schneider, Valuations on convex bodies, in *Convexity and Its Applications*, ed. by P.M. Gruber, J.M. Wills (Birkhäuser, Basel, 1983), pp. 170–247

[MU03] R.D. Mauldin, M. Urbanski, *Graph Directed Markov Systems. Geometry and Dynamics of Limit Sets* (Cambridge University Press, Cambridge, 2003)

[PR13] D. Pokorný, J. Rataj, Normal cycles and curvature measures of sets with d.c. boundary. Adv. Math. **248**, 963–985 (2013)

[PRZar] D. Pokorný, J. Rataj, L. Zajíček, On the structure of WDC sets. Math. Nachr. (2019, to appear). https://doi.org/10.1002/mana.201700253

[PW12] E. Pearse, S. Winter, Geometry of canonical self-similar tilings. Rocky Mountain J. Math. **42**, 1327–1357 (2012)

[PW14] D. Pokorný, S. Winter, Scaling exponents of curvature measures. J. Fractal Geom. **1**, 177–219 (2014)

[Rat96] J. Rataj, The iterative version of a translative integral formula for sets with positive reach. Suppl. Rend. Circ. Mat. Palermo II **46**, 13–20 (1996)

[Rat99] J. Rataj, Translative and kinematic formulae for curvature measures of flat sections. Math. Nachr. **197**, 89–101 (1999)

[Rat02] J. Rataj, Absolute curvature measures for unions of sets with positive reach. Mathematika **49**, 33–44 (2002)

[Reg61] T. Regge, General relativity without coordinates. Nuovo Cimento **19**, 551–571 (1961)

[Roc70] R.T. Rockafellar, *Convex Analysis* (Princeton University Press, Princeton, 1970)

[RW04] R.T. Rockafellar, R.J.-B. Wets, *Variational Analysis* (Springer, Berlin, 2004)

[RW10] J. Rataj, S. Winter, On volume and surface area of parallel sets. Indiana Math. J. **59**, 1661–1686 (2010)

[RW13] J. Rataj, S. Winter, Characterization of minkowski measurability in terms of surface area. J. Math. Anal. Appl. **400**, 120–132 (2013)

[RZ90a] W. Rother, M. Zähle, Absolute curvature measures, II. Trans. Am. Math. Soc. **321**, 547–558 (1990)

[RZ90b] W. Rother, M. Zähle, A short proof of a principal kinematic formula and extensions. Trans. AMS **321**, 547–558 (1990)

[RZ95] J. Rataj, M. Zähle, Mixed curvature measures for sets of positive reach and a translative integral formula. Geom. Dedicata **57**, 259–283 (1995)

[RZ01] J. Rataj, M. Zähle, Curvatures and currents for unions of sets with positive reach, II. Ann. Glob. Anal. Geom. **20**, 1–21 (2001)

[RZ02] J. Rataj, M. Zähle, A remark on mixed curvature measures for sets with positive reach. Beiträge Alg. Geom. **43**, 171–179 (2002)

[RZ05] J. Rataj, M. Zähle, General normal cycles and lipschitz manifolds of bounded curvature. Ann. Global Anal. Geom. **27**, 135–156 (2005)

[RZ12] J. Rataj, M. Zähle, Curvature densities of self-similar sets. Indiana Univ. Math. J. **61**, 1425–1449 (2012)

[RZ17] J. Rataj, L. Zajíček, On the structure of sets with positive reach. Math. Nachr. **290**, 1806–1829 (2017)

[San74] L.A. Santaló, Mean values and curvatures, in *Stochastic Geometry*, ed. by E.F. Harding, D.G. Kedall (Wiley, London, 1974), pp. 165–175

[San76] L.A. Santaló, *Integral Geometry and Geometric Probability* (Addison Wesley, Reading, 1976)

[Sch75] R. Schneider, Kinematische Berührmaße für konvexe Körper. Abh. Math. Sem. Univ. Hamburg **44**, 12–23 (1975)

[Sch78] R. Schneider, Curvature measures of convex bodies. Ann. Mat. Pura Appl. **116**, 101–134 (1978)

[Sch80] R. Schneider, Parallelmengen mit Vielfachheit und Steiner-Formeln. Geom. Dedicata **9**, 111–127 (1980)

[Sch94] A. Schief, Separation properties of self-similar sets. Proc. Am. Math. Soc. **122**, 111–115 (1994)

[Sch14] R. Schneider, *Convex Bodies: The Brunn-Minkowski Theory (Second Expanded Edition)* (Cambridge University Press, Cambridge, 2014)

[SW72] R. Sulanke, P. Wintgen, *Differentialgeometrie und Faserbündel* (Birkhäuser Verlag, Basel/Stuttgart, 1972)

[SW86] R. Schneider, W. Weil, Translative and kinematic integral formulae for curvature measures. Math. Nachrichten **129**, 67–80 (1986)

[Vol57] W. Volland, Ein Fortsetzungssatz für additive Eipolyederfunktionale im euklidischen Raum. Arch. Math. **8**, 144–149 (1957)

[Wal82] P. Walters, *An Introduction to Ergodic Theory* (Springer, New York, 1982)

[Wei90] W. Weil, Iterations of translative integral formulae and non-isotropic poisson processes of particles. Math. Z. **205**, 531–549 (1990)

[Wey39] H. Weyl, On the volume of tubes. Am. J. Math. **1939**, 461–472 (1939)

[Win82] P. Wintgen, Normal cycle and integral curvature for polyhedra in Riemannian manifolds, in *Differential Geometry, Colloq. Math. Soc. János Bolyai 31, 1979*, ed. by Gy. Soos, J. Szenthe (North-Holland, Amsterdam, 1982), pp. 805–816

[Win08] S. Winter, Curvature measures and fractals. Diss. Math. **453**, 1–66 (2008)

[Win11] S. Winter, Curvature bounds for neighborhoods of self-similar sets. Comment. Math. Univ. Carolin. **52**, 205–226 (2011)

[Win15] S. Winter, Minkowski content and fractal curvatures of self-similar tilings and generator formulas for self-similar sets. Adv. Math. **274**, 285–322 (2015)

[WZ13] S. Winter, M. Zähle, Fractal curvature measures of self-similar sets. Adv. Geom. **52**, 229–244 (2013)

[Zäh86a] M. Zähle, Curvature measures and random sets II. Probab. Th. Rel. Fields **71**, 37–58 (1986)

[Zäh86b] M. Zähle, Integral and current representation of Federer's curvature measures. Arch. Math. **46**, 557–567 (1986)

[Zäh87] M. Zähle, Curvatures and currents for unions of sets with positive reach. Geom. Dedicata **23**, 155–171 (1987)

[Zäh89] M. Zähle, Absolute curvature measures. Math. Nachr. **140**, 83–90 (1989)

[Zäh11] M. Zähle, Lipschitz-Killing curvatures of self-similar random fractals. Trans. Am. Math. Soc. **363**, 2663–2684 (2011)

[Zäh13] M. Zähle, Curvature measures of fractal sets. Contemp. Math. **600**, 381–399 (2013)

List of Symbols

J. Rataj, M. Zähle, *Curvature Measures of Singular Sets*, Springer Monographs in Mathematics, https://doi.org/10.1007/978-3-030-18183-3

$\mathcal{LD}_d$	a family of closed Lipschitz domains page 195
$\mathbf{M}(T)$	mass of a current T page 21
$\mathbf{M}_K(T)$	local variant page 21
$\dim_M F$	Minkowski dimension page 209
$\mathcal{M}^D(F)$	Minkowski content page 210
$\mathbf{N}(T)$	norm of a normal current page 22
$\mathbf{N}_K(T)$	local variant page 22
N_X	normal cycle of X page 78
$\mathrm{Nor}(X, a)$	normal cone of X at a page 58
$\mathrm{nor}\, X$	unit normal bundle of a set X with positive reach page 60
$\mathrm{nor}\, X$	unit normal bundle of a $\mathcal{U}_{\mathcal{PR}}$-set X page 90
$N(X, x)$	Clarke normal cone of X at x page 186
$\mathrm{nor}^{\mathrm{Cl}}\, X$	Clarke normal bundle of X page 186
$\mathcal{O}_q$	Hausdorff measure of the unit q-sphere page 112
p_L	orthogonal projection onto a subspace L page 1
$\mathcal{PR}$	the family of sets with positive reach page 87
$s_k(X; x, n)$	symmetric function of principal curvatures page 71
$\mathrm{SO}(d)$	special orthogonal group page 41
$\mathrm{spt}\, T$	support of a current T page 19
$\\|T\\|$	Radon measure associated with a current T representable by integration page 19
$\mathrm{Tan}^k(A, a)$	$(\mathcal{H}^k, k)$-approximate tangent cone of A at a page 7
$\mathcal{U}_{\mathcal{PR}}$	the family of generic unions of sets with positive reach page 87
$\chi(A)$	Euler-Poincaré characteristic page 50
ι_X	Fu's index function of X page 92
$\kappa_i(x, n)$	principal curvatures page 68
μ_j^d	invariant measure on $\mathcal{A}(d, j)$ page 42
ω_s	volume of the unit ball in $\mathbb{R}^s$ if s is an integer page 2
$\varphi_k(x, n)$	Lipschitz-Killing curvature form of order k page 79
π_L	normalized orthogonal projection onto L page 32
Π_X	metric projection to X page 55
Σ_k	set of all permutations of $\{1, \dots, k\}$ page 11
$\Sigma(k, m)$	shuffle symbol page 11
$\Theta^s(A, a)$	s-dimensional density of A at a page 3
ϑ_d	normalized Haar measure on $\mathrm{SO}(d)$ page 41
$\bigwedge^k V$	space of k-covectors of V page 11
$\bigwedge_k V$	space of k-vectors of V page 12
$\bigwedge^k \mathbf{L}$	pullback of k-forms under a linear mapping $\mathbf{L}$ page 13
$\bigwedge_k \mathbf{L}$	pushforward of k-vectors under a linear mapping $\mathbf{L}$ page 13
$[L, M]$	projection volume related to L and M page 16
$\bullet$	scalar product in $\bigwedge_k \mathbb{R}^d$ or $\bigwedge^k \mathbb{R}^d$ page 14
$\wedge$	exterior product page 11
$\xi \,\lrcorner\, \alpha$	interior product of $\xi \in \bigwedge_k V$ and $\alpha \in \bigwedge^m V$ if $k \le m$ page 14
$\xi \,\llcorner\, \alpha$	interior product of $\xi \in \bigwedge_k V$ and $\alpha \in \bigwedge^m V$ if $k \ge m$ page 14

$\|\cdot\|$	comass norm in $\bigwedge^k \mathbb{R}^d$ page 14
$\langle \xi, \alpha \rangle$	dual pairing of a k-vector ξ and a k-form α page 13
∂T	boundary of the current T page 20
$f^\# \phi$	pullback of a k-form ϕ page 18
$F_\# T$	push forward of a current T page 23
$T \llcorner \phi$	restriction of a current by means of a differential form page 20
$T \wedge \xi$	extension of a current by means of a multivector field page 20
$\langle T, f, y \rangle$	slice of T through f at y page 29
$\star$	Hodge star operator page 15
$S \times T$	Cartesian product of currents page 22

Index

J. Rataj, M. Zähle, *Curvature Measures of Singular Sets*, Springer Monographs in Mathematics, https://doi.org/10.1007/978-3-030-18183-3

GPSR Compliance
The European Union's (EU) General Product Safety Regulation (GPSR) is a set of rules that requires consumer products to be safe and our obligations to ensure this.

If you have any concerns about our products, you can contact us on

ProductSafety@springernature.com

In case Publisher is established outside the EU, the EU authorized representative is:

Springer Nature Customer Service Center GmbH
Europaplatz 3
69115 Heidelberg, Germany

www.ingramcontent.com/pod-product-compliance
Ingram Content Group UK Ltd.
Pitfield, Milton Keynes, MK11 3LW, UK
UKHW021840270726
14058UKWH00002B/256

* 9 7 8 3 0 3 0 1 8 1 8 5 7 *